# *Role of the Unsaturated Zone In Radioactive and Hazardous Waste Disposal*

Edited by

James W. Mercer
P. S. C. Rao
I. Wendell Marine

ANN ARBOR SCIENCE
THE BUTTERWORTH GROUP

230 Collingwood, P.O. Box 1425, Ann Arbor, Michigan 48106

Library of Congress Catalog Card Number 82-46067
ISBN 0-250-40620-9

Manufactured in the United States of America


Butterworths, Ltd., Borough Green, Sevenoaks
Kent TN15 8PH, England

## PREFACE

The majority of hazardous and low-level radioactive wastes that are placed in the subsurface are affected by the physical and chemical processes active in the unsaturated zone. This book deals with problems associated with waste disposal practices, and focuses on the uses of laboratory analyses, field observations, and numerical and analytical calculations. Topics include policy, modeling, statistical techniques, liners, and field applications.

During the week of May 28 - June 1, 1979, the Committee on Water in the Unsaturated Zone of the American Geophysical Union (AGU) sponsored a special session as part of the Annual Meeting of the AGU on "The Unsaturated Zone as a Barrier in Waste Disposal." There was sufficient interest in this topic, and research developments were such that, three years later, the committee sponsored another special session as part of the AGU Annual Meeting. This latest session took place during the week of May 31 - June 4, 1982, in Philadelphia, and was titled, "Role of the Unsaturated Zone in Radioactive and Hazardous Waste Disposal." There were 23 abstracts contributed; as a result, the backgrounds of the speakers were quite varied, and the topics ranged from disposal of high-level radioactive waste to the role of partially saturated soil in hazardous waste disposal. Field sites included the Nevada Test Site, Barnwell, South Carolina, and the Price Landfill, near Atlantic City, New Jersey. There was considerable interest in this session, and Ann Arbor Science Publishers asked the speakers to expand their abstracts into papers and publish the result, with the editing being performed by the session conveners. With the permission of AGU and the consent of the conveners and 17 of the speakers, this book was written.

James W. Mercer
P. S. C. Rao
I. Wendell Marine

## EDITORS' BIOGRAPHIES

JAMES W. MERCER

James W. Mercer attended Florida State University and the University of Illinois, where he received his Ph.D. in Geology in 1973. Prior to graduation, he worked for Exxon Oil Corporation and the Desert Research Institute, University of Nevada. Most recently, he worked for the U.S. Geological Survey, and is currently President of GeoTrans, Inc. Dr. Mercer has taught ground-water courses at the USGS, and is currently involved in the programs at the Holcomb Research Institute (Butler University) and at George Washington University, where he is an Associate Professorial Lecturer. He has co-authored several articles on modeling transport problems in ground water, including work performed at the Love Canal site.

P. SURESH CHANDRA RAO

P. Suresh Chandra Rao is Associate Professor of Physics in the Soil Science Department at the University of Florida. He received his Ph.D. in Soil Physics from the University of Hawaii, M.S. in Soil Science from the Colorado State University, and B.Sc. in Agriculture from the A.P. Agricultural University in India. Dr. Rao's research interests include experimental and theoretical investigations of the fate of agrochemicals and toxic substances in soil-water systems. He is an author or co-author of over 50 papers that deal with adsorption-desorption, transformations, and transport of fertilizers, pesticides, and toxic organic substances in soils. He is an Associate Editor of the Journal of Environmental Quality published by the American Society of Agronomy. He has served as a consultant to USEPA and several environmental consulting firms. He teaches a graduate course in Soil Physics at the University of Florida that deals with water and solute transport in porous media.

I. WENDELL MARINE

I. Wendell Marine received his B.A. from St. John's College, Annapolis, Maryland, and his Ph.D. in Geology from the University of Utah in 1960. He worked for the Water Resources Division of the U.S. Geological Survey in Maryland, Delaware, Utah, Oklahoma and South Carolina, and has published numerous papers on the water resources in these areas. He has worked on the geologic storage of radioactive waste both with the USGS and in his current position with the Du Pont Company. His recent work and publications have been on earth tides, seismic response of subsurface facilities, use of tracers in fractured rock, dating water using helium, osmotic properties of geologic materials, and flow of contaminants in the unsaturated and shallow saturated zones.

ACKNOWLEDGMENTS

The editors would like to thank all of the contributors for making this proceedings possible. We also acknowledge the American Geophysical Union for permission to publish this book. The completion of this book in a timely fashion was made possible by the diligent efforts of Mrs. Carol Roberts and Mr. Geoff Jones, both of GeoTrans, Inc.

CONTENTS

NRC-FUNDED STUDIES ON WASTE DISPOSAL IN PARTIALLY SATURATED MEDIA - David L. Siefken and R. John Starmer . . . . . . . 1

MODELING OF MOISTURE MOVEMENT THROUGH LAYERED TRENCH COVERS - T.M. Johnson, K. Cartwright, B.L. Herzog, and T.H. Larson . . . . . . . . . . . 11

A MODEL TO SIMULATE INFILTRATION OF RAINWATER THROUGH THE COVER OF A RADIOACTIVE WASTE TRENCH UNDER SATURATED AND UNSATURATED CONDITIONS - Cheng Y. Hung . . . . . . . . . . . . 27

SIMULATION OF THE EFFECTS OF ORGANIC SOLUTES ON THE HYDRAULIC CONDUCTIVITY OF VARIABLY SATURATED, LAYERED MEDIA - E.G. Lappala . . . . . 49

UNSATURATED FLOW IN HETEROGENEOUS SOILS - T.-C. Jim Yeh and Lynn W. Gelhar . . . . . . . . . 71

ROLE OF PARTIALLY SATURATED SOIL IN LINER DESIGN FOR HAZARDOUS WASTE DISPOSAL SITES - D.B. McWhorter, J.D. Nelson, T.A. Shepherd, and R.E. Wardwell . . . . . . . . . . . . . . . . 81

MOISTURE CHARACTERISTICS OF COMPACTED SOILS FOR USE IN TRENCH COVERS - S. Klein, T.M. Johnson, and K. Cartwright . . . . . . . . . 101

FIELD EXPERIMENTS TO DETERMINE SATURATED HYDRAULIC CONDUCTIVITY IN THE VADOSE ZONE - Daniel B. Stephens, Scott Tyler, Kevin Lambert, and Scott Yates . . . . . . . . . . . . . . . . . 113

COMPOSITION, DENSITY AND FABRIC EFFECTS ON BULKY WASTE CAPILLARY RETENTION CHARACTERISTICS - George E. Veyera and Joseph P. Martin . . . . . . 127

LABORATORY ANALYSIS OF WATER RETENTION IN UNSATURATED ZONE MATERIALS AT HIGH TEMPERATURE - Jim Constantz . . . . . . . . . . . . . . . . . . 147

ROLE OF DESATURATION ON TRANSPORT THROUGH FRACTURED ROCK - Daniel D. Evans and Chi-hua Huang . . . . . 165

POTENTIAL USE OF GPR IN ASSESSING GROUNDWATER POLLUTION IN PARTIALLY AND FULLY SATURATED SOILS - J.J. Bowders, Jr., R.M. Koerner, and A.E. Lord, Jr. . . . . . . . . . . . . . . . . . . 179

NUCLEAR WASTE ISOLATION IN THE UNSATURATED ZONE OF ARID REGIONS - H.A. Wollenberg, J.S.Y. Wang, and G. Korbin . . . . . . . . . . . . . . . . . . . 195

HYDROGEOLOGICAL INVESTIGATION HAZARDOUS WASTE SITE ATLANTIC CITY, NEW JERSEY - Dennis Pennington . . 211

CASE HISTORY STUDY OF WATER FLOW THROUGH UNSATURATED SOIL - Stephen J. Trautwein, David E. Daniel, and Michael W. Cooper . . . . . . . . . . . . . . 229

A HYDROLOGIC STUDY OF THE UNSATURATED ZONE ADJACENT TO A RADIOACTIVE WASTE DISPOSAL SITE AT THE SAVANNAH RIVER PLANT, AIKEN, SOUTH CAROLINA - Paul Gruber . . . . . . . . . . . . . . . . . . . 255

GEOLOGIC CHARACTER OF TUFFS IN THE UNSATURATED ZONE AT YUCCA MOUNTAIN, SOUTHERN NEVADA - Robert B. Scott, Richard W. Spengler, Sharon Diehl, A.R. Lappin, and Michael P. Chornack . . . . . . . . . . . . . . 289

INDEX . . . . . . . . . . . . . . . . . . . . . . 337

FIGURES

MODELING OF MOISTURE MOVEMENT THROUGH LAYERED TRENCH COVER


1. Soil water content as a function of pressure head for soils used in simulations. 14

2. Samples of compacted Peoria Loess. 14

3. Total moisture entry into compacted Peoria Loess under differing initial conditions. 15

4. Predicted advance of wetting front in columns of Peoria Loess compacted wet and dry of optimum under constant surface ponding. Initial moisture content = θ; saturated moisture content = θ sat. 15

5. Predicted effect of a coarse-grained layer of gravel on moisture retention in the overlying fine-grained material as compared to moisture movement through a single layer of fine-grained material. 16

6. Predicted effect of a layer of medium sand on moisture retention in the overlying fine-grained material as compared to moisture movement through a single layer of fine-grained material. 16

7. Hydraulic conductivity as a function of pressure head for soils used in simulations. Calculations of hydraulic conductivity based on method of Green and Corey [14]. 18

8. Predicted distribution of pressure head at selected times in the 3-layer column under conditions of constant ponding at the surface and a seepage face at the bottom. 19

9. Predicted distribution of water content at selected times in a 3-layer column under conditions of constant ponding at the surface and a seepage face at the bottom. 19

10. Cross section of hypothetical cover design used in two-dimensional simulations. 21

11. Comparison of predicted distributions of pressure head at selected times in a 1-layer cover of Peoria Loess (P-2) and in a 3-layer cover of Peoria Loess (P-2), gravel (B), and Peoria Loess (P-2) at x = 100 cm under conditions of constant ponding at the surface and constant pressure head = -z at the bottom. 21

12. Predicted distribution of pressure head at selected times in the 3-layer cover at x = 300 cm under conditions of constant ponding at the surface and a seepage face at the bottom. 22

13. Predicted distribution of soil water content at selected times in the 3-layer cover at x = 300 cm under conditions of constant ponding at the surface and a seepage face at the bottom. 22

14. Initial distribution of total hydraulic head (cm) in 3-layer cover corresponding to a constant initial pressure head of -100 cm. 24

15. Predicted distribution of total head (cm) after 171 hr in the 3-layer cover under conditions of constant ponding at the surface and a seepage face at the bottom. 24

16. Predicted distribution of total head (cm) after 289 hr in the 3-layer cover under conditions of constant ponding at the surface and a seepage face at the bottom. 24

A MODEL TO SIMULATE INFILTRATION OF RAINWATER THROUGH THE COVER OF A RADIOACTIVE WASTE TRENCH UNDER SATURATED AND UNSATURATED CONDITIONS

1. Comparison of a schematic natural wetness distribution curve and its corresponding step-wise wetness distribution curve. 34

2. Comparison of the results of simulation for gravity drainage of a typical sand using the proposed model and the results interpreted from Hillel and Van Bavel's study. 40

3. Comparison of the results of simulation for the steady state evaporation of Chino clay by using the proposed model and that obtained by Ripple. 41

4. Transitions in pellicular water transport from liquid phase to vapor phase calculated from Equations 19, 25, and 26 for sandy and clay soils. 43

5. The results of simulations using the proposed model for loamy sand and clay soil, Barnwell radioactive waste disposal site. 44

6. The results of simulating the variation of pellicular water deficit, loamy sand, Barnwell radioactive waste disposal site. 45

SIMULATION OF THE EFFECTS OF ORGANIC SOLUTES ON THE HYDRAULIC CONDUCTIVITY OF VARIABLY SATURATED, LAYERED MEDIA

1. Hydraulic conductivity as a function of pore volumes for several organic fluids (after Anderson and Brown [1]). 52

2. General flow chart of the simulation program. 56

3. Saturation profiles for a = 0. 59

4. Time history of solvent penetration depth for a = 0.0, 0.5, 1.0, 1.5, and 2.0. 61

5. Time required for solvent to break through the clay layer as a function of the value of a. 62

6. Rate of solvent inflow at the top of the column. 63

7. Saturation history at a depth of 134 cm. 64

UNSATURATED FLOW IN HETEROGENEOUS SOILS

1a. Permeability variation of Mt. Simon Sandstone, Illinois. 72

1b. Porosity variation of Mt. Simon Sandstone, Illinois. 72

2. Variation of saturated hydraulic conductivity of Sevilleta dune sand. 73

3a. Some representative K-$\psi$ curves of Maddock sandy loam. 75

3b. Some representative K-$\psi$ curves of Panoche silty clay loam. 75

4. Anisotropy ratios of Maddock sandy loam and Panoche silty clay loam as a function of mean soil tension. 76

5. Soil tension distribution beneath a disposal basin. 77

6. Configuration of the shallow underground storage facility. 78

ROLE OF PARTIALLY SATURATED SOIL IN LINER DESIGN FOR HAZARDOUS WASTE DISPOSAL SITES

1. Schematic water impoundment disposal area. 83

2. Stages of seepage from waste disposal area. 84

3. Typical relationship between capillary pressure head and saturation. 87

4. Typical relationship between hydraulic conductivity ($K_e$) and pore water pressure head. 87

5. Liner thickness and conductivity for partially saturated wetting front. 94

6. Stage I flow: Effects of liner conductivity on seepage flow. 94

7. Stage I flow: Effects of liner thickness on seepage flow. 96

MOISTURE CHARACTERISTICS OF COMPACTED SOILS FOR USE IN TRENCH COVERS

1. Effects of compaction on structure [Lambe, 1980]. 102

2. Results of compaction and hydraulic conductivity tests on Peoria Loess. 104

3. Results of compaction and hydraulic conductivity tests on Fairgrange Till. 104

4. Soil moisture retention curves for various soils. 106

5. Desaturation curves for compacted Peoria Loess samples. 106

6. Desaturation curves for compacted Fairgrange Till samples. 106

7. Hydraulic conductivity vs degree of saturation for Peoria Loess. 108

8. Hydraulic conductivity vs degree of saturation for Fairgrange Till. 108


FIELD EXPERIMENTS TO DETERMINE SATURATED HYDRAULIC CONDUCTIVITY IN THE VADOSE ZONE


1. Classical free surface concept of flow from a borehole above a deep water table. 114

2a. Water retention curve. 116

2b. Relative hydraulic conductivity versus pressure. 116

3. Schematic of borehole infiltration test equipment and instrumentation. 116

4. Total hydraulic head as measured by tensiometers during borehole infiltration test. 118

5. Total hydraulic head throughout flow field at steady state, S3T4. 119

6. Pressure head throughout flow field at steady state, S3T4. 121

7. Infiltration rate into boreholes showing effects of $CO_2$ flooding. 123

COMPOSITION, DENSITY AND FABRIC EFFECTS ON BULKY WASTE CAPILLARY RETENTION CHARACTERISTICS


1. Gradations of uranium tailings, Grand Junction sites. 132

2. Gradations of uranium tailings, Gunnison site. 132

3. Retention curves of intact samples, Grand Junction. 133

4. Retention curves of intact samples, Gunnison. 133

5. Laboratory hydraulic fabric curves, Grand Junction site tailings. 135

6. Dense compaction fabric curves, Grand Junction site tailings. 135

7. Experimental fabric and density effects, Grand Junction site (beach sand). 136

8. Experimental fabric and density effects, Grand Junction site (mixture). 136

9. Experimental fabric and density effects, Grand Junction site (slimes). 136

10. Comparison of compacted and intact sample fabrics, Gunnison plastic slimes (saturation measure). 137

11. Comparison of compacted and intact sample fabrics, Gunnison plastic slimes (water content measure). 137

12. Stress history effects on compacted slimes, Gunnison site. 139

13. Leaching effects on retention, Grand Junction tailings. 139

14. Density effects on retention, Tosco spent oil shale. 139

15. Effects of fabric and fines content on residual moisture content, Grand Junction site tailings. 141

16. Effects of fabric and fines content on displacement pressure head, Grand Junction site tailings. 141

17. Comparison of optimum and residual moisture tent, Grand Junction compacted tailings. 141

18. Schematic of apparatus for testing remolded samples. 143

19. Schematic of apparatus for testing intact samples. 143

20. Comparison of laboratory hydraulic deposition models, screened concrete sand. 144

21. Compacted density effects on retention for screened concrete sand. 144

LABORATORY ANALYSIS OF WATER RETENTION IN UNSATURATED ZONE MATERIALS AT HIGH TEMPERATURE

1. Cross-sectional and exploded views of a submergible pressure outflow cell (SPOC). Four bolts hold the SPOC firmly together and two o-rings form top and bottom gland seals to contain the applied air pressure. 149

2. The complete experimental assembly shown during the weighing of a submergible pressure outflow cell (SPOC). The data acquisition and equipment control unit is shown as an option which was not used for this experiment. 150

3. A comparison of S vs. $\psi$ at 25°, 95°, and 25°C in Tipperary Sand for Experiment One. 153

4. A comparison of S vs. $\psi$ at 25°, 75°, and 25°C during draining in Tipperary Sand for Experiment Two. 154

5. A comparison of S vs. $\psi$ at 25°, 75°, and 25°C during wetting in Tipperary Sand for Experiment Two. 155

6. A comparison of S vs. $\psi$ at 95°C for experimentally measured values and theoretically calculated values from Eq.[2] in Tipperary Sand for Experiment One. 159

7. A comparison of S vs. $\psi$ at 75°C for experimentally measured values and theoretically calculated values from Eq.[2] during draining in Tipperary Sand for Experiment Two. 161

8. A comparison of S vs. $\psi$ at 75°C for experimentally measured values and theoretically calculated values from Eq.[2] during wetting in Tipperary Sand for Experiment Two. 162

ROLE OF DESATURATION ON TRANSPORT THROUGH FRACTURED ROCK

1. A conceptual drawing of a repository. 165

2. Air entry potential versus aperture size. 168

3. Film thickness versus water potential. 169

4. Lognormal distribution of aperture size with a variation k=0.5, and varying mean aperture. 171

5. Lognormal distribution of aperture size with a mean aperture of 4μm and varying coefficients of variation. 172

6. The effect of fracture drainage on flux or area of the flow for a set of parallel fractures with mean aperture 4 μm and a coefficient of variance 1.0. 173

7. The effect of fracture drainage on flux or area of the flow for a set of parallel fractures with mean aperture 4 μm and a coefficient of variance 0.25. 174

8. Moisture and salt movement under thermal gradient (results of Gurr et al. [7]). 176

9. Mechanism of moisture transport in an unsaturated, finite medium under thermal gradients. 177

10. Hypothetical water potential vs. distance from a repository at three times. 178

POTENTIAL USE OF GPR IN ASSESSING GROUNDWATER POLLUTION IN PARTIALLY AND FULLY SATURATED SOILS


1. GP trace of a stationary water table surface located within a uniform sandy soil with measured water table locations superimposed on the trace. 182

2. Photographs of laboratory test setup showing filled container and container with GPR test equipment. 184

3. GPR traces of well graded sand at varying water contents using fresh water with steel plate beneath box. 185

4. GPR traces of well graded sand at varying water contents using salt saturated water with steel plate beneath box. 187

5. GPR traces of well graded sand at varying water contents using fresh water with "Ecosorb" beneath box. 188

6. GPR traces of well graded sand at varying water contents using salt saturated water with "Ecosorb" beneath box. 189

7. Pulse/reflection bands showing differences between fresh water and salt water tests at all saturation states from dry to saturated. 190

8. GPR traces of uniform well graded sand in other possible situations. 192


NUCLEAR WASTE ISOLATION IN THE UNSATURATED ZONE OF ARID REGIONS


1. The northwestern portion of the Basin and Range province, showing highland areas that contain tuffaceous rocks. Valleys with appreciable components of tuffaceous alluvium lie between these highlands. 197

2. Schematic geologic cross section through an idealized Basin and Range valley, showing alluvium (Qal) overlying tuff (Tt) and Paleozoic carbonate rock (Pzc), with the water table at depth in the Paleozoic rock (based on [1]). 199

3. Thermal conductivities measured in situ in the southern Black Rock Desert (average value = 1.05 ± 0.15 W/m°C, 71 samples over 12 holes, data from [14]). 201

4. Effects of thermal conductivity on alluvial repository temperature, spent fuel, 10 years out of the reactor. 202

5. Variation of temperature rise with water saturation. 203

6. Histogram on the thermal conductivity of tuff; data from [15]. 204

7. Effects of thermal conductivities on tuff repository temperature, spent fuel, 10 years out of the reactor. 205

HYDROGEOLOGICAL INVESTIGATION
HAZARDOUS WASTE SITE
ATLANTIC CITY, NEW JERSEY

1. Site location map. 213

2. Leachate plume at Price Landfill. 225

CASE HISTORY OF WATER FLOW THROUGH UNSATURATED SOIL

1. Plan view of site. 231

2. Generalized stratigraphy at the site. 232

3a. Monitor well installation. 234

3b. Lysimeter installation. 235

4. Field permeameter installation. 236

5. Moisture-characteristic curves used in the analyses. 242

6. Degree of saturation profiles for the various analyses. 244

7. Relative conductivity versus degree of saturation relationships used in the analyses. 245

8. Comparison of measured values of saturated hydraulic conductivity ($k_{sat}$) with those used in the model. 246

9. Predicted advancement of the wetting front. 247

10. Degree of saturation profile after ponding is eliminated. 249

11. Comparison of hydraulic conductivity measured in the laboratory and in field tests for upper layer of clay. 252

A HYDROLOGIC STUDY OF THE UNSATURATED ZONE ADJACENT TO A RADIOACTIVE WASTE DISPOSAL SITE AT THE SAVANNAH RIVER PLANT, AIKEN, SOUTH CAROLINA

1. Location map of the Savannah River plant. 257

2. Location map of burial grounds, production facilities, and major streams of the Savannah River plant. 258

3. Outcrop map of selected geologic units at the Savannah River plant. 262

4. Location map of plots in relation to burial grounds and F area. Drill core sites are also indicated on map. 263

5. Laboratory derived soil-water characteristic curves for the soil cores. 273

6. Measured values and soil-water storage as a function of time after steady-state infiltration has ceased, values given are the mean of all plots. 275

7. Soil-water content profiles for the study area following the cessation of steady-state infiltration. 277

8. Hydraulic conductivity as a function of water saturation over the study area. 278

9. Average hydraulic conductivity as a function of soil-water content for the entire study area assuming a homogeneous soil to a depth of 120 inches. 280

10a. Field versus laboratory derived hydraulic conductivity. Values are matched at the saturated hydraulic conductivity value. 281

10b. Field versus laboratory derived hydraulic conductivity. Values are matched at the saturated hydraulic conductivity value. 282

GEOLOGIC CHARACTER OF TUFFS IN THE UNSATURATED ZONE AT YUCCA MOUNTAIN, SOUTHERN NEVADA

1. Location map. Index map at lower left shows the location of the Nevada Test Site (NTS) within Nevada, and the center shows an expansion of the southwest portion of the NTS. 291

2. Geologic map of central Yucca Mountain generalized from the work of Christiansen and Lipman [3] and Lipman and McKay [4]; and modified based on results of recent detailed mapping. Exact boundaries of the area of the potential repository within this map area are not drawn because data are still being collected that may affect the most favorable boundaries. 292

3. Generalized north-south geologic section through volcanic strata under Yucca Mountain. SWL marks the static water level. Horizontal and vertical scales are the same. The zero depth is mean sea level. 295

4. Comparison between stratigraphic units defined by petrologic criteria and those defined by physical-property criteria. The conventional stratigraphic nomenclature based upon petrologic criteria is given for drill hole USW-GU3/G3 on the left and the physical-property stratigraphies based on degree of welding for drill holes USW-GU3/G3, -G1, and -G2 are drawn in order toward the right. The conventional stratigraphic names of major units are also listed on the far right for general reference but exact boundaries are not shown. The shaded zones are moderately welded to densely welded; unshaded are nonwelded to partially welded. BT = bedded tuff. 301

5. Generalized zonation within the Tiva Canyon Member. The rounded step zone grades laterally into the clinkstone zone. Measured section numbers are located on Figure 2. 302

6. Relations between dry bulk density, matrix porosity, and degree of welding of the Tiva Canyon Member. The central portion includes the devitrified zones: the hackly, lower lithophysal, clinkstone, and upper lithophysal zones. Brackets show the general regions typical of rocks of different degrees of welding. 305

7. Grain density (g/cm$^3$) correlation with stratigraphic zones in the Tiva Canyon member, Paintbrush Tuff. 306

8. Photographs of thin sections of devitrified textures in the Tiva Canyon Member in the (a) hackly zone and (b) clinkstone zone. The glass shard texture is more distinct in the hackly zone but partially destroyed in the clinkstone zone. 308

9. Stratigraphy of devitrification textures in the Tiva Canyon Member, Paintbrush Tuff. The dashed lines mark contacts between devitrification zonations that do not coincide with zone boundaries. 309

10. Lithophysal zonation in the Topopah Spring Member, Paintbrush Tuff. Locations of drill holes USW-G2, -G1, and -GU3/G3 and the measured sections 3, 7, and 8 are shown on Figure 2. 310

11. Detailed fault pattern at Yucca Mountain. Faults in the Abandoned Wash area are mapped only on south-facing slopes where exposures are favorable. Attitudes of primary foliation and detailed zones within the Tiva Canyon and Topopah Spring Members are not shown on this figure because of space limitations. Examples of foliation attitudes in the Abandoned Wash area are given in Figure 12. 314

12. Structure section across Yucca Crest and Abandoned Wash showing the nature of the increase in foliation attitudes toward a north-northeast-striking normal fault and apparent dip-slip movement on representative north-northwest-striking faults. 315

13. Rose diagrams of the fracture strikes encountered along traverses essentially perpendicular to the average fracture attitude. Five traverses were made in the Tiva Canyon Member. One of the two longer traverses (A) was made south of Drill Hole Wash and the other (B) was made just north of Abandoned Wash. The northernmost of the traverses was made in the tuffaceous beds of Calico Hills. The outline of Yucca Mountain is the Tertiary volcanic rock contact with Quaternary alluvium and colluvium. 316

14. Density of fractures compared to physical-property stratigraphy expressed by differences in degree of welding in drill hole USW-GU3/G3 shown in Figure 4. These fractures include only mineralized cracks or cracks that have evidence of shear such as breccia, gouge, or slickensides. The density of fractures has been calculated per unit $m^3$. The density of fractures at drill holes USW-G1 and -G2 is somewhat lower than at -GU3/G3, but the overall correlation with degree of welding is similar. 319

15. Lower hemisphere poles to attitudes of average fractures and faults measured in oriented core from USW-GU3/G3 and corrected for drilling deviation (R.B. Scott, USGS, and Mayra Castellanos, F&S, written commun., 1982). The average fracture plane (N.15°W., 81°SW.) was calculated in the upper diagram by averaging the stereonet maxima for all the poles to fracture planes from each stratigraphic unit. Also in the upper diagram, the averaged direction of drill hole deviation is shown by the X. The stereonet maximum for the fault attitude (N.21°W., 73°SW.) is shown in the lower diagram. The Tiva Canyon Member (Tpc) and Topopah Spring Member (Tpt) are part of the Paintbrush Tuff. The Prow Pass Member (Tcp), Bullfrog Member (Tcb), and Tram unit (Tct) from the Crater Flat Tuff. Tlr symbolizes the tuff of Lithic Ridge and Tt symbolizes the older unnamed tuffs. 321

16. Drill-hole deviation directions on Yucca Mountain. Note that no horizontal scale is given because the amount of lateral deviation varies. Line lengths are normalized to represent percent of horizontal deviation. 322

17. Location of the Las Vegas Valley shear zone [21] relative to the Yucca Mountain region. Note the parallelism of the shear zone and the elongation of Yucca Wash. 323

18. Conceptual hydrologic model of Yucca Mountain, assuming a positive net recharge for the mountain. Determination of vertical and possible horizontal rates of transmission of recharge through densely and moderately welded Tiva Canyon and Topopah Spring Members and through less welded vitiric, argillic and zeolitized tuffs awaits hydrologic testing specifically designed for the unsaturated zone. If faults are locally sealed, they may create local perched water tables, but if they are open, the faults may rapidly transmit water. 326

TABLES

NRC-FUNDED STUDIES ON WASTE DISPOSAL IN PARTIALLY SATURATED MEDIA

I. High-Level Waste Projects 3

II. Low-Level Waste Projects 6

III. Uranium Recovery Projects 9

A MODEL TO SIMULATE INFILTRATION OF RAINWATER THROUGH THE COVER OF A RADIOACTIVE WASTE TRENCH UNDER SATURATED AND UNSATURATED CONDITIONS

I. Soil Characteristics of the Trench Cover Used in Simulating the Annual Rainfall Infiltration, Barnwell Radioactive Waste Disposal Site 44

SIMULATION OF THE EFFECTS OF ORGANIC SOLUTES ON THE HYDRAULIC CONDUCTIVITY OF VARIABLY SATURATED, LAYERED MEDIA

I. Hydraulic Conductivity for Several Organic Fluids [1] 53

II. Values Used for Parameters in the Brooks and Corey Equations for Saturation and Relative Liquid Conductivity 57

III. Time and Iteration Data 66

ROLE OF PARTIALLY SATURATED SOIL IN LINER DESIGN FOR HAZARDOUS WASTE DISPOSAL SITES

I. Impoundment and Strata Properties 92

MOISTURE CHARACTERISTICS OF COMPACTED SOILS FOR USE IN TRENCH COVERS

I. Results of Classification Tests 103

FIELD EXPERIMENTS TO DETERMINE SATURATED HYDRAULIC CONDUCTIVITY IN THE VADOSE ZONE

I. Borehole Infiltration Results 120

II. Field and Lab Permeameter Results 122

COMPOSITION, DENSITY AND FABRIC EFFECTS ON BULKY WASTE CAPILLARY RETENTION CHARACTERISTICS

I. Index Properties of Uranium Tailings Samples 132

II. Volumetric Data for Undisturbed (Intact) Uranium Tailings Field Samples 133

III. Unsaturated Material Parameters Grand Junction Site Retention Tests Brooks-Corey Method 140

LABORATORY ANALYSIS OF WATER RETENTION IN UNSATURATED ZONE MATERIALS AT HIGH TEMPERATURE

I. Comparison of the Matric Potential, $\psi$, Versus the Percent Volume Saturation, S, in Tipperary Sand for Experiment One 152

II. Comparison of the Matric Potential, $\psi$, Versus the Percent Volume Saturation, S, in Tipperary Sand for Experiment Two 156

III. Comparison of Experimental and Theoretical Values of the Matric Potential for a Given Water Content at 95°C for Experiment One 157

IV. Comparison of Experimental and Theoretical Values of the Matric Potential for a Given Water Content at 75°C for Experiment Two 158

ROLE OF DESATURATION ON TRANSPORT THROUGH FRACTURED ROCK

I. Numerical Example of Data Extracted From Figure 6 for a Mean Aperture of 4 m and a Coefficient of Variation of 1.0 175

II. Numerical Example of Data Extracted From Figure 7 for a Mean Aperture of 4 m and a Coefficient of Variation of 0.25 175

POTENTIAL USE OF GPR IN ASSESSING GROUNDWATER POLLUTION IN PARTIALLY AND FULLY SATURATED SOILS

I. Details of Pulsed Radio Frequency Methods, after Lord and Koerner [2] 181

NUCLEAR WASTE ISOLATION IN THE UNSATURATED ZONE OF ARID REGIONS

I. Major Advantages and Concerns of Waste Isolation in Unsaturated and Saturated Regimes 207

HYDROGEOLOGICAL INVESTIGATION HAZARDOUS WASTE SITE ATLANTIC CITY, NEW JERSEY

I. Contaminants in Surface Waters (ppb) 217

II. Range of Concentrations in Water Samples 218

III. Selected Contaminants at Local Pollution Sources (ppb) 220

IV. Concentrations Found at an Illegal Dump Site (S1) 221

GEOLOGIC CHARACTER OF TUFFS IN THE UNSATURATED ZONE AT YUCCA MOUNTAIN, SOUTHERN NEVADA

I. Hydraulic Conductivities of Tuffaceous Aquifers and Aquitards, NTS 299

II. Physical Properties of the Tiva Canyon Member of the Paintbrush Tuff, Yucca Mountain (Analyses by Terra Tek, Inc., Salt Lake City) 303

# NRC-FUNDED STUDIES ON WASTE DISPOSAL IN PARTIALLY SATURATED MEDIA

David L. Siefken and R. John Starmer
U.S. Nuclear Regulatory Commission
Low-Level Waste Licensing Branch
Washington, D.C. 20555

ABSTRACT

The Division of Waste Management within the Office of Nuclear Material Safety and Safeguards encompasses three distinct types of commercial radioactive waste disposal: high-level, low-level, and uranium recovery wastes. Disposal of the latter two types of wastes is generally at or near the ground surface in partially saturated media above fluctuations of the water table. Use of unsaturated media for a geologic repository for high-level wastes, which is subject to NRC licensing, is being considered by the Department of Energy in their national siting program.

Technical assistance projects, through the Office of Nuclear Material Safety and Safeguards, and research projects, through the Office of Nuclear Regulatory Research, provide technical support to NRC staff in a wide range of areas related to waste disposal in partially saturated media. These studies include such diverse areas as: field studies and modeling of flow and transport in partially saturated porous media or fractured rock; laboratory studies and modeling of changes in unsaturated hydraulic conductivity due to consolidation or changes in moisture content and temperature; assessment of methodologies for paleohydrologic evaluation; laboratory and field testing of capillary (wick-effect) barriers to infiltration; field studies of the relationship between wetting fronts and leaching; field studies and modeling of geochemical processes affecting the transport of reactive radioactive solutes; field studies involving hydraulic monitoring of flow using tracers; field studies of the effects of subsidence on infiltration-limiting trench cap covers; field studies of vapor phase releases; and assessment of unsaturated media at the Nevada

Test Site as alternatives for geologic disposal of high-level wastes.

## INTRODUCTION

The Division of Waste Management within the Office of Nuclear Material Safety and Safeguards of the U.S. Nuclear Regulatory Commission regulates three distinct types of commercial radioactive waste disposal: high-level, low-level, and uranium recovery (mill tailings). The NRC staff receives technical support in these three types of radioactive waste disposal through technical assistance projects funded by the Division of Waste Management and through research projects funded by the Office of Nuclear Regulatory Research.

The subject of this paper is those NRC-funded technical assistance and research projects that deal in part or in whole with the movement of moisture and/or solute transport in partially-saturated media. Projects dealing with paleoclimatology and paleohydrology are also included because of the need to evaluate past and future variations in the water table. The majority of these projects are currently in progress and have no formal publications available at this time. Reports which are available have been indicated in the following text.

The purpose of this paper is to make the studies more useful by increasing the awareness of the technical community to the studies and by more broadly disseminating the results of the studies. Critical review and application of the results by others, in disposal of radioactive or other types of waste, can further refine the results of the studies, improve the methodologies developed, and lead to more environmentally-acceptable disposal sites.

For program planning and budgetary purposes, the technical assistance and research projects are grouped under the type of radioactive waste. It should be emphasized, however, that the studies are typically applicable, at least in part, to disposal of all three types of radioactive wastes.

## HIGH-LEVEL WASTES

The Department of Energy (DOE) is responsible for the disposal of commercial high-level radioactive wastes. The Environmental Protection Agency (EPA) is responsible for setting environmental standards for such disposal; and, the Nuclear Regulatory Commission (NRC) is responsible for licensing the disposal and enforcing the environmental standards.

DOE is currently considering disposal of commercial high-level wastes in a geologic repository situated in partially-saturated media above fluctuations of the water table. At present, these studies have focused on tuff units in Yucca Mountain at the Nevada Test Site; however, consideration of unsaturated basalt and alluvium has been recommended by others (Panel on Hanford Wastes [1], Winograd [2]).

The proposed regulation for geologic disposal of high-level wastes (10 CFR Part 60, published July 8, 1981 in draft form for public comments) primarily addressed disposal at great depths in saturated media; however, it indicated that this reflected current DOE siting studies rather than a preference by the NRC. In the final regulation currently under preparation, siting criteria for geologic disposal in unsaturated media will be specifically addressed.

High-level waste projects related to partially-saturated media are listed on Table I. Contractors are identified parenthetically after the project titles.

Table I. High-Level Waste Projects

| |
|---|
| Division of Waste Management |
| Development of Nuclear Waste Management Criteria (LLNL) |
| Geotechnical Services Program (LBL) |
| Benchmarking of Computer Codes and Licensing Assistance (Teknekron) |
| Office of Nuclear Regulatory Research |
| Unsaturated Flow and Transport Through Fractured Rock (U. of Arizona) |
| Climates of Holocene: Methodologies for Paleoclimatic Reconstruction (NSF) |

The first waste management project was performed at Lawrence Livermore National Laboratory and has been completed. Two reports related to disposal of high-level waste in partially-saturated media were prepared and are available from NTIS. These reports are UCRL-13992, Preliminary Study of Radioactive Waste Disposal in the Vadose Zone, and UCID-18118, Past Climate Reconstruction: A Tool for Assessing Site Suitability.

The first report, prepared under subcontract by Geotechnical Engineers, Inc., is a substantial, comprehensive volume covering the mechanics of flow in partially-saturated media, the hydrogeology of four regions having a deep water table, characteristics and modeling of a vadose zone repository, and appendices summarizing available literature, soil-water relationships, infiltration, measurement of unsaturated flow parameters, and effects of thermal gradients. The second report is primarily a summary of ongoing (1979) work in paleoclimatic reconstruction, with examples of reconstructed maximum, mean, and minimum precipitation in 25-year intervals for the last 300 years in various regions of the United States.

The Geotechnical Services Program being performed at Lawrence Berkeley National Laboratory includes, as a small part of the study, an assessment of radioactive waste disposal in unsaturated media. In this task, partially-saturated tuff and alluvium at the Nevada Test Site have been evaluated on the basis of the geologic, hydrogeologic and geomechanical properties of the media; geologic hazards; construction techniques; and, the presence of mineral resources. The paleohydrology of the Nevada Test Site was also considered in the evaluation. This study is the subject of a paper presented later in this volume by Harold A. Wollenberg.

The benchmarking project, currently being performed by Teknekron Research, Inc., is a multiple-component study of which computer codes applicable to disposal in the unsaturated zone is a small part. The work on unsaturated codes is being performed on subcontract by GeoTrans and includes codes such as FEMWATER, FEMWASTE, and TRUST. The work consists of preparing (1) detailed code summaries, (2) benchmark problems to test the codes against published analytical solutions and field simulations and to evaluate hypothetical repository sites, (3) a data base on unsaturated media for use in solving the benchmarking problems, and (4) an evaluation of the codes based upon the results of the preceeding work. The project will also provide assistance to the NRC staff in the review of performance assessment portions of site characterization reports and license applications submitted by DOE.

The first research project, dealing with unsaturated flow and transport through fractured rock, is being performed at the University of Arizona. A progress report covering the period September 21, 1981 to April 1, 1982 has been completed and is available from the contractor. That report covers the mechanisms of water flow in unsaturated fractured rock and work completed to-date on methods of measurement and instrumentation for characterizing (1) water (liquid/vapor phases) movement, (2) contaminant transport, and (3) fracture geometry. In addition, the planned approach for modeling of unsaturated flow and transport in fractured rock is outlined. This project is the subject of a paper presented later in this volume by Daniel Evans and Chi-hua Huang.

The project on paleoclimatic reconstruction consists of a review and application of several methodologies, primarily pollen studies. This project is a small portion of an ongoing National Science Foundation study involving the University of Minnesota, University of Wisconsin, and Brown University. One major goal of the study is to reconstruct North American climates in 3000-year intervals back to 12,000 years B.P. Three major products expected by the NRC include (1) an analysis of methodologies for reconstructing quantitative paleoclimatic conditions, (2) establishment of a data base for such reconstructions, and (3) refinement of the modeling capability to simulate paleoclimatic (and project future climatic) conditions.

## LOW-LEVEL WASTE

The NRC is responsible for licensing the disposal of commercial low-level radioactive wastes in conformance with the environmental standards developed by EPA for such disposal. The proposed regulation for near-surface disposal of low-level wastes (10 CFR Part 61, published July 24, 1981 in draft form for public comment) requires that, with one exception, disposal will be in the unsaturated zone above fluctuations of the water table. In addition, the proposed regulation requires covers over the disposal units which are designed to minimize infiltration, direct percolating water away from the disposed waste, and resist degradation by surface geologic processes and biotic activity. Further, the disposal site must be designed to minimize the contact of percolating or standing water with the wastes after disposal. In light of these requirements, NRC-funded low-level waste projects related to partially-saturated media, listed in Table II, encompass performance assessment, site characteristics, and facility design.

The first waste management project, being performed at Battelle Pacific Northwest Laboratories, covers two subjects, (1) a review of the state-of-the-art of modeling moisture movement and solute transport in partially-saturated media, and (2) a review of available methodologies for paleohydrologic evaluations. The review of models included a symposium on unsaturated flow and transport modeling in Seattle, Washington on March 23-24, 1982. Proceedings of the symposium were published as NUREG/CP-0030, PNL-SA10325, 1982. The review of models will conclude with detailed code summaries for models recommended for NRC use in review of low-level waste disposal facilities. The recommendations will be primarily based on the degree of documentation and previous applications of the codes. The review of methodologies for paleohydrologic evaluations will include the applicability of, data requirements for, and results of field studies using the various methodologies.

Table II. Low-Level Waste Projects

| |
|---|
| **Division of Waste Management** |
| Unsaturated Zone Hydrology (PNL) |
| Model Documentation and In-House Training for Unsaturated Flow and Transport Modeling (Undesignated) |
| Field Studies and Modeling of Chemical Processes in the Unsaturated Zone (LANL) |
| Trench Cap Study (ISGS) |
| Trench Cap Hydraulic Monitoring by Tracers (U. of Arizona) |
| **Office of Nuclear Regulatory Research** |
| Burial Ground Design, Construction, and Operating Practice (U. of Arizona) |
| Unsaturated Zone Hydrology - Maxey Flats, KY (UCLA) |

The project on model documentation and in-house training is planned for FY83. A request for proposals was advertised in Fall, 1982 in Commerce Business Daily; and, the Contractor will be selected by the competitive process. The purpose of this project is to develop staff expertise with a group of prequalified codes (centered on those identified in the previous project) and to make these codes, complete with documentation, user's manual, and sample problems, available to potential licensees, reviewers, or interested parties. The codes will be set up on NRC-leased computer facilities using the CDC 6600/7600 at Brookhaven National Laboratory. Magnetic tapes of the codes will be available through the NRC's Office of Management and Program Analysis.

The project on chemical processes in the unsaturated zone is just beginning and will use clustered lysimeters in the experimental Engineered Test Facility at Los Alamos National Laboratory (DePoorter [3], DePoorter and Hakonson [4]). This study consists of two major tasks, field and laboratory column studies of (1) the relationship between wetting fronts and leaching of wastes, and (2) chemical processes affecting solute transport in partially-saturated media. Specific questions to be addressed in the first task include: (1) Do significant releases occur only during the passage of wetting fronts or are there continual releases of significance due to moisture movement in partially-saturated soil? (2) Does a wetting front become contaminated primarily by flushing contaminated, capillary-held moisture from the surface of the wastes or by leaching of the wastes during the time of contact of the wetting front with the wastes? (3) Does an increased frequency and/or volume of water passing through successive wetting fronts result in a dilution of the leachate? and (4) Does the assumption frequently used in modeling of continuous infiltration distributed evenly throughout the year result in higher or lower estimated leaching of the wastes than that estimated by sporadic infiltration events (wetting fronts) with the same cumulative amount of infiltration? The second task attempts to quantify the effects of chemical processes, such as precipitation/coprecipitation, on the transport of reactive solutes in the unsaturated zone, identify the relative affect of chemical processes when compared to adsorptive processes, and improve the basis for estimating the combined effects of processes which remove radionuclides from the liquid phase to the solid phase during migration (frequently input to modeling as the distribution coefficient).

The trench cap study, currently being performed by the Illinois State Geological Survey (ISGS), consists of a systematic evaluation of infiltration-limiting trench cap covers, especially multiple-layered systems employing the wick effect. The first task report, NUREG/CR-2478, Volume 1, A Study of Trench Covers to Minimize Infiltration at Waste Disposal Sites: Review of Present Practices and Annotated Bibliography, has been completed. Studies in progress include computer simulations of trench cap covers and laboratory testing of soil columns representing various trench cap covers. Full-scale field testing of the most promising covers should be underway this summer. This study is the subject to three papers presented later in the volume by Thomas Johnson, Stephen Klein, and Henry Harris.

The project on hydraulic monitoring of trench caps is planned for FY83 and consists of using fluorinated organic tracers to monitor moisture movement through the field-scale trench cap covers in the previous project. The work will be performed by the University of Arizona to confirm measurements of moisture movement using tensiometers and to provide a basis to compare the two techniques for monitoring moisture movement.

The first research project, dealing with burial design, construction and operating practice, is also being performed at the University of Arizona. This project compliments the ISGS trench cap study by investigating the soil beam or arching capability of various trench cap designs. Hydraulic monitoring by tracers and settlement monitoring are being used to define the failure of the trench cap covers in terms of increased infiltration and deflection as the trench caps settle over catastrophically-generated or progressively-enlarging voids under the trench cap covers.

The project on unsaturated zone hydrology at Maxey Flats is part of a group of studies being performed at the closed Maxey Flats, Kentucky low-level waste disposal site. The purpose of these studies is to assist the State of Kentucky in developing the technical data necessary to evaluate site performance and to decommission the site. The unsaturated zone hydrology study has determined the method and rate of moisture movement through the trench caps and has provided sampling locations for the study of tritium migration upward through the trench caps. The study is now being used to demonstrate the effectiveness of a plastic cover in limiting infiltration. In addition, lysimeter studies are underway to evaluate the effectiveness of vegetative cover and management practices in limiting infiltration. Studies of tritium translocation by evaporation and plant transpiration are also in progress.

## URANIUM RECOVERY PROJECTS

The NRC is responsible for licensing source and byproduct material. The licensing requirements are contained in 10 CFR Part 40, which defines byproduct material to be the tailings or wastes produced by the extraction of uranium or thorium from any ore processed primarily for its source material content. This includes discrete surface wastes resulting from uranium solution extraction processes.

Only one uranium-recovery research project has been listed in Table III as being applicable to partially-saturated media. As part of that project, being performed at Battelle Pacific Northwest Laboratories, NUREG/CR-2360, TRUST: A Computer Program for Variably Saturated Flow in Multidimensional Deformable Media, was published.

Table III. Uranium Recovery Projects

| |
|---|
| Division of Waste Management |
| None |
| Office of Nuclear Regulatory Research |
| Assessment of Leachate Movement from Ponded Tailings (PNL) |

## SUMMARY

The NRC is currently funding, through the Division of Waste Management and the Office of Nuclear Regulatory Research, a diverse group of studies related to moisture movement and solute transport in partially-saturated media. Few reports have been finalized for public distribution at this time; however, final task and/or project reports will be forthcoming as NUREG documents for each project. Notices of the availability of these reports can be received by sending a request for such notification to Docket Control Center, Division of Waste Management, U.S. Nuclear Regulatory Commission, Washington, D.C. 20555. Once available, copies of the NUREG reports can be purchased from the NRC/GPO Sales Program Office, Division of Technical Information and Document Control, U.S. Nuclear Regulatory Commission, Washington, D.C. 20555.

Additional projects related to disposal of radioactive waste in the unsaturated zone have been and will continue to be identified by the NRC staff. It is the intent of the NRC staff to make the technical community aware of the current, and future, technical assistance and research projects and to broadly disseminate the results of the projects in order to assure more environmentally-acceptable waste disposal sites.

## REFERENCES

1. Panel on Hanford wastes, "Radioactive Wastes at the Hanford Reservation - A Technical Review," National Research Council Committee on Radioactive Waste Management (1978).
2. Winograd, I., "Radioactive Waste Disposal in Thick Unsaturated Zones," *Science*, 212(4502): 1457-1464(1981).

3. DePoorter, G. L., "The LASL Experimental Engineered Waste Burial Facility: Design Considerations and Preliminary Plans," in *Scientific Basis for Nuclear Waste Management, Vol. 3*, J. Moore, Ed. (Plenum Publishing Corporation, 1981), pp. 259-266.
4. DePoorter, G. L. and Hakonson, T. E., "Novel Experiments for Understanding the Shallow Land Burial of Low-Level Radioactive Wastes," Los Alamos Scientific Laboratory Technical Report LA-UR 81-3411 (1981).

MODELING OF MOISTURE MOVEMENT THROUGH LAYERED TRENCH COVERS

T. M. Johnson, K. Cartwright, B. L. Herzog, and T. H. Larson
Illinois State Geological Survey
Champaign, Illinois 61820

## INTRODUCTION

Low-level radioactive wastes in the United States are currently buried in trenches in 11 major shallow land-burial grounds. Six of these sites—two federal and four commercial—are located in the relatively wet eastern part of the country. Tritium migration has been observed at five of these sites, and is thought to be caused by infiltration of precipitation through trench covers.

Trench covers at all 11 sites consist of material excavated from the trenches. The material is typically mounded to a depth of 1-3 m and planted with grass. At sites in relatively wet regions, covers are compacted by earthmoving equipment, or material excavated from an adjacent trench is temporarily placed on the cover. At some sites, infiltration through trench covers has resulted in accumulation of water within the waste; at other sites, infiltration has caused migration of contaminants, although no free water has been observed in the trench.

In order to further reduce infiltration, two general types of infiltration-limiting covers have been proposed:

- Manmade "impermeable" covers (concrete, asphalt, membranes)
- Natural materials (clay-layered designs).

The use of layered covers consisting of soils of highly contrasting texture (utilizing what is often called the "wick effect") has been proposed. Several researchers [1,2] have shown that a coarse-grained layer underlying a fine-grained material acts as a barrier to infiltration. Experimental studies of infiltration into a column of dry loam or fine sand overlying dry gravel show that water entry into the gravel will not occur until the moisture content above the interface approaches saturation. These studies indicate that the greater the contrast in texture, the more effective the barrier. It has been proposed that a sloping interface between the fine- and coarse-grained material be used to produce lateral flow and prevent excessive moisture buildup in the upper layer. A second fine-grained layer would be used below the gravel to direct water away from the waste through drains, under saturated conditions.

Several researchers have proposed cover designs utilizing the wick effect [3,4,5]. These designs have been based on laboratory experiments and numerical simulations. Previous studies on the application of the wick effect to trench cover design have also often assumed that moisture movement into the coarse layer will not occur until the pressure head at the interface reaches 0.

The Illinois State Geological Survey, under contract to the U.S. Nuclear Regulatory Commission, is currently studying the use of layered covers to limit infiltration at shallow land-burial waste disposal sites. A review of present trench cover practices with an annotated bibliography dealing with trench covers, was published previously [6]. The present paper discusses the preliminary results of computer modeling of infiltration through various types of cover designs.

## MATHEMATICAL MODELS

Numerous mathematical models have been developed to describe the flow of moisture through variably saturated porous media [7]. These models are based either on analytical solutions or on approximate numerical solutions to the equations governing moisture flow. The selection of materials for use in alternative cover designs in this study is based on the results of laboratory column experiments and numerical simulations. One-dimensional models were used initially to evaluate many different layered systems. Simulations of one-dimensional, vertical flow indicate the response of a variety of materials under simplified conditions (such as those encountered in laboratory column experiments). More sophisticated multi-dimensional models were used subsequently to test selected field-scale cover designs under a variety of simulated field conditions.

## ONE-DIMENSIONAL SIMULATIONS

Vertical infiltration into layered covers was studied initially by using a one-dimensional model of moisture flow developed by Rojstaczer [8] that incorporates a fully implicit, finite-difference approximation of Richard's equation. The computer program is entitled MOMOLS (Moisture Movement in Layered Soils). Required data in MOMOLS include data from the pressure head, $\psi$, versus water content, $\theta$, curve for each material.

The measurement of $\theta$ versus $\psi$ is a straightforward procedure; however, the $K(\psi)$ relationship is difficult to measure directly. Soil-moisture characteristic curves of $\theta$ versus $\psi$ were obtained for each material during desaturation by using pressure cell, pressure plate, and pressure membrane tests. From these curves, MOMOLS calculated the hydraulic conductivity function, $K(\psi)$, using the Mualem method [9]. It is assumed that $\psi$ and K are single-valued functions of $\theta$ for this analysis; therefore, hysteresis is ignored. Philip's quasi-analytical method [10] was used to check the accuracy of the model.

We observed, during laboratory experiments, that soils compacted wet of optimum moisture content exhibited much different saturated hydraulic conductivities and soil moisture characteristic curves than did soils compacted dry of optimum. The physical characteristics of soils compacted wet and dry of optimum were included in the simulations to determine the influence on infiltration and percolation. Soil-moisture characteristic curves for samples of compacted loess, sand, and gravel used in the one-dimensional simulations are shown in Figures 1 & 2 (from Klein et al, this volume).

The MOMOLS program allows for either constant head or constant flux boundary conditions at the surface. The factors influencing moisture entry or infiltration into the surface layer under constant ponding ($\psi = 0$) at the surface are shown in Figure 3. More negative values of initial pressure head (tension) increase moisture entry because of the available pore space and significant pressure-head gradient. Moisture entry is also influenced by the compaction moisture content of fine-grained materials. Soils compacted wet of optimum exhibited much lower moisture entry and a much slower rate of moisture front advance (Figure 4) than did soils compacted dry of optimum.

If the texture of the intermediate layer in a layered cover is much coarser than the overlying layer, moisture will build up in the upper layer before breakthrough occurs into the coarse layer. The effects of a coarse layer of differing texture on moisture retention in a fine-grained upper layer are shown in Figures 5 and 6. The water content in a 30-cm layer of compacted Peoria Loess overlying a coarse gravel with a hydraulic conductivity of 1 cm/sec (Figure 5) builds up significantly prior to breakthrough after 2600 hours of continuous surface ponding. For comparison, the moisture content in a single layer of Peoria Loess is also shown. Figure 6 illustrates the more limited moisture retention that occurs in compacted Peoria Loess overlying a medium sand with

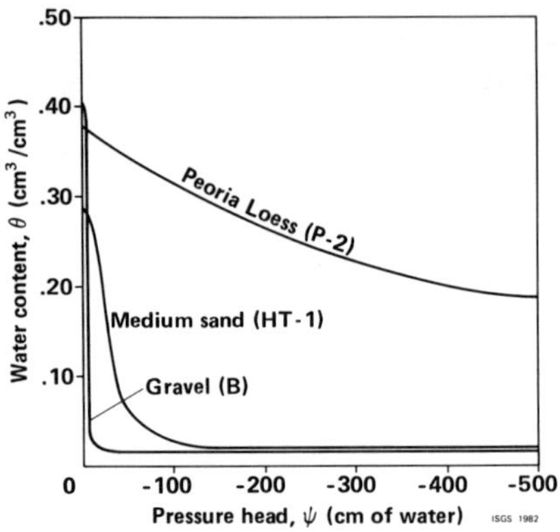


Figure 1. Soil water content as a function of pressure head for soils used in simulations.

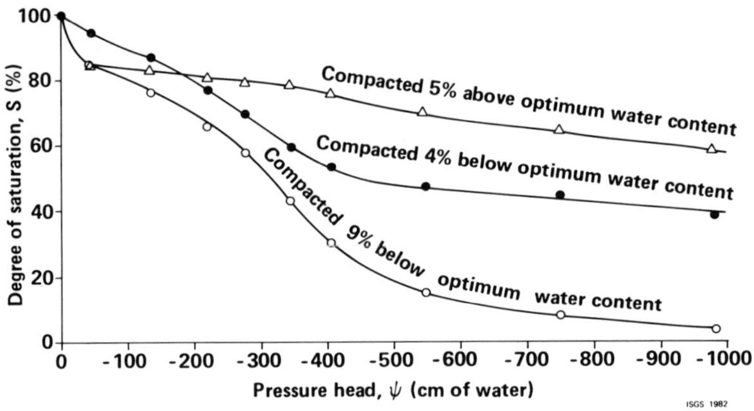


Figure 2. Samples of compacted Peoria Loess.

a hydraulic conductivity much less than that of the gravel. In this case, predicted moisture breakthrough occurs after only 940 hours; comparison with the single layer example indicates that there is little effect due to layering.

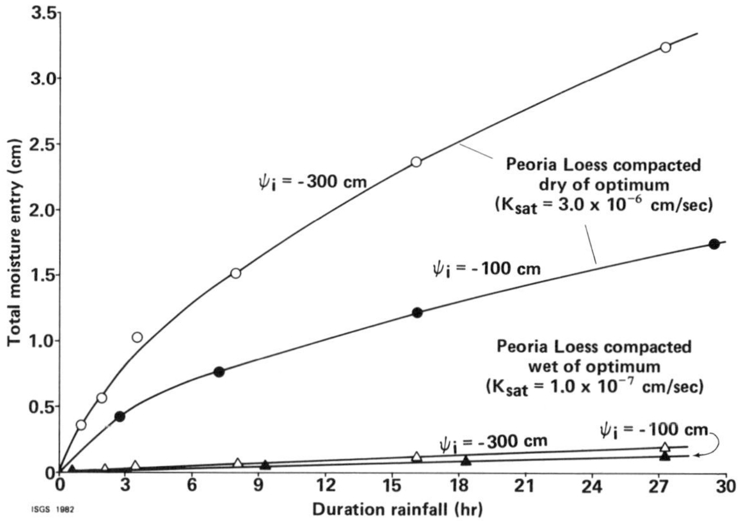


Figure 3. Total moisture entry into compacted Peoria Loess under differing initial conditions.

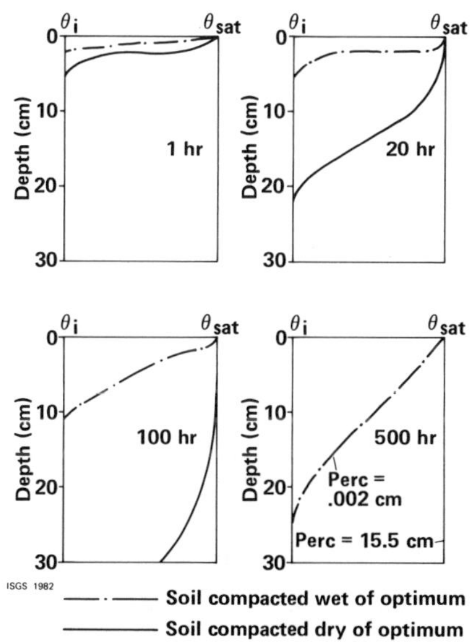


Figure 4. Predicted advance of wetting front in columns of Peoria Loess compacted wet and dry of optimum under constant surface ponding. Initial moisture content = θ; saturated moisture content = θ sat.

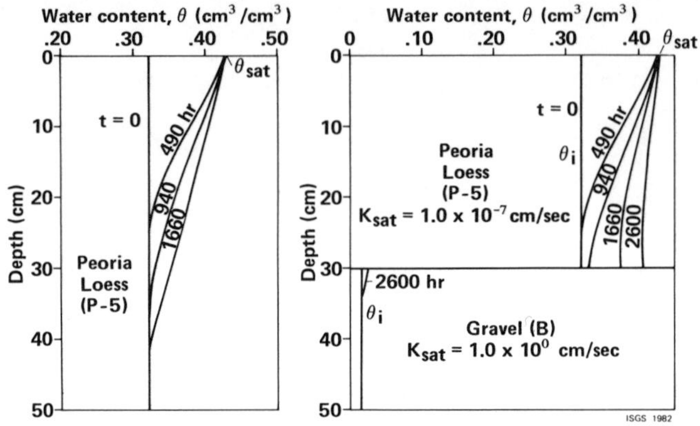


Figure 5. Predicted effect of a coarse-grained layer of gravel on moisture retention in the overlying fine-grained material as compared to moisture movement through a single layer of fine-grained material.

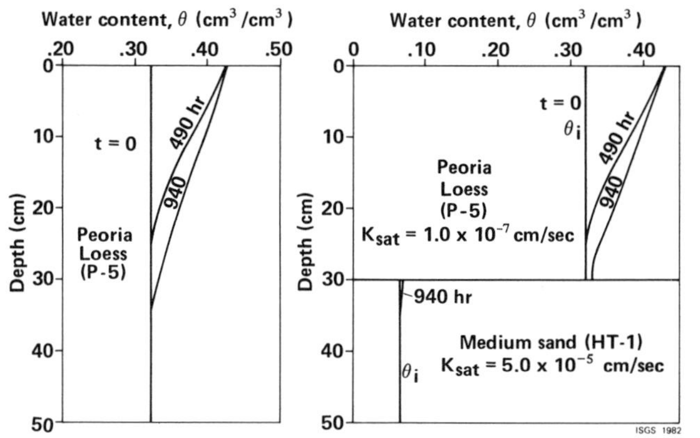


Figure 6. Predicted effect of a layer of medium sand on moisture retention in the overlying fine-grained material as compared to moisture movement through a single layer of fine-grained material.

To summarize, the one-dimensional simulations indicate that coarse, unsaturated material in layered systems can impede moisture movement. In all simulations, progress of the wetting front temporarily or permanently ceased once the interface between the gravel and the overlying fine-grained material was reached. The lack of moisture movement across the interface caused the pressure head and moisture content to build up in the overlying layer. Pressure head and moisture content increased until either a static profile was established or the interfacial pressure head reached a critical maximum value, $\psi_c$. If $\psi_c$ was reached at the interface, the moisture front moved rapidly into the coarse layer. The critical pressure ($\psi_c$) at which moisture breakthrough occurred was still negative (unsaturated) in all simulations, indicating that moisture movement through such layered systems can occur when the moisture content of the overlying layer is less than saturation.

## TWO-DIMENSIONAL SIMULATIONS

For the simulation of moisture movement through laboratory columns and field-scale cover designs, the finite-element model, UNSAT2, developed by Neuman [11] and Neuman et al. [12] was used. The model uses a Galerkin-type finite element method to simulate two-dimensional, nonsteady flow of water in incompressible, saturated-unsaturated soils; evaporation, water uptake by roots, and all types of boundary conditions are considered in this model. The theoretical development of the model is presented by Neuman et al. [12], and the results of verification of the model are described in Feddes et al. [13].

UNSAT2 incorporates several types of boundary conditions, including seepage faces. In the simulations described here, boundary conditions at the surface consisted of continuous rainfall (constant surface flux) and continuous ponding (constant pressure head, $\psi = 0$). At the bottom, boundary conditions included either a constant pressure head, $\psi = 0$ (water table), a prescribed constant negative pressure head, or a seepage face. The sides of both the column and field scale simulations are no-flow boundaries.

A seepage face is an atmospheric pressure boundary along which water seeps out from the saturated portion of the exposed face. A seepage face could occur at the bottom of a landfill cover if the waste remains unsaturated and if drains within the waste maintain atmospheric pressures.

Soil moisture characteristic curves of $\theta$ versus $\psi$ determined from laboratory tests for the materials in these simulations are presented in Figure 2. Hydraulic conductivity, K, as a function of water content, $\theta$, was calculated for each material, using the method of Green and Corey [14] (Figure 7).

## Laboratory Column-Scale Simulations

Simulations of a soil column 29 cm long and 8 cm wide were conducted to compare the predicted response with actual measurements of moisture content in a laboratory soil column determined by gamma-ray attenuation. The soil column consisted of 11 cm of Peoria Loess (P-2), overlain by 9 cm of medium sand (HT-1), in turn overlain by 9 cm of Peoria Loess (P-2).

The results of one laboratory column simulation are presented here. In this case, constant ponding occurs at the surface and a seepage face is presumed to be present at the base of the column. The predicted vertical distributions of pressure head and water content, at selected times, in the central portion of the column are shown in Figures 8 and 9. Pressure head and moisture content increase somewhat in the upper layer and at the interface with the underlying sand; however, moisture breakthrough into the sand occurs after only 3 hours of continuous ponding. Predicted saturation of the bottom layer and the lower portion of the sand occurs after only 130 hours. The effectiveness of the sand as a barrier to moisture movement in the column simulations is limited because of its relatively low hydraulic conductivity and consequently low contrast with the hydraulic conductivity of the overlying loess.

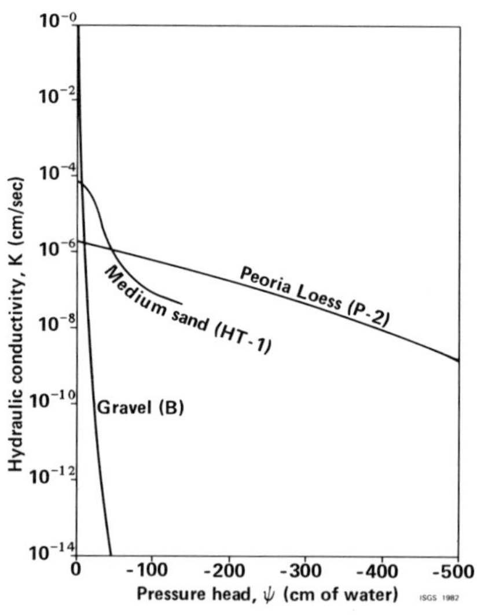


Figure 7. Hydraulic conductivity as a function of pressure head for soils used in simulations. Calculations of hydraulic conductivity based on method of Green and Corey [14].

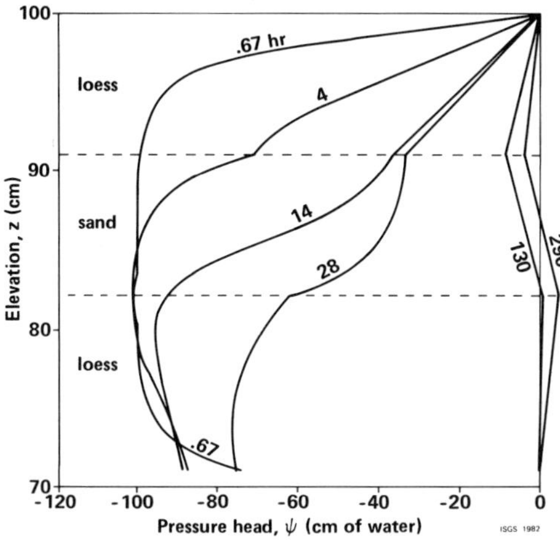


Figure 8. Predicted distribution of pressure head at selected times in the 3-layer column under conditions of constant ponding at the surface and a seepage face at the bottom.

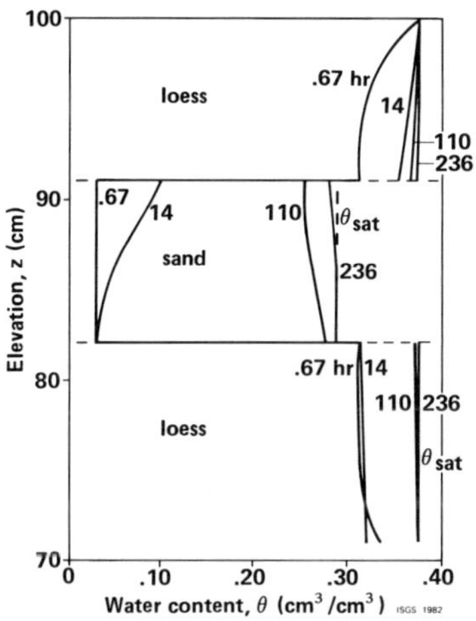


Figure 9. Predicted distribution of water content at selected times in a 3-layer column under conditions of constant ponding at the surface and a seepage face at the bottom.

## Field-Scale Cover Simulation

Infiltration and moisture movement through a hypothetical field-scale layered cover was simulated, using the preliminary design shown in Figure 10. This initial design consists of three layers, each 60 cm thick. For comparison, the behavior of a single layer cover 180 cm thick was also simulated. The cover is assumed to be 6 m wide with each layer sloping 5 percent (2.9°) from the center to the sides. (We anticipate that this will be the approximate configuration of pilot-scale covers to be constructed in the fall of 1982.) Since the hypothetical cover design in Figure 10 is symmetrical, moisture movement was modeled through only half of the cover. The region of interest is divided into 108 quadrilateral elements and 133 nodes. The sides of the cover are assumed to be no-flow boundaries. A constant pressure head of zero, indicating continual ponding, is imposed on the surface of the cover, while the boundary condition at the bottom varies in each case. Initial conditions within the cover consist of a constant pressure head of -100 cm. The magnitude of the first time step is 1 hour; subsequent time steps increase to a maximum of 3 hours.

The materials used in the field-scale cover simulations consist of compacted Peoria Loess (P-2) and gravel (B). The saturated hydraulic conductivities of the loess and gravel are $2.0 \times 10^{-6}$ cm/sec and 1.0 cm/sec, respectively. The soil-moisture characteristic curves for each material are presented in Figure 2; Figure 7 illustrates the relationship between K and $\psi$, calculated by using the method of Green and Corey [14].

When a 180-cm single layer cover of loess is simulated, the predicted wetting front propagates steadily downward as expected to a depth of more than 140 cm after 297 hours of ponding. The introduction of a 60-cm intermediate layer of unsaturated gravel (B) has a significant effect on moisture through the cover. As was predicted by the one-dimensional simulations, moisture movement through the loess overlying the gravel temporarily ceases once the interface is reached. Moisture content and pressure head increase in the overlying layer until the interfacial pressure reaches a critical maximum value, $\psi_c$. If $\psi_c$ is reached at the interface, the wetting front progresses rapidly into and through the gravel layer. After moisture breakthrough, the pressure head at the interface and in the underlying gravel layer remains approximately equal to $\psi_c$ for the remainder of the simulation.

The effects of a layered cover on infiltration are graphically illustrated in Figure 11, in which the predicted vertical distribution of pressure head for the column of nodes at selected times at x = 100 cm in the 3-layer cover is the predicted increase in pressure head (and water content)

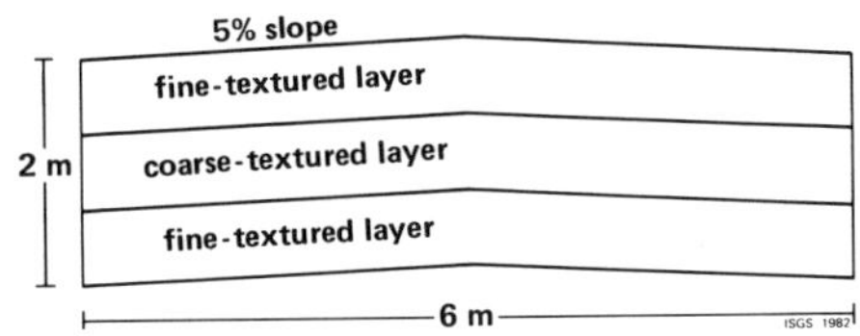


Figure 10. Cross section of hypothetical cover design used in two-dimensional simulations.

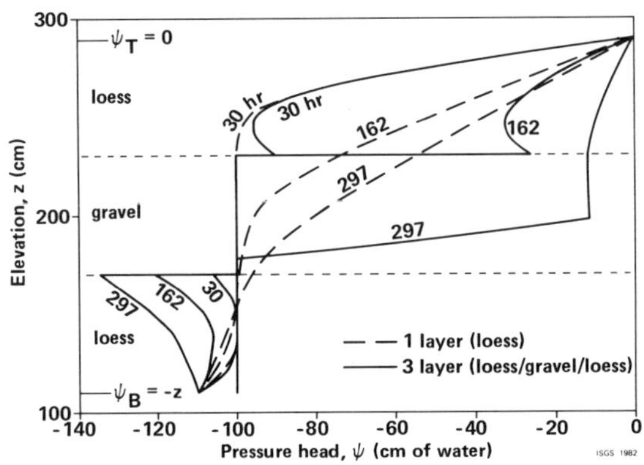


Figure 11. Comparison of predicted distributions of pressure head at selected times in a 1-layer cover of Peoria Loess (P-2) and in a 3-layer cover of Peoria Loess (P-2), gravel (B), and Peoria Loess (P-2) at x = 100 cm under conditions of constant ponding at the surface and constant pressure head = -z at the bottom.

above the gravel layer and lack of movement into the gravel for more than 162 hours of ponding. It should be noted, however, that following breakthrough, the wetting front progresses much more rapidly through the gravel than in the 1-layer model.

Figures 12 and 13 show the predicted vertical distribution of pressure head and water content, at selected times, with a seepage face at the bottom for the column of nodes at the center of the cover at x = 300 cm. The seepage face behaves as a no-flow boundary as long as the boundary remains unsaturated; however, once saturated, it allows seepage in response to hydraulic gradients and atmospheric conditions.

As in the other simulations involving layered covers, pressure head and moisture content increase in the upper layer until relatively sudden breakthrough into the gravel occurs after approximately 180 hours of ponding. The pressure head in the gravel then attains a relatively constant value of approximately -12 cm ($\psi_c$), equal to that at the interface with the overlying loess. The moisture content of the gravel remains quite low, however, and increases significantly only after the wetting front has penetrated the entire system. Pressure head in the bottom layer of loess, which initially was assumed to be -100 cm throughout, attains a steady-state distribution until the wetting front reaches the bottom layer after 286 hours.

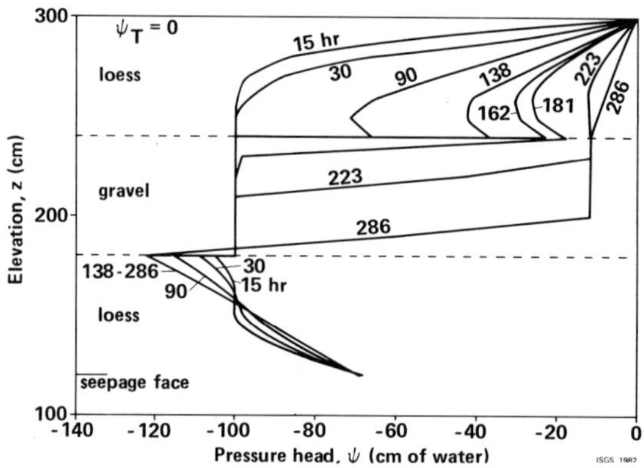


Figure 12. Predicted distribution of pressure head at selected times in the 3-layer cover at x = 300 cm under conditions of constant ponding at the surface and a seepage face at the bottom.

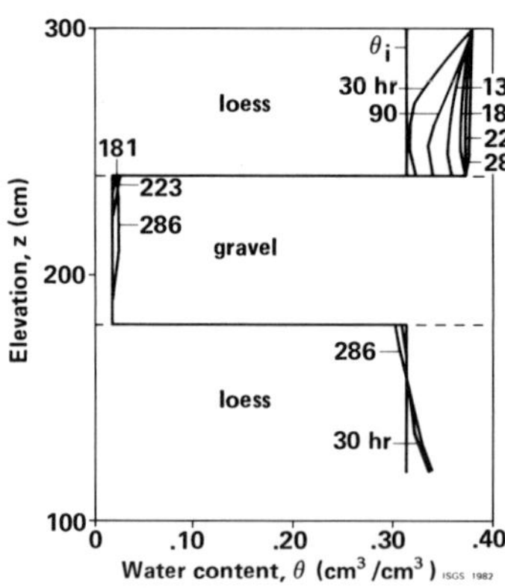


Figure 13. Predicted distribution of soil water content at selected times in the 3-layer cover at x = 300 cm under conditions of constant ponding at the sur- and a seepage face at the bottom.

The layered cover system has a slope of 5 percent (2.9°). The effects of this slope on moisture flow are illustrated in Figures 14, 15, and 16, which show the predicted distribution of total hydraulic head in the cover. Initial hydraulic head conditions at the onset of ponding at the surface are shown in Figure 14. The initial gradient associated with the assumption of a constant initial pressure head for the entire system is apparent. The total head distribution after 171 hours (Figure 15) shows conditions immediately prior to moisture breakthrough into the gravel layer. The vertical gradient at the interface of the upper loess and gravel is very sharp. The presence of a lateral component of flow in response to the slope of the cover is also apparent in both the upper and lower layers. Although still unsaturated, a portion of the moisture which builds up in the upper layer of loess begins to move laterally through the loess above the gravel. The no-flow boundary imposed along the edge of the cover imposed by the model stops this lateral flow; however, in practice, it may be possible to direct this lateral movement of water away from the waste through the edge of the cover. Future modeling efforts and field studies will investigate this phenomenon. The lateral gradient in the bottom layer is due to the slope and no-flow nature of the unsaturated seepage face boundary.

Figure 16 shows the predicted distribution of total head after 289 hours. The wetting front has penetrated the gravel layer and there is no longer any noticeable lateral component of flow in loess overlying the gravel. There is, however, some apparent lateral flow within the gravel layer and in the underlying loess.

## CONCLUSIONS

The results of laboratory experiments and computer simulations of several preliminary cover designs indicate that a layer of coarse-textured, unsaturated material overlain by fine-grained material serves as a barrier to moisture movement. The effectiveness of the barrier is related to the contrast in saturated hydraulic conductivity and texture between the two layers. Our investigations also indicate that prior to breakthrough, moisture in the fine-grained layer overlying a coarse-textured layer in a sloping cover flows laterally downslope above the interface.

It has been suggested previously that saturation of the overlying layer is required before moisture breakthrough will occur in such layered systems. However, our results and those of Rojstaczer [8] and others indicate that moisture movement through layered systems of highly contrasting texture can occur when the moisture content of the overlying

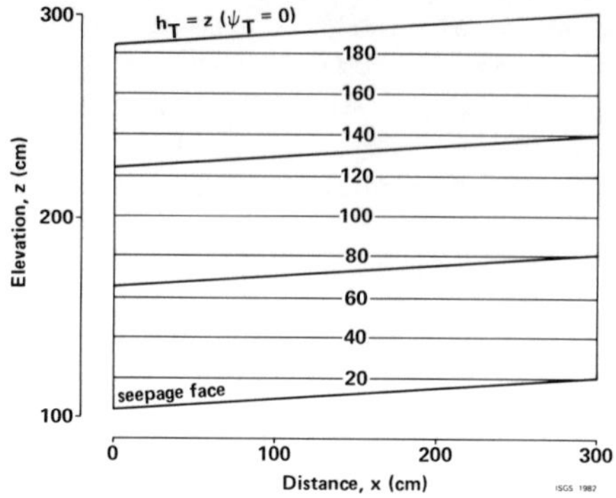


Figure 14. Initial distribution of total hydraulic head (cm) in 3-layer cover corresponding to a constant initial pressure head of -100 cm.

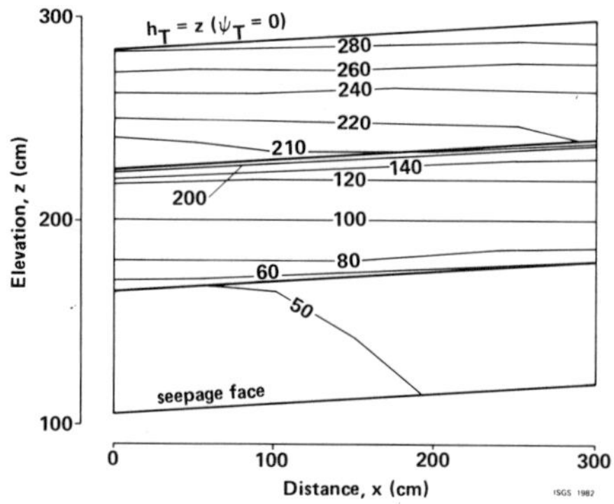


Figure 15. Predicted distribution of total head (cm) after 171 hr in the 3-layer cover under conditions of constant ponding at the surface and a seepage face at the bottom.

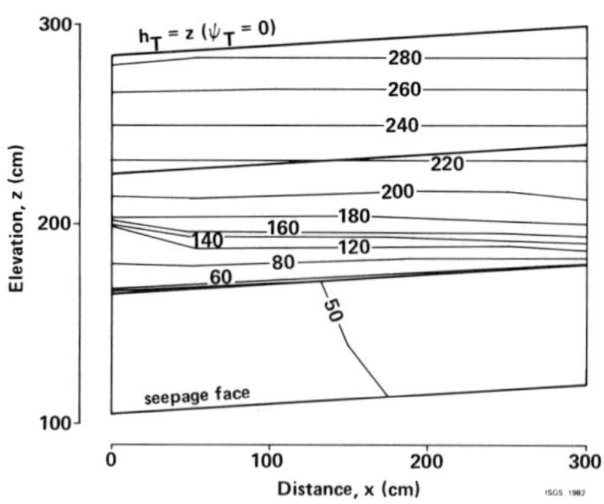


Figure 16. Predicted distribution of total head (cm) after 289 hr in the 3-layer cover under conditions of constant ponding at the surface and a seepage face at the bottom.

layer is less than saturation and the pressure head at the interface is less than zero. The significance of these results must be evaluated in terms of observed behavior in laboratory columns, field experiments, and present capability of instrument measurement of the parameters of interest.

## ACKNOWLEDGEMENTS

This study was supported, in part, by funds provided through the University of Illinois by the U.S. Nuclear Regulatory Commission, contract no. NRC 02-80-074.

## REFERENCES

1. Horton, J. H., and R. H. Hawkins. "Flow Path of Rain from the Soil Surface to the Water Table." *Soil Science* 100(6):377-383 (1965).
2. Corey, J. C., and J. H. Horton. "Influence of Gravel Layers on Soil Moisture Content and Flow." DuPont Savannah River Laboratory, DP-1160, 23 p. (1969).
3. Frind, E. O., R. W. Gilham, and J. F. Pickens. "Application of Unsaturated Flow Properties in the Design of Geologic Environments for Radioactive Waste Storage Facilities." in *Finite Element Simulation in Surface and Subsurface Hydrology*, G. Pinder and W. Gray, Ed. (Academic Press, New York, N.Y., 1977), pp. 3.133-3.163.
4. Rancon, D. "Application of the Capillary Barrier Technique to Storage in Trenches." in *Underground Disposal of Radioactive Wastes*, v. 1, (International Atomic Energy Agency, Vienna, Austria, 1980), pp. 241-265.
5. Lu, A. H., S. J. Phillips, and M. R. Adams. "Finite Element Model Evaluation of Barrier Configurations to Reduce Infiltration into Waste Storage Disposal Structures: Preliminary Results and Design Considerations." Rockwell Hanford Operations, RHO-SA-255, Richland, WA, 20 p. (1982).
6. Herzog, B. L., K. Cartwright, T. M. Johnson, and H.J.H. Harris. "A Study of Trench Covers to Minimize Infiltration at Waste Disposal Sites: Review of Present Practices and Annotated Bibliography." U.S. Nuclear Regulatory Commission, NUREG/CR-2478, 245 p. (1982).
7. Lappala, E. G. "Modeling of Water and Solute Transport Under Variably Saturated Conditions: State of the Art." in *Proceedings of the Interagency Workshop on Radioactive Waste Modeling*, Denver, CO, 57 p. (1980).
8. Rojstaczer, S. A. "Moisture Movement Through Layered Soils of Highly Contrasting Texture." Univ. of Illinois, Unpub. M.S. Th., 86 p. (1981).

9. Mualem, Y. "Hydraulic Conductivity of Unsaturated Porous Media: Generalized Macroscopic Approach." *Water Resources Research* 14(2):325-334 (1978).
10. Philip, J. R. "The Theory of Infiltration: 1. The Infiltration Equation and Its Solution." *Soil Science* 83:354-357 (1957).
11. Neuman, S. P. "Saturated-Unsaturated Seepage by Finite Elements." *Jour. Hydraul. Div. Am. Soc. Civ. Eng.*, 99(HY12):2233-2250 (1973).
12. Neuman, S. P., R. A. Feddes, and E. Bresler. "Finite Element Analysis of Two-Dimensional Flow in Soils Considering Water Uptake by Roots: I. Theory." *Soil Sci. Soc. Amer. Proc.*, 39:244-230 (1975).
13. Feddes, R. A., S. P. Neuman, and E. Bresler. "Finite Element Analysis of Two-Dimensional Flow in Soils Considering Water Uptake by Roots: II. Field Applications." *Soil Sci. Soc. Amer. Proc.*, 39:231-237 (1975).
14. Green, R. E., and J. C. Corey. "Calculation of Hydraulic Conductivity: A Further Evaluation of Some Predictive Methods." *Soil Sci. Soc. Amer. Proc.*, 35:3-8 (1971).

# A MODEL TO SIMULATE INFILTRATION OF RAINWATER THROUGH THE COVER OF A RADIOACTIVE WASTE TRENCH UNDER SATURATED AND UNSATURATED CONDITIONS

CHENG Y. HUNG
OFFICE OF RADIATION PROGRAMS
U.S. ENVIRONMENTAL PROTECTION AGENCY
WASHINGTON, D.C. 20460

## ABSTRACT

This paper presents a mathematical model which simulates the infiltration of rainwater through a homogeneous earth cover over radioactive, chemical, or municipal waste disposal trenches under saturated and unsaturated conditions. The infiltration model includes overland flow, subsurface flow, and atmospheric diffusion systems. Space dependent variables of the basic differential equations which govern the movement of water or vapor through these three systems were transformed into space independent variables by introducing some engineering assumptions. This paper emphasizes the transformation of the subsurface flow system; the other two systems have been investigated by previous investigators.

Transformation of the subsurface flow system involves the concept of dividing soil moisture into three components: gravity, pellicular, and hygroscopic waters. The proposed model was tested against the studies coducted by other investigators by applying the model to the Barnwell radioactive waste disposal site. The results indicated close agreement between the results obtained from the proposed model and those obtained from other investigators.

## INTRODUCTION

The disposal of low-level radioactive wastes by the shallow land burial method has been used for decades and has proven to be an effective disposal method if the waste treatment, packaging, site selection, and disposal operations are managed properly. An important consideration in evaluating the performance of a waste disposal site is the potential effects on the health of downstream populations. A dynamic simulation model has been commonly used to predict this impact because limited experimental or field data are available. Since one of the major driving forces in accelerating the release of radionuclides is the rainwater which infiltrates through the cover of a disposal trench, the simulation of the infiltration process is an important part of the model for evaluating health effects.

An accurate model for simulating the infiltration of rainwater through a trench cover involves the modeling of three flow systems: an overland flow system, a subsurface flow system, and an atmospheric diffusion system. The overland flow system receives the rainwater and diverts the excess water from percolation and evaporation into the receiving drainage system. The subsurface flow system receives the percolated water from the overland flow system and transports the water either downward as infiltration into the trench and/or upward as evaporation into the atmospheric diffusion system. The atmospheric diffusion system receives water/vapor from the overland flow system or subsurface flow system and transports the vapor to the atmosphere. One of the early efforts in simulating this system for storm runoff was the Stanford Watershed Model conceived by Crawford and Linsley [1]. The same model was significantly improved by Moore and Claborn [2], Hung and Keifer [3], and others. However, these models were developed for water resources planning purposes with emphasis on the overland flow system. Significant improvements are required to apply the same models to the simulation of trench cover infiltration. This study presents an efficient and yet accurate infiltration model based on the dynamic equations governing the above systems.

## BASIC EQUATIONS

The simulation of rainwater movement through a homogeneous trench cover, defined here as the "total infiltration system," involves analyzing the hydrologic processes of overland flow, subsurface flow, and atmospheric diffusion systems. The basic momentum and continuity equations governing these three systems have been studied thoroughly by other investigators. They are detailed as follows:

### Overland Flow System

The one-dimensional momentum and continuity equations governing an overland flow system are expressed by (4):

$$\frac{1}{g}\frac{\partial u}{\partial t} + \frac{u}{g}\frac{\partial u}{\partial x} + \frac{\partial h}{\partial x} - \sin\alpha + \frac{n^2u^2}{h^{4/3}} = 0 \qquad (1)$$

and

$$\frac{\partial h}{\partial t} + h\frac{\partial u}{\partial x} + u\frac{\partial h}{\partial x} = P - E_o - q_o \qquad (2)$$

where $u$ = velocity of overland flow
$h$ = depth of flow
$\alpha$ = average inclination of the trench cover
$n$ = Manning's coefficient of roughness
$P$ = rate of precipitation
$E_o$ = rate of evaporation from the overland flow system
$q_o$ = rate of percolation from the overland flow system
$g$ = acceleration potential due to gravitational force
$x$ = space coordinate along the slope of the trench cover
$t$ = time

The above equation is derived from the assumptions that Corioli's energy correction factor, Boussinesq's momentum correction factor, and Jaeger's hydrostatic pressure correction factor are all unity; the flow is one dimensional, i.e., the velocity component in the longitudinal direction predominates the velocity component in lateral directions, and the flow is a gradually varied flow. The above system equations cannot be solved independently and must be coupled with the basic equations governing the other two systems.

Because Manning's law of resistance is an empirical formula developed for open-channel flow, it is conceivable that the application of the same law to a thin overland flow may be questionable. However, comprehensive laboratory and field studies conducted by Takasao et al., [5-8] indicated that the Manning Law can also reasonably be applied to an overland flow system. This finding was confirmed by Foster et al. [9], Hjelmfelt [10], and others.

### Subsurface Flow System

The movement of soil moisture in a subsurface flow system can be either upward or downward depending on the direction of potential head gradient. In general, the moisture in the soil is simultaneously transported in both liquid and vapor phases. The basic equation governing this system has been derived by Currie [11] and Hillel [12]. The results of their studies for the momentum and continuity equations are summarized as follows:

$$q = -D_L(\theta)\frac{\partial \theta}{\partial z} + K(\theta) - \gamma D_V \frac{\partial \rho_V}{\partial z} \tag{3}$$

and

$$\frac{\partial \theta}{\partial t} = -\frac{\partial q}{\partial z} \tag{4}$$

where $q$ = flux of moisture in the vertical direction
$\theta$ = volumetric wetness of soil
$K$ = hydraulic conductivity
$D_L$ = hydraulic diffusivity
$\gamma$ = conversion factor for transforming the vapor flux into liquid water flux
$D_V$ = diffusivity for water vapor
$\rho_V$ = concentration of water vapor in the air-filled void

The above equations were based on an isothermal condition and assumed that both viscous flow in the liquid phase and diffusion of vapor are impelled by the force field of capillarity and gravity. No explicit account has been taken of osmotic or solute effects on vapor pressure. In reality, the upper layer of the soil surface is warmer during the daytime and is cooler than the deeper layer during nighttime. The movement of vapor due to the thermal gradient effect tends to be downward during the day and upward during the night. Therefore, the error due to the assumption that the system is isothermal may compensate each other within the same day and will not significantly affect the accuracy of simulation.

## Atmospheric Diffusion System

The atmospheric diffusion system transports the water vapor through the turbulent boundary layer into the atmosphere. The vapor transport through an unsteady turbulent boundary layer involves the analysis of time-averaged boundary layer equations for the air stream and the continuity equation for the water vapor. This analysis is extremely complicated. However, by assuming the system is quasi-steady, the solution of the system equation becomes obtainable. The "quasi-steady" technique has been commonly used in solving unsteady mass transport systems. It assumes that within a small time increment the flow of the carrying fluid is steady and that the effect of the change from one flow state to the other due to the change in time step is negligible. By imposing the above assumption, the water vapor flux within the fully developed boundary layer was expressed by employing Fick's law for diffusion [13] as:

$$J = -k^2 y^2 \left| \frac{dv}{dy} \right| \frac{d\rho}{dy} \tag{5}$$

where J = water vapor flux
k = Prandtl's mixing length coefficient
v = time averaged wind speed
ρ = concentration of water vapor in the boundary layer
y = distance from ground surface.

## BASIC EQUATIONS FOR THE PRACTICAL MODEL

Theoretically, the basic equations compiled in the previous section can be solved numerically. However, the processes of analysis are complicated and consume excessive computation time. Several attempts have been made to solve a simplified system. For example, a conjunctive overland-subsurface flow model without atmospheric diffusion system has been developed by Akan and Yen [14], and a subsurface flow-atmospheric diffusion model without overland flow system was reported by Hillel [15, 16]. The above models solved the dependent variables as functions of space and time and involve long and costly calculations if the real time of simulation is prolonged. For example, the time increment used in Akan and Yen's overland flow system was 30 seconds which would require excessive computation costs if the simulation were for a year or more.

One may simplify the above system equations by transforming all of the space dependent variables into space independent variables to avoid time-consuming simulation. The transformation for each system is described in the following sections.

### Atmospheric Diffusion System

The solution of Equation 5 depends on many complicated fluid dynamic and boundary conditions. One of the simplest solutions is obtained by assuming that the vapor flux will not be limited by the availability of vapor transmitted from the subsurface system or the overland flow system. The vapor flux under this condition is known as the evaporation potential and is normally used as the upper bound of the evaporation rate from the overland flow and subsurface flow systems. The actual evaporation rate from these systems may then be calculated from the conjunctive system to be discussed latere. The solution of Equation 5 under the above condition was proposed by Rohwer [17] as:

$$E_p = 0.372(1-0.000374p_a)(1. + 0.6V_w)(e_s - e_a) \tag{6}$$

where $E_p$ = evaporation potential in mm/day
$p_a$ = atmospheric pressure in mb
$V_w$ = wind speed in m/sec
$e_s$ = saturated vapor pressure in mb
$e_a$ = vapor pressure in the atmosphere in mb

Equation 6 requires time dependent input for atmospheric pressure, wind velocity, and vapor pressure to compute the diurnal variation of evaporation potential. The above information is normally not readily available from existing records. Therefore, from a practical viewpoint, the computation of evaporation potential may be reduced to a daily average level by using the Hamon Equation [18]:

$$E_p = \eta r^2 s \tag{7}$$

where $\eta$ = coefficient of daily evaporation potential
$r$ = duration of daylight in 12-hour unit/day
$s$ = saturated absolute humidity corresponding to the daily average temperature

Equation 7 is used to calculate the evaporation potential which is the upper bound of the evaporation rate.

## Overland Flow System

It is well known that the local and convective acceleration terms in Equation 1 govern the wave form and the celerity of flood waves. Kichikawa [19] indicated that these terms are predominated by the friction term and suggested that they may be ignored. This conclusion was also confirmed by Takasao [5,6] in his comprehensive study of overland flow systems through laboratory and field observations. Takasao further simplified the system equations by replacing the space dependent depth of flow h by an average flow depth. After this modification, Equations 1 and 2 become:

$$Q_o = \frac{\sin^{1/2}\alpha}{n} H^{5/3} \tag{8}$$

and

$$\frac{dH}{dt} = P - E_o - q_o - \frac{Q_o}{L} \tag{9}$$

where $Q_o$ = rate of overland flow per unit width of trench cover
$H$ = average depth of overland flow over the entire trench cover
$L$ = length of slope (or half of trench width)

Furthermore, it is assumed that the rate of evaporation from the total infiltration system will be preferentially obtained from the overland flow system; then the component of evaporation rate may be written as:

$$\begin{aligned} E_o &= E_p, && \text{when } P + H/\Delta t > E_p \\ E_o &= P + H/\Delta t, && \text{when } E_p > P + H/\Delta t > 0 \\ E_o &= 0, && \text{when } P + H/\Delta t = 0 \end{aligned} \tag{10}$$

where $E_o$ = component of evaporation rate contributed by the overland flow system

On the other hand, since the maximum rate of percolation from the overland flow system can not exceed the saturated hydraulic conductivity of the trench cap, the percolation rate can be written as:

$$\begin{aligned} q_o &= K_s, && \text{when } P - E_o + H/\Delta t > K_s \\ q_o &= P - E_o + H/\Delta t, && \text{when } K_s > P - E_o + H/\Delta t > 0 \\ q_o &= 0, && \text{when } P - E_o + H/\Delta t = 0 \end{aligned} \tag{11}$$

The above expressions show that there are four equations available for solving four dependent variables, $Q_o$, $H$, $E_o$ and $q_o$. Therefore the system can be solved independently.

Subsurface Flow System

For the purpose of transforming the space dependent variables into space independent variables, one may subdivide the moisture contained in the system into three components: gravity, pellicular, and hygroscopic waters. Gravity water is the moisture in a soil which can be drained by gravity force; pellicular water is the moisture in a soil which cannot be drained by gravity force but can be lost to the atmosphere through natural evaporation; and hygroscopic water is the moisture which will never be lost through the above natural processes. Furthermore, it is assumed that soil wetness can be mathematically approximated by a step-wise distribution composed of these three components. Figure 1 shows a comparison of the step-wise wetness distribution and its corresponding original wetness distribution.

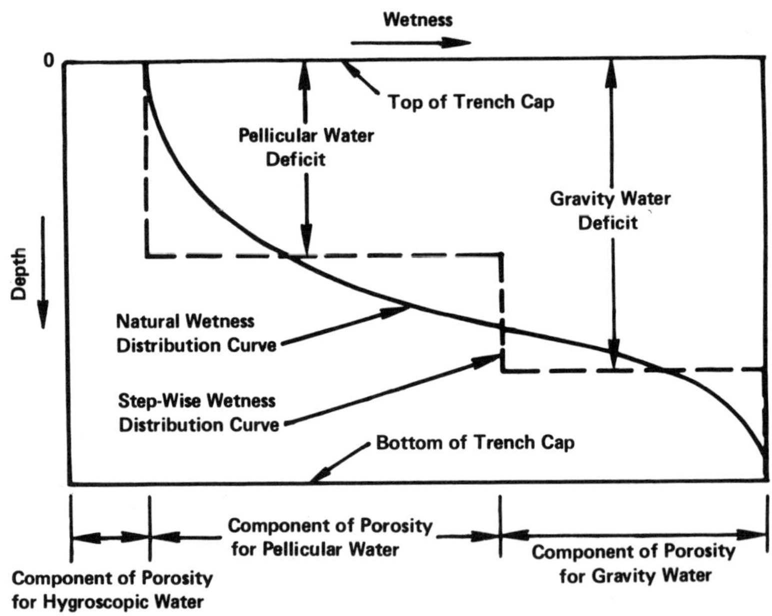


Figure 1. Comparison of a schematic natural wetness distribution curve and its corresponding step-wise wetness distribution curve.

Based on the above concept, one may subdivide the dependent variables appearing in the basic differential equations into three components, i.e.,

$$\theta = \theta_1 + \theta_2 + \theta_3$$
$$q = q_1 + (q_2 + q_t) + (q_3 - q_t) \qquad (12)$$

where $\theta_1$ = component of wetness for the hygroscopic water
$\theta_2$ = component of wetness for the pellicular water
$\theta_3$ = component of wetness for the gravity water
$q_1$ = component of flux for the hygroscopic water
$q_2$ = component of flux for the pellicular water
$q_3$ = component of flux for the gravity water
$q_t$ = flux of moisture being transformed from gravity water to pellicular water

Substituting Equation 12 into Equations 3 and 4, one obtains

$$q_3 = -D_L(\theta) \frac{\partial \theta_3}{\partial z} + K(\theta) \qquad (13)$$

$$(K(\theta) = 0 \text{ when } \theta_3 = 0)$$

$$\frac{\partial \theta_3}{\partial t} = - \frac{\partial (q_3 - q_t)}{\partial z} \qquad (14)$$

for the component of gravity water and

$$q_2 = -D_L(\theta) \frac{\partial \theta_2}{\partial z} + K(\theta) - \gamma D_v \frac{\partial \rho_v}{\partial z} \qquad (15)$$

$$(K(\theta) = 0 \text{ when } \theta_3 > 0)$$

$$\frac{\partial \theta_2}{\partial t} = - \frac{\partial (q_2 + q_t)}{\partial z} \qquad (16)$$

for the component of pellicular water.

Applying the step-wise wetness distribution concept to the above system equations, Equations 13 and 14 may be re-written by substituting the proper difference forms of the dependent variables. The result is

$$q_i = K_s \qquad \text{when } Z_g < Z_{max}$$
$$q_i = 0 \qquad \text{when } Z_g = Z_{max} \qquad (17)$$

and

$$\frac{dZ_g}{dt} = (q_i - q_o + q_t)/\theta_g \tag{18}$$

where $q_i$ = flux of moisture infiltrating into the trench
$K_s$ = saturated hydraulic conductivity of the soil
$Z_g$ = deficit of gravity water
$Z_{max}$ = maximum deficit of gravity water (equivalent to the thickness of the trench cover)
$\theta_g$ = component of wetness for the gravity water under a fully saturated condition and is numerically identical to the porosity for the gravity water

To simplify the solution of Equation 15, one may compute the component of moisture flux for the pellicular water $q_2$ based on the assumptions that (1) the flow is predominated by liquid phase transport, or (2) the flow is predominated by vapor phase transport. The moisture flux for the pellicular water at any instant is then chosen from the larger flux calculated from the above two assumptions.

When the flow is predominated by a liquid phase transport, the third term in Equation 15 vanishes. By substituting the proper difference form of the dependent variables, one obtains:

$$q_L = - D_e \theta_p / Z_p + K_e$$
$$q_L < E_p - E_o \tag{19}$$

where $q_L$ = flux of pellicular water transported in the liquid phase
$\theta_p$ = component of wetness for the pellicular water under fully saturated condition and is numerically identical to the porosity for pellicular water
$Z_p$ = deficit of the pellicular water
$D_e$ = hydraulic diffusivity at equivalent wetness
$K_e$ = hydraulic conductivity at equivalent wetness

The equivalent wetness for the purpose of this study is calculated by:

$$\theta_e = \theta_h + \sigma\theta_p \tag{20}$$

where $\theta_h$ = component of wetness for the hygroscopic water under fully saturated condition and is numerically identical to the porosity for hygroscopic water

$\sigma$ = coefficient of equivalent wetness for pellicular water which should be greater than 0.5 and smaller than 1.0

When the flow is predominated by vapor phase transport, the first and second terms on the righthand side of Equation 15 drop out. Substituting the proper difference form of the dependent variables, Equation 15 becomes

$$q_v = -\gamma D_v(\rho_s - \rho_o)/Z_p \tag{21}$$

where $q_v$ = flux of moisture being transported in the vapor phase

$D_v$ = diffusivity of water vapor in the trench cap

$Z_p$ = deficit of pellicular water

$\rho_s$ = water vapor concentration at the front of pellicular water

$\rho_o$ = water vapor concentrations at the top of trench cover

The diffusion of water vapor from the top of the trench cover to the atmosphere, on the other hand, can be written based on Fick's Law as:

$$q_v' = -\gamma D_a(\rho_o - \rho_a)/\delta \tag{22}$$

where $q_v'$ = flux of water vapor in the boundary layer of diffusion

$D_a$ = diffusivity of vapor in the boundary layer

$\delta$ = equivalent thickness of boundary layer

$\rho_a$ = water vapor concentration in the atmosphere outside the boundary layer

Since the vapor flux in the trench cover and in the boundary layer should be identical, Equations 21 and 20 may be combined to yield:

$$q_v = -\frac{\gamma D_a(\rho_s - \rho_o)/\delta}{1 + \dfrac{D_a}{\delta D_v} Z_p} \tag{23}$$

Furthermore, the ratio of diffusivities, $D_a/D_v$, may be obtained by introducing Penman's study [20] on the diffusion of vapor through a porous solid and Rohwer's study [17] on evaporation potentials. Substituting their results into Equation 23 and replacing its numerator with the evaporation potential, one obtains

$$q_v = - \frac{E_p - E_o}{1 + \frac{1 + 0.6V_w}{\delta} \frac{Z_p}{0.66(\theta_p + \theta_g)}} \qquad (24)$$

For the purpose of this study, the coefficient, $(1 + 0.6V_w)/\delta$, is assumed to be a constant and equal to $0.5\ m^{-1}$ (see next section on model testing). When this coefficient is used, Equation 24 becomes

$$q_v = - \frac{E_p - E_o}{1 + \frac{0.5Z_p}{0.66(\theta_p + \theta_g)}} \qquad (25)$$

Now, the flux of moisture being lost through evaporation $q_p$ can be computed by

$$q_p = - \text{Max}\ (|q_L|,\ |q_v|) \qquad (26)$$

where "Max" = a function designed to select the largest number among those values designated in the parentheses.

Again, by substituting the proper difference form of the dependent variables, Equation 16 becomes

$$dZ_p/dt = - (q_p + q_t)/\theta_p \qquad (27)$$

where $q_t$ may be computed by

$$q_t = q_o, \qquad \text{when } Z_p > 0$$
$$q_t = 0, \qquad \text{when } Z_p = 0 \qquad (28)$$

Equations 8-10, 11, 17-19, and 25-28 are the basic equations for the practical infiltration model. There are eleven equations and eleven dependent variables in the system equations. Since all of the dependent variables are space independent and each pair of momentum and continuity equations can be solved in sequence, the computer coding for the mathematical model is greatly simplified from that of the original system equations. A computer model was developed for testing the proposed system equations and for the application of the model to the evaluation of a low-level waste disposal site.

## MODEL TESTING

The overland flow system, using Equations 8 and 9 as basic equations, has been studied through laboratory studies and field observations by Takasao et al. [5-8]. They concluded that the space independent system equations, as represented by Equations 8 and 9, can reasonably be used to characterize the nature of overland flow for flood routing purposes. Since the primary purpose of this study is the simulation of infiltration rates, the proposed overland flow system equations are judged to be adequate for the infiltration model. Testing of the model was therefore concentrated on the subsurface flow system.

To test the subsurface flow system submodel, three special cases with simplified boundary conditions were selected. The results of simulation were compared with existing studies having the similar boundary conditions. They are described in the following sections.

### Gravity Water Drainage without Evaporation

A computer simulation of gravity water drainage by gravitational force without evaporation loss was conducted by Hillel and van Bavel [15]. This simulation was conducted by solving equations similar to the basic differential equations expressed in Equations 3 and 4 without the term representing the vapor phase transport. Their results of simulation for a sandy soil (1.16 m thick) were used to calculate the cumulative water volume being drained from the soil (see Figure 2).

By applying the same boundary conditions used in Hillel and van Bavel's study to the model developed for this study, the cumulative water volume drained from the soil is simulated and plotted in Figure 2. The comparison of the two results indicates that the proposed model does not simulate the time variation of the drainage precisely, but the calculation of the cumulative water volume drained

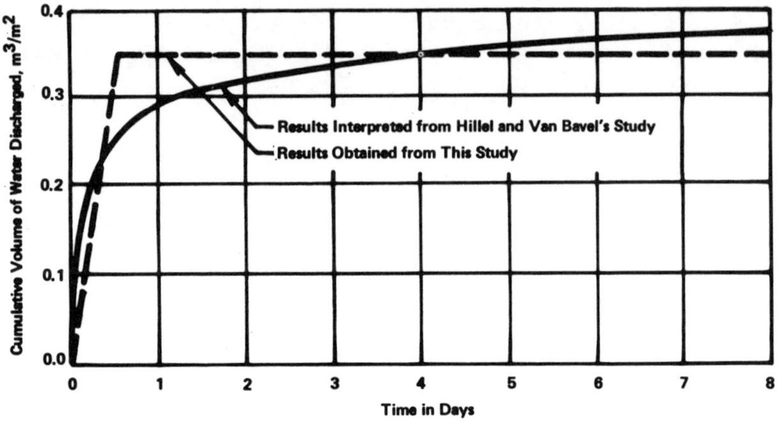


Figure 2. Comparison of the results of simulation for gravity drainage of a typical sand using the proposed model and the results interpreted from Hillel and Van Bavel's study.

from the soil is reasonably close to the results obtained from Hillel and van Bavel's simulation. Since the main purpose of the simulation described in this study is to obtain the cumulative volume of water infiltrated into the waste trench, the proposed model simulates the rate of infiltration with reasonable accuracy.

## Steady State Evaporation Rate

The steady state evaporation rate from a clay soil system with a relatively high groundwater table has been analyzed by Ripple [21]. This study analyzed the evaporation rate by solving the system equation characterizing the water flux transported in the subsurface flow system and the vapor flux diffused into the atmosphere. The results of his analyses for the groundwater table at depths of 0.6, 0.9, 1.2 and 1.8 meters are shown in Figure 3.

By applying similar parameters and boundary conditions to the proposed model and reorganizing the results into the form corresponding to Ripple's results, one obtains the results as shown in Figure 3. The results of simulation

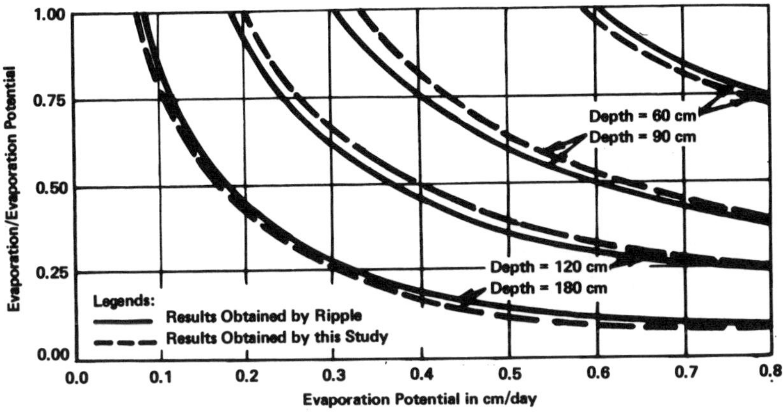


Figure 3. Comparison of the results of simulation for the steady state evaporation of Chino clay by using the proposed model and that obtained by Ripple.

from the proposed model fit reasonably well with those obtained by Ripple.

The above evaporation rate was simulated for a relatively shallow groundwater table. Therefore, the transport of soil moisture was predominated by liquid phase transport. However, when the groundwater table is very deep, the steady state rate of evaporation may be predominated by the vapor phase transport. The computation of vapor transport in Equation 25 becomes important under this condition.

## Vapor Phase Transport

There are no data available for evaluating the transport of pellicular water in a vapor phase. To compare the characteristics of vapor phase transport and liquid phase transport, the flux of moisture being transported by these two phases are computed for various depths of pellicular water deficit based on Equations 19, 25, and 26, using evaporation potentials of 0.3 and 0.6 cm/day. The results of analyses for both sandy and clay soils are plotted and presented in Figure 4. The author selected the coefficient represented by $(1. + 0.6V_w)/\hat{\delta}$ as 0.5 $m^{-1}$ to obtain the best transition from liquid phase transport to vapor phase transport for both soils and for the evaporation potentials which ranged from 0.3 to 0.6 cm/day. The trend of pellicular water flux variations with the increase in the pellicular water deficit agrees, in general, with the trend described by Philip [22].

## APPLICATION OF MODEL TO AN ACTUAL WASTE DIPOSAL SITE

The low-level radioactive waste disposal site located at Barnwell, South Carolina, was selected for an application of the model to an actual waste disposal site. The trench cover is constructed in two soil layers, loamy sand (120 cm thick) at the top and clay soil (80 cm thick) at the bottom. First, the analyses simulated the infiltration rates for a homogeneous trench cover for each of the soil types constituting the trench cover; then, the rate of infiltration for the composite trench section was interpreted from the results.

The soil characteristics for both layers were interpreted from basic data reported by Chem-Nuclear Systems, Inc. [23], and the results used for the computer simulation are listed in Table I.

Using the annual rainfall of 118 cm/year and the rainfall distribution pattern observed in Augusta, Geogia, the simulation was conducted for the loamy sand and the clay soil. The results of simulation, presented in Figure 5, indicate that the infiltration rates are 45 cm/year for the loamy sand and 7 cm/year for the clay soil. It is obvious that the rate of infiltration will not be affected by the soil characteristics below the level of annual maximum deficit for pellicular water plus the allowance for gravity water storage. The pellicular water deficit simulated for the loamy sand (Figure 6) indicated that the maximum pellicular water deficit reached the level of approximately 50 cm which is far above the upper boundary of the clay soil layer (120 cm below the top of trench cap). Thus, one may conclude that the infiltration rate for the Barnwell site is controlled by characteristics of the top layer, and the rate of infiltration is 45 cm/year or 17.7 in/year.

The above analyses assumed a uniform rainfall distribution within the hourly period when the rainfall was

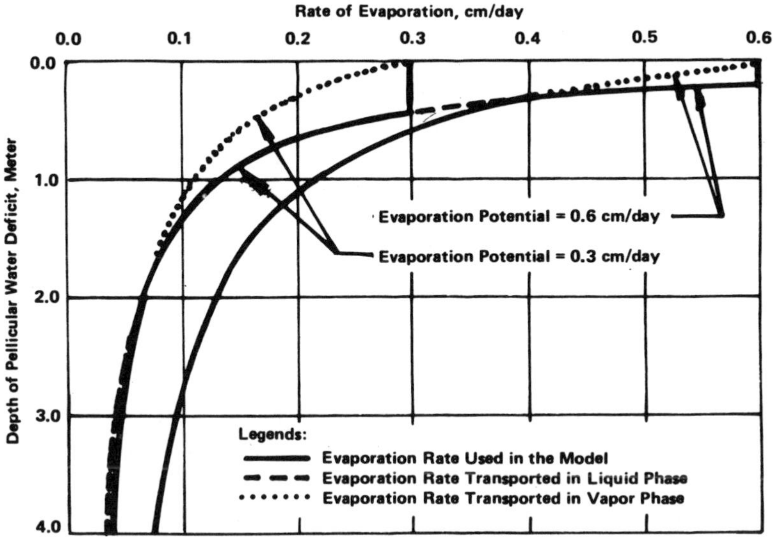


Sandy Soil

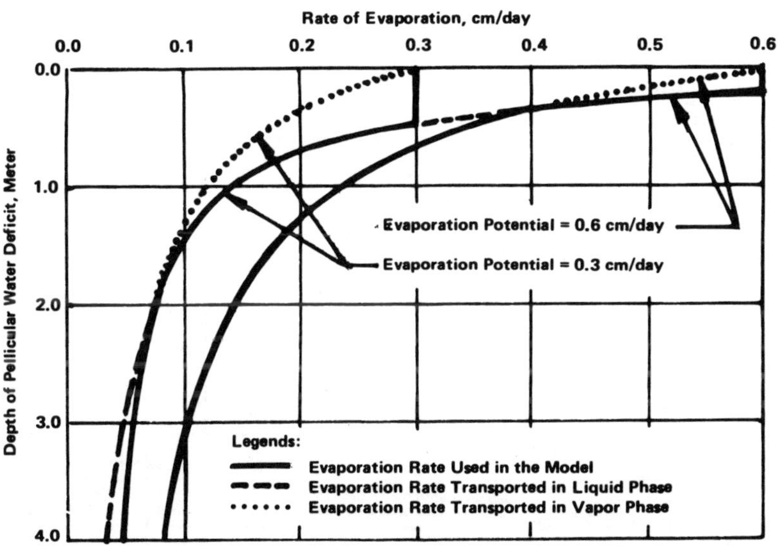


Clay Soil

Figure 4. Transitions in pellicular water transport from liquid phase to vapor phase calculated from Equations 19, 25, and 26 for sandy and clay soils.

Table I. Soil Characteristics of the Trench Cover Used in Simulating the Annual Rainfall Infiltration, Barnwell Radioactive Waste Disposal Site

| Type | Soil Characteristics | | | | | |
|---|---|---|---|---|---|---|
| | Saturated Hydraulic Conductivity (m/hr) | Equivalent Upward Diffusivity ($m^2$/hr) | Equivalent Upward Conductivity (m/hr) | Porosity | | |
| | | | | Total | Pellicular Water | Gravity Water |
| Top Layer Loamy Sand | 0.02 | $0.02 \times 10^{-2}$ | $0.12 \times 10^{-5}$ | 0.51 | 0.24 | 0.25 |
| Bottom Layer Clay Soil | 0.005 | $0.13 \times 10^{-3}$ | $0.01 \times 10^{-4}$ | 0.54 | 0.44 | 0.08 |

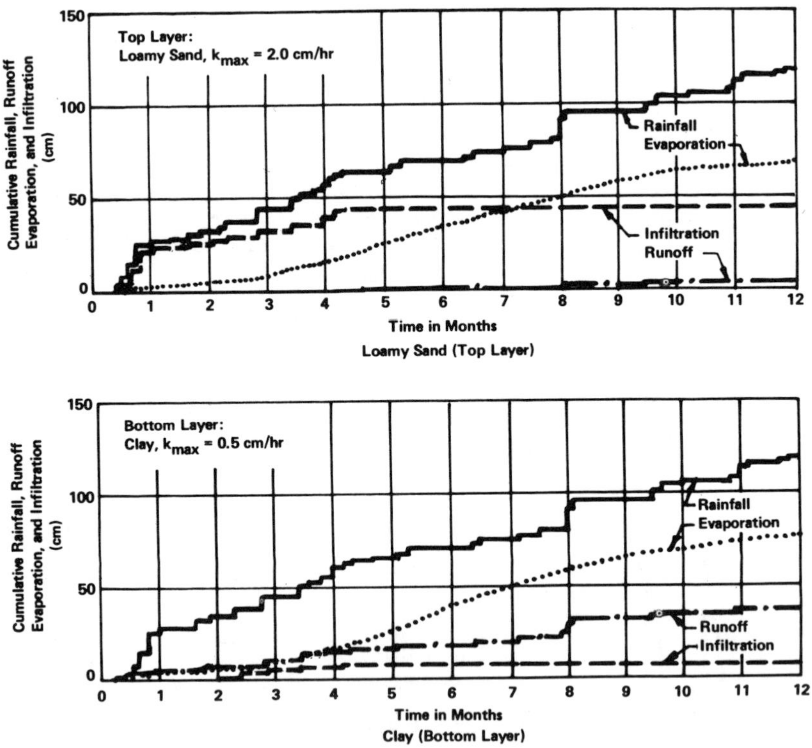


Figure 5. The results of simulations using the proposed model for loamy sand and clay soil, Barnwell radioactive waste disposal site.

recorded. However, the rainfall distribution within any hour in a real case is not necessarily uniform. By assuming a peak factor of two, i.e., all of the rainfall being recorded hourly is assumed to be concentrated in the second half of the recorded period, and reapplying the same input data to the infiltration model to rerun the case for the loamy sand (top layer), the rate of infiltration is 35 cm/year or 13.8 in/year. Therefore, one may conclude that the infiltation rate for the Barnwell site may vary between 13.8 and 17.7 in/year due to the non-uniform hourly rainfall distribution.

Independent studies have been conducted by the U.S. Nuclear Regulatory Commission (NRC) [24] and the U.S. Geological Survey (USGS) [25] on the rate of infiltration for the Barnwell site, based on an analysis of groundwater being discharged to a nearby creek. NRC's analyses indicated a rate of 14 in/year; USGS's, 14 to 17 in/year. The results of simulation using the infiltration model are in agreement with these two studies.

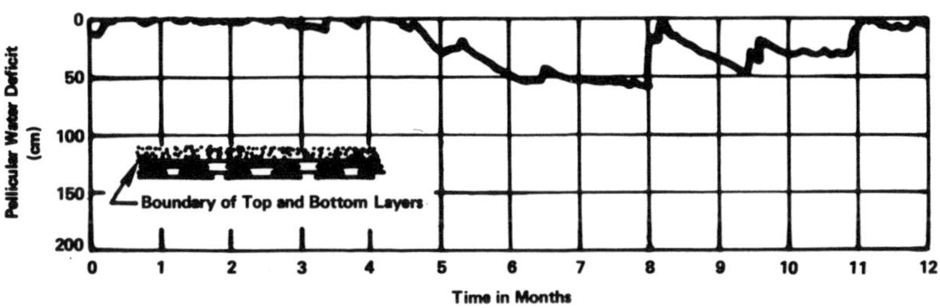


Figure 6. The results of simulating the variation of pellicular water deficit, loamy sand, Barnwell radioactive waste disposal site.

CONCLUSIONS

1. The space dependent variables in the momentum and continuity equations of a subsurface flow system can be transformed into space independent variables by breaking down the soil moisture into gravity water, pellicular water, and hygroscopic water components.

2. The results of transformations for the overland flow system, the subsurface flow system, and the atmospheric diffusion system have greatly simplified the computational procedures for simulating these systems and have improved the stability and efficiency of the numerical analysis.

3. Although the transformed system equations cannot simulate the dynamic response of soil moisture as precisely as the original partial differential equations, the derived system equations can be used to simulate the long-term infiltration rates with reasonable accuracy.

4. When the model was applied to the Barnwell radioactive waste disposal site, the results of simulation fit very well with the results of analyses conducted by other investigators using the other methods. This fact implies that the proposed model can be employed to simulate the infiltration of rainwater through the cap of a waste disposal trench.

5. The proposed practical infiltration model is suitable for integrating into a larger system model for risk assessment of a low-level radioactive waste disposal site.

## ACKNOWLEDGEMENTS

The author would like to express his sincere gratitude to Mr. Floyd L. Galpin and Mr. G. Lewis Meyer of the U.S. Environmental Protection Agency for their valuable advice and suggestions in developing this work.

## REFERENCES

1. Crawford, N. H., Linsley, R. K., "Digital Simulation in Hydrology: Stanford Watershed Model IV, Tech. Rep. 39, Dept. of Civil Engr., Stanford Univ., July 1966.
2. Moore, W. L. and Claborn, B. J., "Numerical Simulation of Watershed Hydrology," *Proceedings of the First Bilateral U.S.-Japan Seminar in Hydrology,* Jan. 1971.
3. Hung, C. Y. and Keifer, C. J., "Storm Runoff Simulation for the Lake Diversion Area," a report prepared for the Illinois Department of Transportation, Division of Water Resources, February 1977.

4. Iwasa, Y., "Fundamental Principles of Open-Channel Flow," published by Hydraulic Committee, Japan Society of Civil Engineers, 1964.
5. Takasao, T. and Kishimoto, S., "An Experimental Study on the Runoff of Rainfall," Bull. Disaster Prev. Research Inst., Japan, No. 4, 1961.
6. Ishihara T. and Takasao T. "A Study on the Subsurface Runoff and Its Effects on Runoff Process," Trans. JSCE No. 79, 1962.
7. Ishihara, T. and Takasao, T., "A Study on the Runoff Pattern and Its Characteristics," Bull. Disaster Prev. Research Inst., Japan, No. 65, 1963.
8. Ishihara, T. and Takasao, T., "Fundamental Research on the Unit Hydrograph Method and Its Application," Trans., JSCE No. 60 Extra paper (3-3), 1959.
9. Foster, G., Huggins, L., and Meyer, L., "Simulation of Overland Flow on Short Field Plots," *Water Resources Research,* Vol. 4, No. 6, 1968.
10. Hjelmfelt, A., "Overland Flow from Time Distributed Rainfall," *Jour. of Hydraulic Division, ASCE,* February, 1981.
11. Currie, J.A., "Gaseous Diffusion in Porous Media, Part 3 - Wet Granular Material," *Brit. J. Appl. Phys.* 12, 275-281, 1961.
12. Hillel, D., "Fundamentals of Soil Physics," Academic Press, New York 1980, pp. 221-223.
13. Bird, R. B., Stewart, W. E., and Lightfoot, E. N., "Transport Phenomena," John Wiley and Sons, Inc., New York, Seventh Edition, 1966.
14. Akan, A. S. and Yen, B. C., "Mathematical Model of Shallow Water Flow over Porous Media," *J. of the Hydr Div., ASCE,* HY4, April 1981.
15. Hillel, D. and van Bavel, C., "Simulation of Profile Water Storage as Related to Soil Hydraulic Properties," *Journal, Soil Science Society of America,* Vol. 40., No. 6, Nov. 1976.
16. Hillel, D., "On the Role of Soil Moisture Hysteresis in the Supression of Evaporation from Bare Soil under Diurnally Cyclic Evaporativity," *Soil Science,* Vol. 122, No. 6, December, 1976.
17. Rohwer, C., "Evaporation from a Free Water Surface," U.S. Dept. of Agr., Tech. Bull. 271, 1931.
18. Hamon, W. R., "Estimating Potential Evapotranspiration," ASCE, HY3, 1961.
19. Kichikawa, H., "A Study on the Underground Flood Control Reservoir," *Proceedings, JSCE,* May 1959.
20. Penman, H. L., "Gas and Vapor Movement in the Soil: I. The Diffusion of Vapor through Porous Solids," *Journal, Agr. Sci.,* 30, 437-461, 1940.

21. Ripple, C. C., Rubin J., and Van Hylkama, T., "Estimating Steady-State Evaporation Rate from Bare Soil under Conditions of High Water Table." U.S. Geological Survey, Water Supply, pp. 2019-A.
22. Philip, J. R., "Water Movement in Soil," Edited by De Varies, D. A. and Afgan, N. H., pp. 29-47, Halsted Press-Wiley, New York, 1974.
23. Chem-Nuclear Systems, Inc., "Environmental Assessments for Barnwell Low-Level Radioactive Waste Disposal Facility," Report, 1980.
24 U.S. Nuclear Regulatory Commission, "Environmental Assessment for the Barnwell Low-Level Waste Disposal Facilities," Report prepared by Division of Waste Management, NUREG-0879, January 1982.
25. Cahill, J., "Movement of Radionuclides in Unconsolidated Burial Site near Barnwell, South Carolina," (in press).

# SIMULATION OF THE EFFECTS OF ORGANIC SOLUTES ON THE HYDRAULIC CONDUCTIVITY OF VARIABLY SATURATED, LAYERED MEDIA

E.G. Lappala
Ertec Western, Inc.
3777 Long Beach Boulevard
Long Beach, CA 90807

## INTRODUCTION

The most common method of attempting to prevent contamination of aquifers by leachate from landfills and disposal ponds is compacted clay layers. Compacted clays are used because they are often locally available and economically attractive. In addition, compaction methods used can often achieve saturated hydraulic conductivities as low as $1x10^{-9}$ cm/sec. Although all clay liners will eventually allow the passage of liquid wastes, conductivities of this magnitude are usually sufficient to contain liquids for periods of tens of years. Such containment, however, requires that the integrity of the clay liner is not compromised by physical, chemical, biological, or human actions. Some chemicals present in primary and secondary leachates may cause significant leakage and potential ground-water contamination ten to 1,000 times faster than their design life (Anderson & Brown [1]). Studies by Montague [2] have demonstrated that leachate has been found beneath the clay liners in four "secure" landfills in the eastern United States.

Numerical analyses of the flux of liquids and solutes under conditions of variable liquid saturation are often used to evaluate design criteria for both surface and shallow subsurface disposal of chemical and radioactive waste. Such models almost always consider water as the solvent fluid, with the solutes present in trace amounts. However, many liquid wastes that have been disposed of in ponds and landfills consist of compounds in sufficient concentrations that may significantly alter their hydraulic conductivity and porosity.

State and federal regulations regulating the surface and landfill disposal of hazardous wastes specify that clay-liners will prevent seepage to ground water during the life of a facility. [3]

Consequently, prediction of the potential for groundwater contamination using numerical analyses should explicitly consider alteration of conductivities by the passage of leachates.

## Objective

The use of numerical methods to predict the flux of liquids and solutes through the unsaturated and saturated zones involves the solution of nonlinear partial differential equations. To our knowledge, no one has investigated the increased difficulty of incorporating dissolution and/or precipitation of the porous media in numerical models of variably saturated flow. The objectives of this investigation were to (1) devise methodology for incorporating alteration of the hydraulic conductivity of porous media by the fluid into an existing variably saturated flow model and (2) to evaluate the increased numerical difficulties, if any, that arise from such incorporation.

This paper presents results of simulating the inclusion of such alterations using finite difference approximations to the combined Cooper-Jacob and Richards equations. The analyses are restricted to increases in hydraulic conductivity, although the methodology is directly applicable to mechanisms that would decrease hydraulic conductivity. Although the example problem chosen considers one-dimensional vertical infiltration of ponded solvent through heterogeneous media, the methodology is applicable to the analysis of multidimensional problems.

## SOLVENT AND CLAY PROPERTIES

The mechanisms that may alter the hydraulic conductivity of clays can be categorized as dissolution, erosion, precipitation, and swelling caused by cation exchange. A discussion of these mechanisms is beyond the scope of this report, but can be found in [4]. Mechanisms that reduce conductivity are not considered in this paper because the primary concern regarding liner design and numerical difficulty is with increases in conductivity. As Anderson and Brown [1] indicate, the mechanisms causing an increase in the conductivity to organic fluids are not well understood. Until models are available to adequately predict the clay-solvent interactions in terms of the physical and chemical properties of both components, empirical relationships based on tests of specific leachate-clay combinations will have to be used.

The work of Anderson and Brown [1] describes methods for measuring leachate conductivity and the results of tests using a wide range of organic fluids and Houston Black and Lufkin clays. These are both smectite clays. Figure 1 presents the results of their tests, which show increases in conductivity as more than 1,000 times after the passage of only a few pore volumes of solvent.

Since the tests were conducted under fully saturated conditions, it is reasonable to consider that the changes in conductivity resulted from changes in intrinsic permeability. For the simulations described in this paper, the results of Anderson and Brown (Figure 1) were approximated by fitting an exponential function to the dimensionless intrinsic permeability expressed as a function of the number of pore volumes of permeant passing through a given volume of clay. The dimensionless permeability was the ratio of the intrinsic permeability obtained using organic solvents to that obtained using water. Based upon inspection, an exponential function of the type:

$$\bar{\bar{K}}_l/\bar{\bar{K}}_w = \exp(\tilde{a}V_p),\ V_p > 0; \tag{1}$$

$$\bar{\bar{K}}_l/\bar{\bar{K}}_w = 1,\ V_p = 0$$

roughly fits the data. All symbols are defined in the list at the end of the paper.

To fit this function to the curves shown in Figure 1, the observed leachate conductivities were corrected using the appropriate fluid viscosities and porosities as follows:

$$\frac{\bar{\bar{K}}_l}{\bar{\bar{K}}_w} = \frac{K_l}{K_w}\ \frac{\rho w \mu_l}{\rho l \mu_w} \tag{2}$$

Table 1 summarizes the values of $\bar{\bar{K}}_l/\bar{\bar{K}}_w$ obtained by applying equation (2) to viscosity and density data given by Anderson and Brown. Table 1 also shows the value of the parameter $\tilde{a}$ in equation (1) that was determined using the computed $\bar{\bar{K}}_l/\bar{\bar{K}}_w$ corresponding to one pore volume.

## MATHEMATICAL MODEL

The equation that describes multidimensional flow of liquids in variably saturated porous media results from combining the continuity equation with a generalized Darcy's Law. In this paper, total hydraulic head is used as

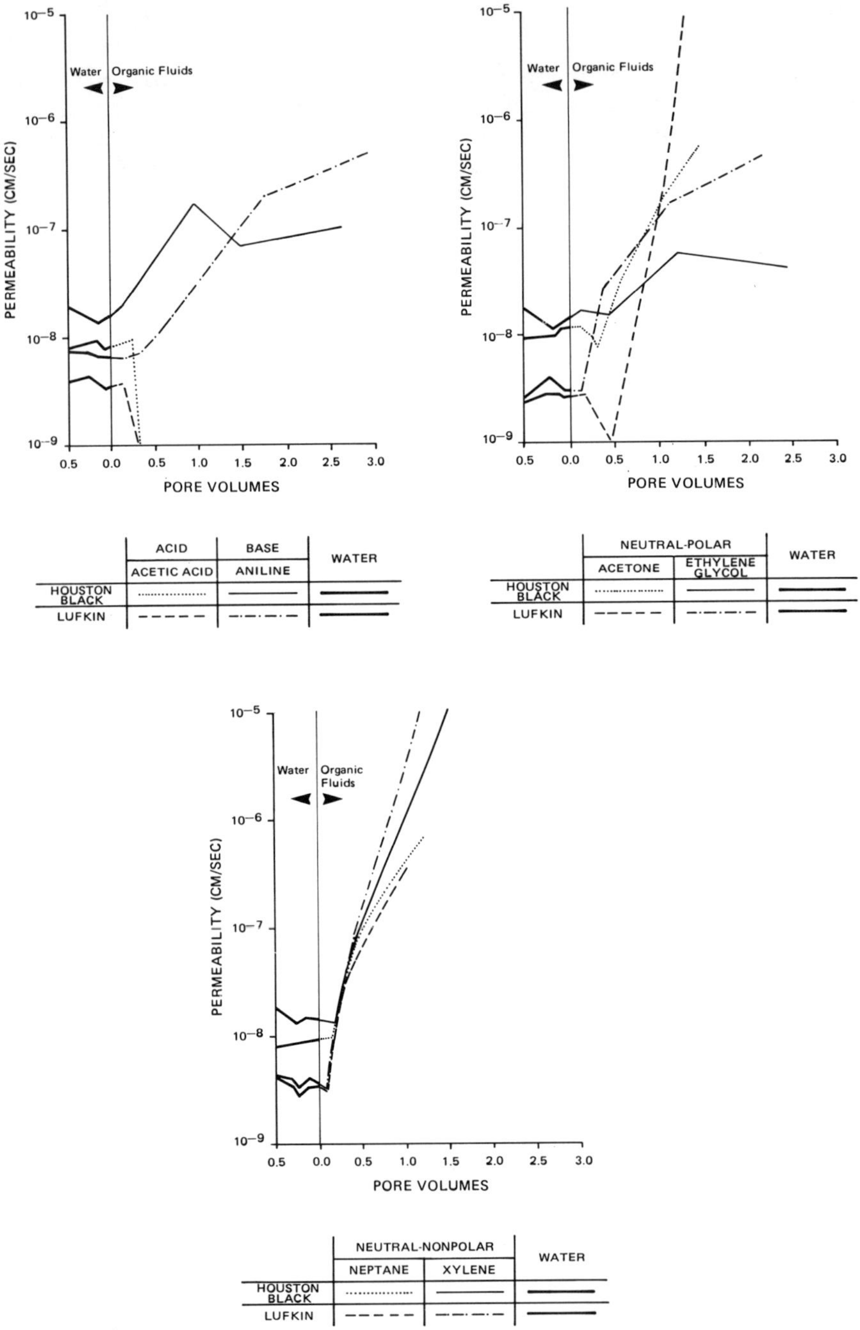


Figure 1. Hydraulic conductivity as a function of pore volumes for several organic fluids (after Anderson and Brown [1]).

the dependent variable. This approach allows a simplified uniform treatment of both saturated and unsaturated conditions since the total head is a continuous variable. Formulations that use saturation or liquid content as the dependent variable cannot simulate both saturated and unsaturated flow in heterogeneous media because these variables are not continuous across saturated-unsaturated interfaces and across boundaries between different materials.

Table I. Hydraulic Conductivity for Several Organic Fluids [1]

| SOLVENT | 1/ $K/K_w$ | 2/ $\frac{\mu_1 \rho_w}{\mu_w \rho_1}$ | $\bar{\bar{K}}/\bar{\bar{K}}_o$ | 3/ $\tilde{a}$ |
|---|---|---|---|---|
| ACID | | | | |
| Acetic Acid | <1 | 1.19 | <1 | - |
| BASE | | | | |
| Aniline | 4-6 | 4.23 | 17-25 | 2.9-3.2 |
| NEUTRAL POLAR | | | | |
| Acetone | 12 | 0.41 | 5 | 1.4 |
| Ethylene Glycol | 3-31 | 18.5 | 56-573 | 4.0-6.4 |
| NEUTRAL NON-POLAR | | | | |
| Heptane | 44-87 | 0.59 | 26-51 | 3.3-3.9 |
| Xylene | 75-500 | 0.91 | 68-455 | 4.2-6.1 |

1/ After 1 Pore Volume
2/ At 20°C
3/ $\bar{\bar{K}}/K_o = \exp(\tilde{a}V_p)$

The descriptive flow equation used can be considered a combination of the Cooper-Jacob and Richards equations as described by Freeze [5] which is written with total head as the dependent variable:

$$\rho\,(S_s + C_s)\,\frac{\partial H}{\partial t} - \frac{\partial}{\partial X_i}\left[\frac{\bar{\bar{K}}\rho^2 g K_r(h)}{\mu}\frac{\partial H}{\partial X_j}\right] + \rho W = 0 \qquad (3)$$

subject to the initial conditions

$$H = H_o(X_{i_o}, t) = 0$$

and the boundary conditions

$$H = H(X_i,\ t) \text{ on specified head boundaries} \qquad (3a)$$

or $$\frac{\bar{\bar{K}}\rho g K_r}{\mu}\,\frac{\partial H}{\partial X_j} = q(X_i, t) \text{ on specified liquid flux boundaries} \qquad (3b)$$

All symbols are defined in the list at the end of the paper.

Equation (3), subject to conditions (3a) and (3b), is a multidimensional nonlinear partial differential equation. The nonlinearities arise from the dependence of the specific moisture capacity ($C_s$), the relative liquid conductivity ($K_r$), boundary conditions, and the source/sink term (W), on the pressure head (h).

The saturation (S), specific moisture capacity ($C_s$), and relative hydraulic conductivity ($K_r$) functions are assumed to be described by equations of the Brooks-Corey type (Laliberte and others [6]):

Saturation: $$\frac{S-S_o}{1-S_o} = \left(\frac{h_b}{h}\right)^{\lambda} \tag{4}$$

Specific Moisture Capacity:
$$C_s = \frac{\partial S}{\partial H} = (S_o-1)\lambda \; [h_b^{\lambda}/h^{(\lambda+1)}] \tag{5}$$

Relative liquid conductivity:
$$K = \left(\frac{h_b}{h}\right)^{(2 + 3\lambda)} \tag{6}$$

For the analyses presented in this paper, the change of liquid conductivity by the passage of solvents is considered to alter the intrinsic permeability alone. Although not documented by experiment, the actual mechanisms undoubtedly involve changes to the relative conductivity and pressure-saturation relationships as well, since dissolution, erosion, and precipitation of the solids would alter the pore size distribution, which determines the shape of both the S(h) and $K_r$(h) functions. The assumption that all changes to liquid conductivity can be described by changes to intrinsic permeability allows the data of Anderson and Brown [1] to be used to analyze unsaturated flow problems. The dependence of the intrinsic permeability on the cumulative volume of solvent passing through the pores introduces an additional nonlinity in equation (1).

Although it is conceivable that the permeability might increase without bound if the entire clay layer was dissolved, it was considered more reasonable to place some upper bound on the increase. This was accomplished by applying equation (1) only until 9 pore volumes had passed. Permeabilities corresponding to $V_p$ greater than 9 were set equal to the value for $V_p$ equal to 9. This upper limit was arbitrary and the results presented should be applicable to other limits.

The use of equation (1) requires $V_p$ to be known. For a given volume (V) of porous medium the pore volume is given by $V_p = \phi V$. Because $\underline{\underline{K}}$ is different from $\underline{\underline{K}}_o$ when $V_p > 0$, the position of the solvent front must be known to determine the portion of the solution domain in which equation (1) should be applied. For this analysis, it is assumed that the solvent and the displaced pore water are immiscible fluids separated by a sharp front. This assumption has been shown to be reasonable in analyzing vertical infiltration problems involving solutes present in low concentrations (deSmedt and Wirenga [7]. For solvents, computation of the position of the solvent front is accomplished by using equation (6) of deSmedt and Wirenga (1978), written in the notation of this paper:

$$\int_0^{\tau} q(0,t)dt = \int_0^{\alpha(t)} \phi S(X_i, t)dX_i \qquad (7)$$

To solve for the position of the solvent front, the lefthand side of equation (7) is assumed to be known by integration of the flux rate at the surface over the duration of simulation. The righthand side is determined by integration along each flow path until equation (6) is satisfied. For one dimensional cases, the flow paths are specified by the geometry of the problem. For multidimensional problems, methods to determine the flow paths such as those described by Nelson [8] would need to be used.

## NUMERICAL MODEL

Equations (1) through (7) were solved using a finite difference model for analyzing multidimensional, nonlinear problems (Lappala [9]). For the analyses in this paper, this model was used with block centered finite differences to approximate the spatial derivatives and implicit difference approximations to the time derivatives. To control oscillations and achieve numerical stability, the model incorporates provisions for automatic control of the maximum allowed time-step used. Linearization of the difference equations was accomplished by using an implicit Newton-Raphson method. Direct solution of the linearized difference equations was used.

Modifications to the model of Lappala [9] were made for the present problem to incorporate both the additional nonlinear functional dependence of the intrinsic permeability on solvent pore volumes using equation (1) and the computation of solvent penetration depth using equation (7). Since the example problem used was one-dimensional, the latter modification was straightforward. Multidimensional problems would require the addition of a flow path or particle tracking routine. Figure 2 shows the general sequence of simulation used by the modified model.

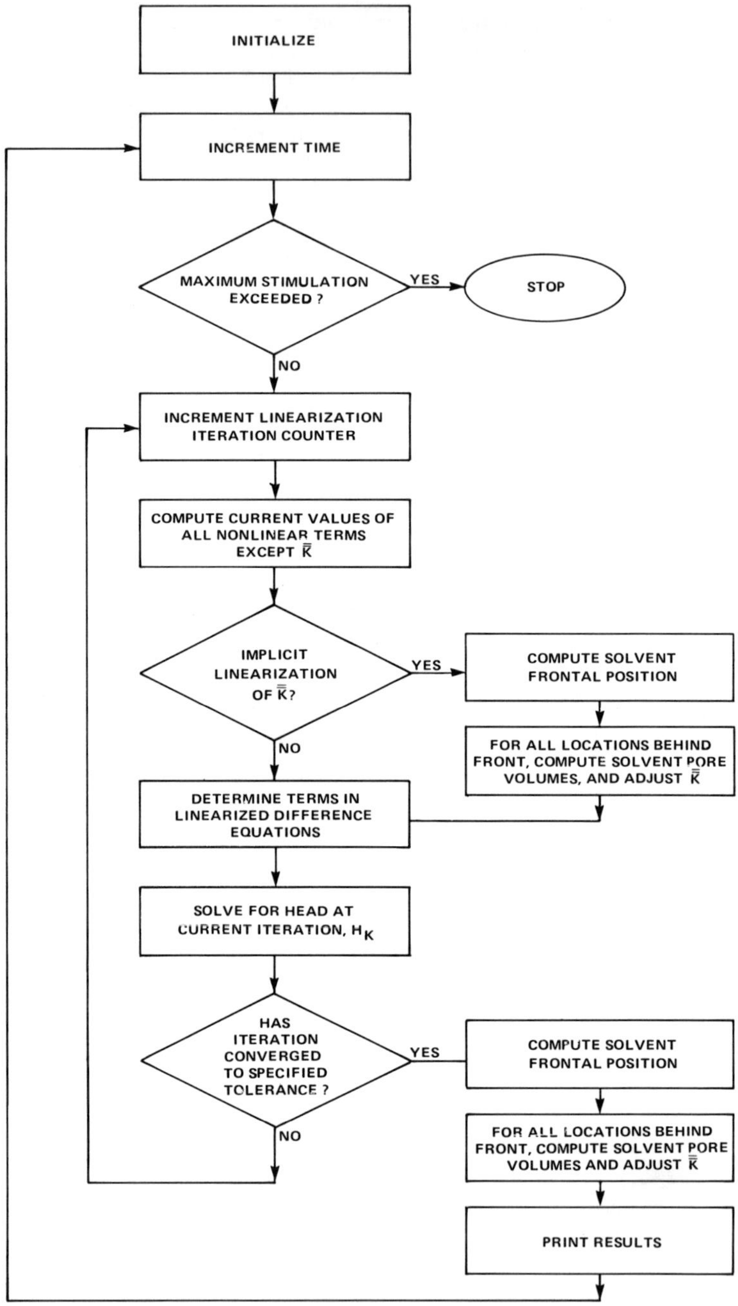


Figure 2. General flow chart of the simulation program.

PROBLEM DESCRIPTION

The problem used to evaluate the numerical difficulties involved in incorporating the intrinsic permeability as a function of pore volumes of solvent utilized a one-dimensional, vertical column. Heterogeneity was included by using the following layers in sequence from the top of the column: 15cm of clay, 123cm of sand, 21cm of silt, and 310cm of sand. The hydraulic properties of these materials are described in Table II. The saturated hydraulic conductivity of the clay was selected near the upper limit of expected values for clays to allow significant flux of solvent in a reasonable simulation time.

The initial conditions simulated a moderately moist subsurface environment, with a constant initial pressure head (h) of -60cm of water for all layers. The initial saturation of each of the materials is listed in Table II and is shown in Figure 3. As can be seen from Figure 3, the clay was considered to be initially saturated with water. Therefore, the simulated displacement of water by solvent in the clay and changes to the intrinsic permeability of the clay occurred under saturated conditions.

The boundary conditions consisted of a constant, ponded solvent depth of 50cm on the top of the column, and zero flux at the base of the column.

Table II. Values Used for Parameters in the Brooks and Corey Equations for Saturation and Relative Liquid Conductivity

| Material | $K_s$ (cm/sec$^{-1}$) | $\bar{\bar{K}}_o$ (cm$^2$) | $\phi$ | $S_r$ | $h_b$ (cm) | $\lambda$ | Initial Saturation $S_o$ |
|---|---|---|---|---|---|---|---|
| Clay | $5.9\times10^{-7}$ | $6.0\times10^{-12}$ | 0.50 | 0.20 | -60 | 1.5 | 1.0 |
| Sand | $6.7\times10^{-3}$ | $6.8\times10^{-8}$ | 0.36 | 0.14 | -15 | 4.5 | .14 |
| Silt | $3.3\times10^{-5}$ | $3.4\times10^{-10}$ | 0.44 | 0.18 | -20 | 3.0 | .21 |

Although the model can simulate fluids of any viscosity and density, it was assumed that these properties were the same for both the solvent and the pore water initially in the column.

Grid spacing used for all simulations varied from 1.0cm to 25cm. Grid spacing in adjacent finite difference cells was not allowed to increase by more than a factor of 1.40.

The model was used to simulate periods of up to $1\times10^5$ minutes (69.4 days). The initial time-step for all simulations was 5 minutes. Time-steps were increased by a factor of 1.3 until a maximum of 1000 minutes was reached or until a maximum saturation change of 10cm occurred within a given time increment.

Implicit linearization was used for terms in the finite difference equations involving the relative hydraulic conductivity. Implicit Newton-Raphson linearization was used for terms involving the specific moisture capacity. The non-linearity caused by the dependence of intrinsic permeability upon pore volumes of solvent as described by equation(1) was linearized by updating the permeability only after each time-step. The alteration of the intrinsic permeability was only applied to the clay layer that occurred at the top of the column. Six simulations were run using values of the parameter $\tilde{a}$ in equation (1) of 0, 0.5, 1.0, 1.5, 2.0 and 3.0. The value of $\tilde{a} = 0.0$ corresponds to no change in intrinsic permeability, and the values of 1.5 and 3.0 fall near the middle of the range of values which were fit to the data of Anderson and Brown (Table 1) [1].

RESULTS

An initial simulation was run without varying the intrinsic permeability, $\bar{\bar{K}}$. The results of this simulation were used as the baseline to compare results from simulations in which $\bar{\bar{K}}$ was increased by the passage of solvent. Figure 3 shows the saturation profiles for the baseline condition after 0, 6.9, 13.9, 34.7, and 69.4 days of simulation. The following features of these profiles that are expected for these boundary conditions for a layered system are evident from Figure 3:

1. The downward advance of a wetting front above the silty clay layer.
2. The establishment of a steady downward flux at a larger rate than the initial condition. This flux is established first above the silt and later below the silt and is shown by establishment of constant saturation in the sands at values higher than the initial condition.
3. The retardation of the downward movement by the perching on silt until the air entry pressure head of the underlying sand is exceeded.

The saturation profiles using values of $\tilde{a}$ equal to 0.5, 1.0, 1.5, 2.0, and 3.0 are also shown in Figure 3. The results for these values of $\tilde{a}$ are essentially identical because the intrinsic permeability increased sufficiently rapidly that the clay ceased to be the control on the downward flux of solvent. By 34.7 days the entire column above the silt was saturated with solvent, and the silt layer was serving as

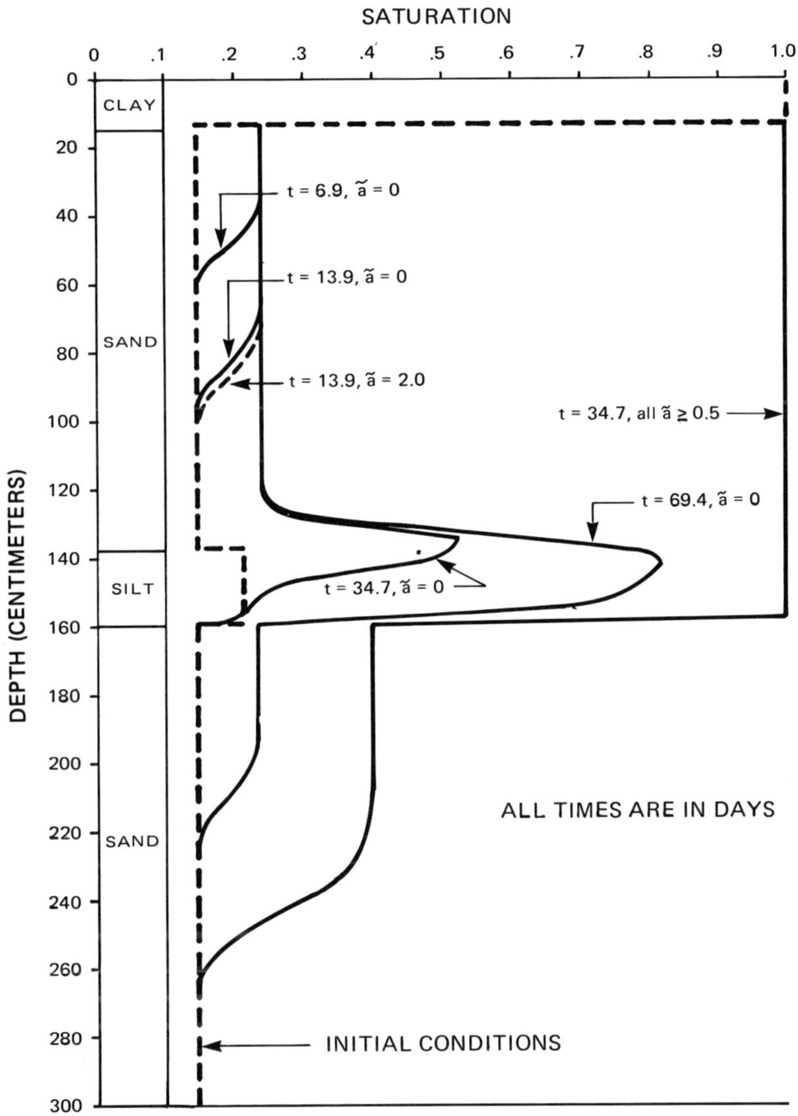


Figure 3. Saturation profiles for $\tilde{a} = 0$.

the control on downward movement. A steady downward flux below this layer was established at a rate of 4.6cm $day^{-1}$.

The solvent penetration depth as computed from Equation (7) for the baseline condition ($\tilde{a}$ = 0) is shown in Figure 4. For this condition the rate of solvent penetration becomes constant within each layer after a short initial transient period corresponding to the time required for passage of the wetting front.

The solvent penetration depths for $\tilde{a}$ = 0.5, 1.0,1.5, 2.0, and 3.0 are also shown on Figure 4. The following features regarding the rate of solvent penetration are evident from these curves:

1. An initial constant penetration rate during which all curves follow the $\tilde{a}$ = 0 curve. The curves for $\tilde{a}>0$ depart from the $\tilde{a}$ = 0 when the permeability has been increased in a sufficient thickness of the clay. The departure begins at about 14 days for all values of $\tilde{a}>0$.
2. A rapid increase in the rate of penetration as the permeability of the remaining clay layer is rapidly increased.
3. A reduction in rate of penetration caused by perching on top of the silt. The rate continues to decline until the solvent breaks through the silt.
4. A constant rate of penetration below the silt layer. The time required for the solvent to break through the base of the clay layer as a function of $\tilde{a}$ is shown in Figure 5.

The effect of increasing the intrinsic permeability is also demonstrated by examining the rate of inflow at the top of the column as shown in Figure 6, and in the saturation history at a depth of 134cm, as shown in Figure 7. The inflow rate for the baseline shows the expected rapid, but minor decrease in the inflow rate before a steady downward flux of 0.19 cm/day is established. For values of $\tilde{a}>0$, the inflow history can be subdivided into four stages as shown in Figure 6.

1. A constant rate of 0.19 cm/day until sufficient volume of solvent has penetrated the clay to begin to increase the permeability noticeably.
2. A rapid increase in the inflow rate as the permeability increase becomes effective in increased thickness of the clay. This stage ends with solvent breakthrough into the sand beneath the clay.
3. A period of constant inflow rate while the solvent moves down through the sand. The permeability of

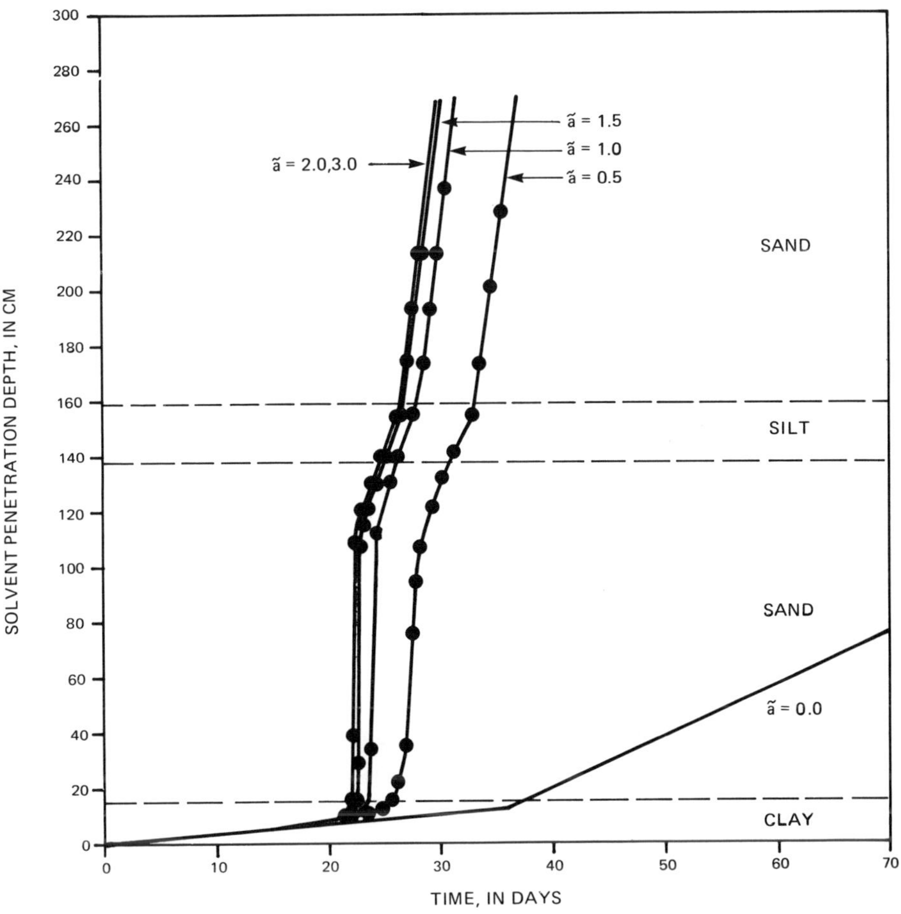


Figure 4. Time history of solvent penetration depth for ã = 0.0, 0.5, 1.0, 1.5, and 2.0.

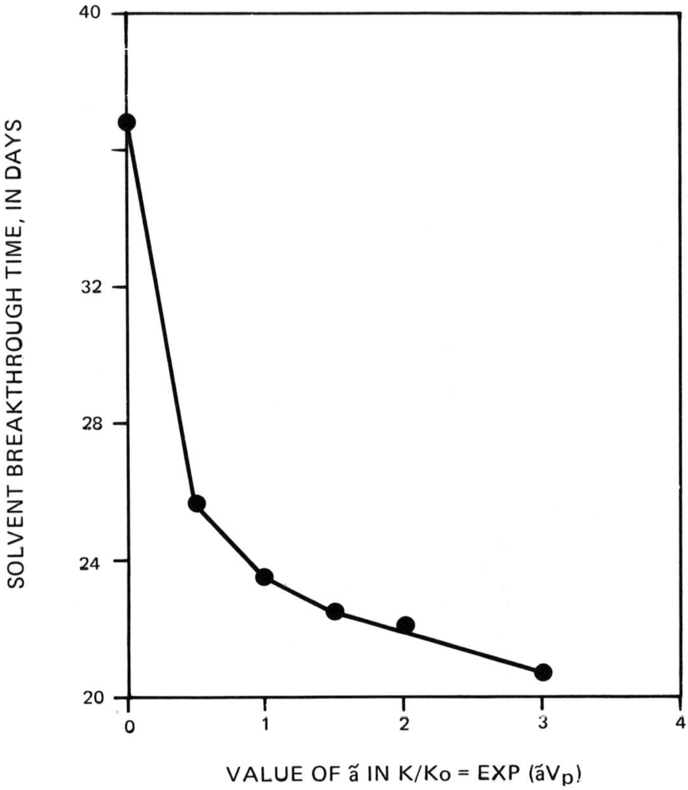


Figure 5. Time required for solvent to break through the clay layer as a function of the value of $\tilde{a}$.

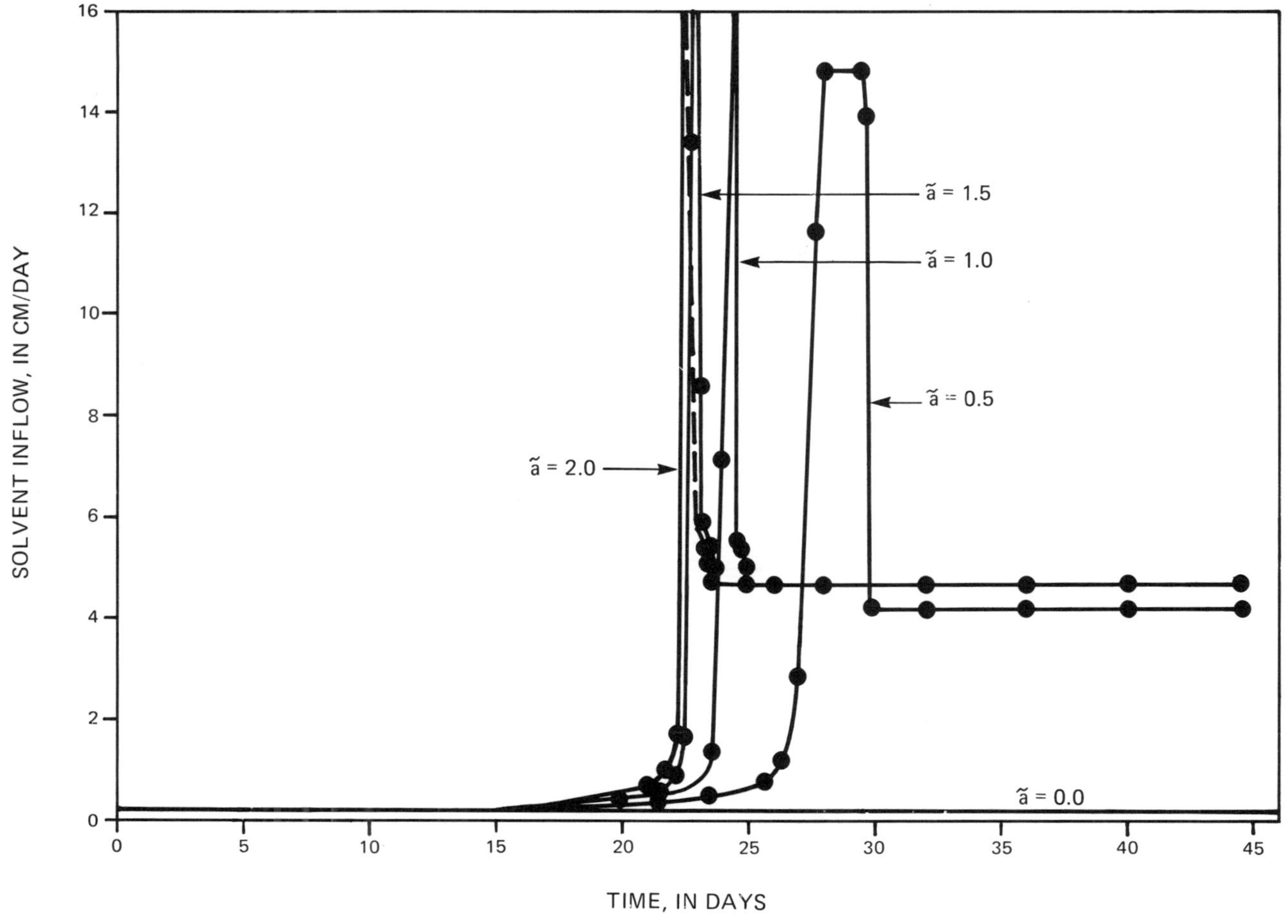


Figure 6. Rate of solvent inflow at the top of the column.

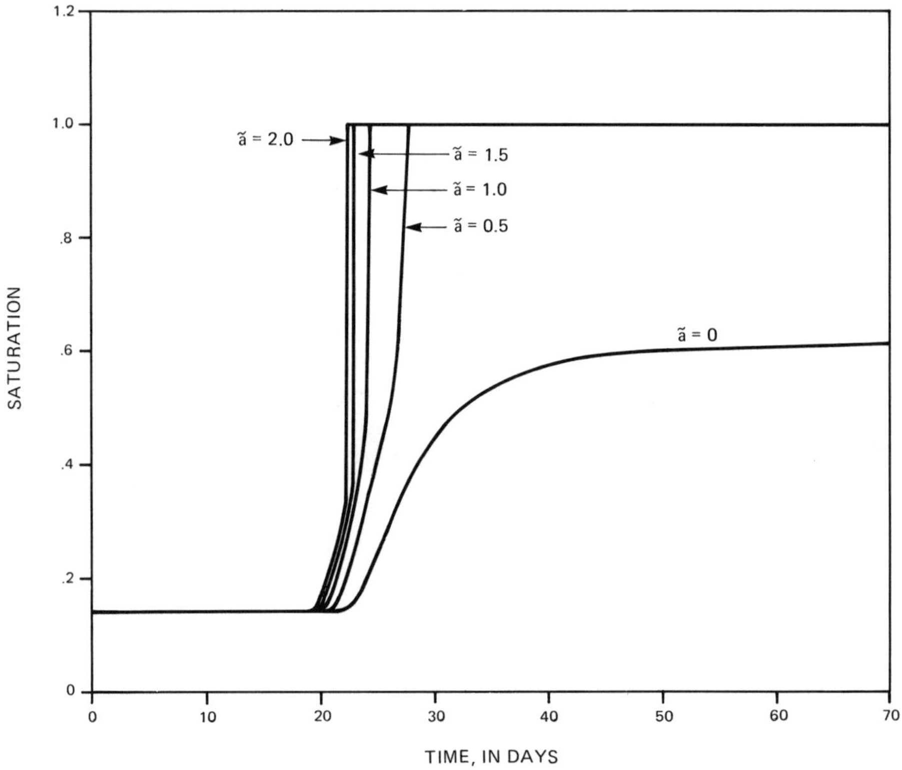


Figure 7. Saturation history at a depth of 134 cm.

the clay has been increased to the point that for values of $\tilde{a}$ greater than 0.5, the conductivity of the sand is the controlling factor in downward movement. For $\tilde{a}$ = 0.5, the permeability of the clay was not increased enough to allow as high a sand conductivity as for $\tilde{a}$> 0.5, and the clay still influenced the rate of downward solvent movement through the sand.

4. A period during which the flow rate rapidly decreases while water is perched on the silt.
5. A period of constant downward flux that is controlled by the conductivity of the silt layer. For an $\tilde{a}$> 0.5, this rate was 4.62cm/day. For $\tilde{a}$ = 0.5, the rate was 4.19cm/day. This lower rate does not make physical sense and is attributed to convergence to a slightly erroneous solution during the rapid decrease in flux rate while the solvent was perching on the silt.

As with many cases involving simulation with the non-linear flow equation, those described in this paper were not without numerical difficulty. However, the problems of oscillation and poor mass balance that are common in numerical solutions of nonlinear problems were not exacerbated by the additional nonlinearity caused by making the intrinsic permeability of the clay a nonlinear parameter. Since the model automatically controls the length of time-step so as to minimize these problems, comparisons of the relative computational effort involved in cases with and without permeability increases are made by using the total number of iterations required to achieve a given simulation time. The times used for comparison were selected to correspond with the time required to achieve steady flow through the silt (Figure 6) for values of $\tilde{a}$ = 0.5, 1.0, 1.5, and 2.0. For $\tilde{a}$ = 3.0, the time required to reach the peak inflow rate was used because the solution did not converge to the correct solution following the rapid rise in the inflow rate.

Table III shows the ratio of the iterations required. For $\tilde{a}$ = 0.5, 1.0, 1.5 and 7.0, the inclusion of permeability increases resulted in 30 to 50% more computational effort. The ratio for $\tilde{a}$ = 3.0 cannot be compared directly to that for the lower values of $\tilde{a}$ because for the lower values of $\tilde{a}$, significantly more iterations were required between the peak and the establishment of steady flow below the silt layer. The maximum number of iterations per time-step and the relative number of time-steps required to achieve the same time used for the iteration comparison are also shown in Table III.

Comparisons of central processor unit (CPU) time are not included in Table III because several simulations for a>0 were not run for the full 69.4 days used for $\tilde{a}$ = 0. However, the iteration ratio should give a direct measure of the increased CPU time required.

The increased computational effort required to include intrinsic permeability as a nonlinear parameter were a result of the rapid variations in flow rates during solvent penetration. However, by automatic control of the time-step length, solutions can be achieved. Values of $\underline{a} > 3$ resulted in convergence to an incorrect solution. Efforts to resolve this problem were not extensively pursued, but use of a smaller value for the maximum allowed pressure head change during one time-step probably would have resolved this problem. These problems are similar to those expected in attempting to simulate the phenomenon of fingering which occurs when liquid perched above a coarse grained layer does not move downward as a uniform front.

Table III. Time and Iteration Data

| VALUE OF $\tilde{a}$ IN $\underline{\underline{K}}/\underline{\underline{K}}_o = \exp(\tilde{a}V_p)$ | TIME USED[1/] IN DAYS | ITERATION[2/] RATIO | TIME[3/] STEP RATIO | MAXIMUM[4/] ITERATIONS |
|---|---|---|---|---|
| 0.5 | 30.32 | 1.30 | 1.25 | 30 |
| 1.0 | 25.46 | 1.41 | 1.40 | 34 |
| 1.5 | 24.12 | 1.36 | 1.34 | 34 |
| 2.0 | 23.85 | 1.50 | 1.57 | 37 |
| 3.0 | 20.78 | 1.18 | 1.21 | 35 |

1/ Times used for comparison (except for $\tilde{a} = 3$) were those required to achieve the steady flow rate through the silty clay or the second steady flow rate shown in Figure 6. For $\tilde{a} = 3$, the time to the peak inflow rate shown in Figure 6 was used.

2/ Ratio of the number of iterations for the given value of $\tilde{a}$ to that required to simulate the same time with $\tilde{a} = 0$.

3/ Ratio of the number of time-steps for the given value of $\tilde{a}$ to that required to simulate the same time with $\tilde{a} = 0$.

4/ Maximum number of iterations per time-step, excluding time-step 1. All occurred just before or just after time of occurence for the peak inflow rates shown on Figure 6.

## CONCLUSIONS

Methodology for simulating the effects of clay dissolution was incorporated into an existing code. The major modification involved adding a method for computing the position of the solvent front and recalculation of the intrinsic conductivity after each time-step. The results of this investigation can be summarized as follows:

1. Results agree with established concepts of variably saturated flow. Establishment of steady, unsaturated flow beneath the fine-grained layers, perching and rapid movement of wetting fronts through coarse materials occurred for both water and the solvent.
2. Increased computational effort was required by the inclusion of an exponential increase in intrinsic permeability as a function of solvent pore volumes. The increased effort, as measured by the number of iterations required to achieve a given simulation time, ranged from 30 to 50 percent.
3. Values of the parameter $\tilde{a}$ in $\bar{\bar{K}}_1/\bar{\bar{K}}_w = \exp(\tilde{a}V_p)$ that was greater than 3.0 caused convergence to an erroneous solution.

As lab and field tests are completed, a compilation of the data should provide a database to establish the functions describing $\bar{\bar{K}}$ changes. These functions can and should be incorporated into models that are used to design and evaluate disposal systems which involve clays subject to dissolution. Use of this feature will cause additional computational difficulty, but should prove to be tractable by resorting a sufficient number of numerical devices used to linearizing the flow equation and to limiting the maximum allowed saturation change during time-steps.

## LIST OF SYMBOLS

$\tilde{a}$ = constant in exponential intrinsic permeability function

$g$ = gravitational acceleration $[LT^{-2}]$

$H$ = total hydraulic head, equal to h+Z, where h is pressure head = P/ρg and Z is elevation head, equal to position relative to an arbitrary datum [L]

$h_b$ = bubbling or air entry pressure head [L]

LIST OF SYMBOLS (continued)

$K_r(h)$ = relative liquid conductivity of the medium. It is a function of the pressure head. $0 \leq K_r \leq 1.0$ [L]

$K_w, K_l$ = Hydraulic conductivity under conditions of full saturation with water, or solvent, $= \rho \bar{\bar{K}} g/\mu$ $[LT^{-1}]$

$\bar{\bar{K}}_w, \bar{\bar{K}}_l$ = intrinsic permeability of the porous medium, without and with alteration by solvent passage $[L^2]$

P = liquid pressure $[ML^{-1}T^{-2}]$

S = liquid saturation, a function of pressure head $[L^o]$

$S_o$ = residual saturation $[L^o]$

t = time [T]

$V_p$ = pore volumes $[L^o]$

W = a general volumetric source/sink term $[L^3T^{-1}]$

$x_i, x_j$ = Cartesian coordinate directions (repeated summation convention is assumed) [L]

$\alpha$ = distance from initial point of contact between the solvent and the medium and the solvent-pore fluid front

$\phi$ = porosity $[L^o]$

$\lambda$ = pore size distribution index

$\mu_w, \mu_l$ = dynamic viscosity of the liquid $[ML^{-1}T^{-1}]$

$\rho$ = liquid density $[ML^{-3}]$

REFERENCES

1. Anderson, D., and K. Brown. "Organic Leachate Effects on the Permeability of Clay Liners," in Procedures of Solid Waste Symposium, U.S. EPA, p. 119-130 (1981).
2. Montague, P. "Four Secure Landfills in New Jersey--A Study of the State of the Art in Shallow Burial Waste Disposal Technology," draft (1982).
3. U.S. Environmental Protection Agency. "Regulations for Owners and Operators of Permitted Hazardous Waste Facilities," 40CFR Part 264, Subparts K and L (1982).
4. Matrecon, Inc. "Lining of Waste Impoundment and Disposal Facilities," Office of Solid Waste, U.S. EPA Report SW870, 385 p. (1980).
5. Freeze, R. A. "Three Dimension, Transient, Saturated-Unsaturated Flow in a Groundwater Basin," Water Resources Research, V.7, p. 347-362 (1972).

REFERENCES (continued)

6. Laliberte, G. E., A. T. Corey, and R. H. Brooks. "Properties of Unsaturated Porous Media," Colorado State University Hydrology Paper 17, 40 p. (1966).
7. deSmedt, F., and P. J. Wierenga. "Approximate Analytical Solution for Solute Flow during Infiltration and Redistribution," Soil Science Society of America Proceedings, V. 42, p. 407-412 (1978).
8. Nelson, R. W. "Evaluating the Environmental Consequences of Groundwater Contamination," Water Resources Research, V.14, No. 3, p. 409-415 (1978).
9. Lappala, E. G. "Documentation of Computer Program VS2D to Solve the Equations of Fluid Flow in Variably Saturated Porous Media," U.S. Geological Survey Open File Report (in review), 165 p. (1982).

# UNSATURATED FLOW IN HETEROGENEOUS SOILS

T.-C. Jim Yeh
New Mexico Institute of Mining and Technology
Socorro, New Mexico 87801

Lynn W. Gelhar+
Professor of Hydrology
New Mexico Institute of Mining and Technology
Socorro, New Mexico 87801

## INTRODUCTION

Hydrologists or soil physicists generally assume the soil is homogeneous and isotropic when they deal with unsaturated flow in most field situations. However, several recent studies (Nielsen et al. [1]; Coelho [2]; and others) show that soil hydraulic parameters could vary significantly in the field. For example, the standard deviation of log saturated hydraulic conductivity has been observed in the range between 0.3 and 3.0. Correspondingly, the coefficient of variation is in the range of 0.20 to 90. This simply implies that the spatial variation of soil properties is substantial in the field situation.

To illustrate the degree of the spatial variation of the soil hydrologic parameters, Figure 1 shows the spatial distribution of the permeability of Mt. Simon sandstone in Illinois (Bakr [3]). As indicated in this figure, the permeabilities can be completely different at distances as small as one foot. The standard deviation of the log saturated hydraulic conductivity is 1.6 in this case. Spatial variation in the porosity also exists. A similar spatial variation in saturated hydraulic conductivity has also been observed in dune sand, which is usually assumed to be homogeneous (see Figure 2). Although the standard deviation of the log-saturated hydraulic conductivity is small, 0.5, the spatial variation of the parameter does exist.

+ Now at MIT.

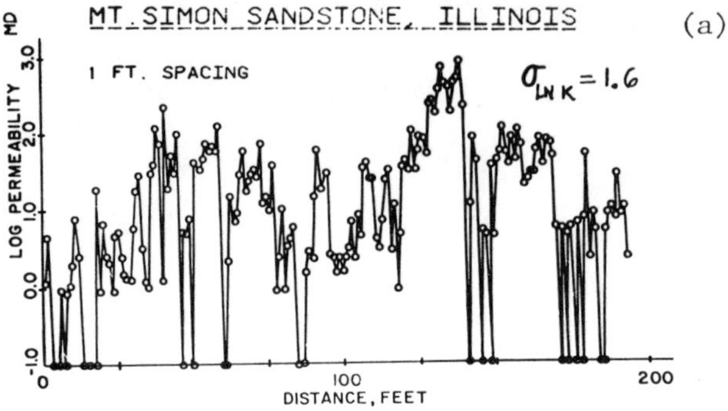


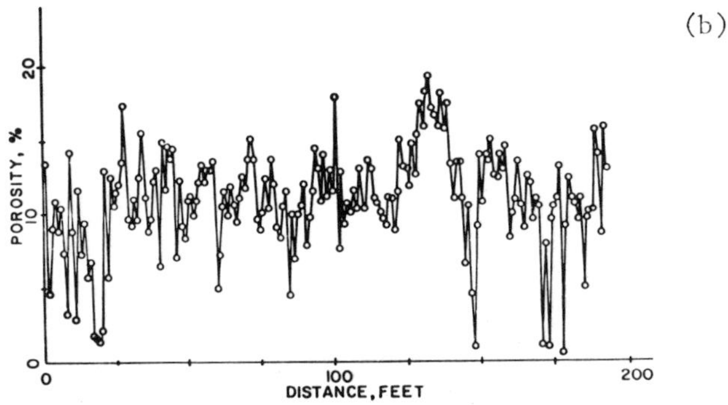


Figure 1a: Permeability variation of Mt. Simon Sandstone, Illinois.

Figure 1b: Porosity variation of Mt. Simon Sandstone, Illinois.

Since soil often exhibits stratification, the variation of the soil parameters are not disordered in space. They are correlated. Such a correlated structure often can be characterized by its correlation scales.

As has been pointed out, the soil presents a large degree of variation in its properties. What can we do if we want to model the flow system in a large scale field problem? Obviously it is not possible to carry out a sampling program to delineate the details of variation of the soil properties because of the physical and the economic constraints. The alternative is to use limited amounts of available data to obtain a mean or average soil property with which a mean behavior of the flow regime can be predicted.

Therefore, the objectives of this study are: a) to determine the average or effective hydraulic conductivity of the unsaturated porous media and b) to determine the variation above the average prediction of the flow regime based on the effective hydraulic conductivity.

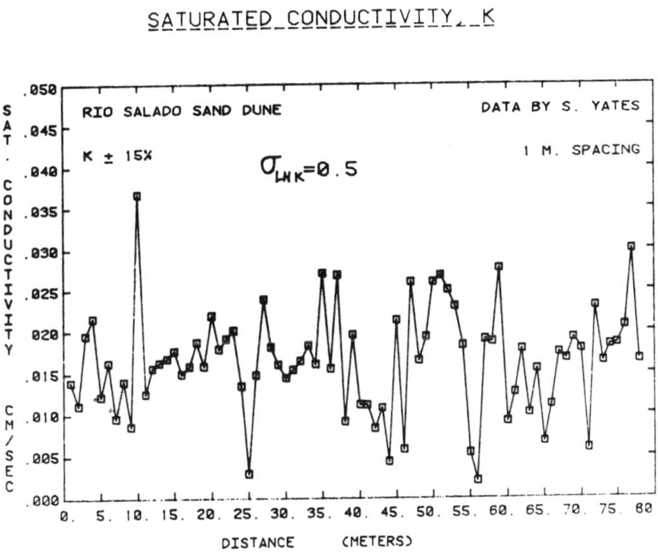


Figure 2: Variation of saturated hydraulic conductivity of Sevilleta dune sand.

APPROACH

In this analysis, a steady-state infiltration under a unit gradient condition is investigated. The porous medium is considered to be infinitely large so that the boundary effects can be neglected. It is also assumed that at the local scale unsaturated hydraulic conductivity can be represented by the exponential function.

$$K(\psi,\vec{x}) = K_s(\vec{x})e^{-\alpha\psi} \qquad (1)$$

where $K_s$ is the saturated hydraulic conductivity, $\alpha$ is the pore-size distribution parameter or the slope of lnK-$\psi$ curve, and $\psi$ is the capillary pressure. The hydraulic conductivity is assumed to be locally isotropic.

Recent studies such as Warrick [5] have shown that the soil parameters, $K_s$ and $\alpha$ , vary significantly in the field. Therefore, to represent the heterogeneous unsaturated media, we assume that log saturated hydraulic conductivity and pore-size distribution parameters are statistically anisotropic stochastic processes. The statistical anisotropy implies that the correlation scales of a process are not the same in all directions. Such a statistical anisotropy is used to characterize the lenticular structure of the sediments, or the stratification of the sediments. Additionally, it is assumed that no correlation exists between $\ln K_s$ and $\alpha$ processes.

Since $\ln K_s$ and $\alpha$ are stochastic processes, the governing steady-state infiltration equation becomes a stochastic partial differential equation where the unsaturated hydraulic conductivity is a stochastic process in space. To solve the stochastic partial differential equation, a perturbation method and spectral representation are used. The details of the formulation and solutions are given by Yeh [5].

The soil tension head variance is derived as a measure of the variation above the average prediction of flow in the unsaturated media. For the case of an arbitrarily oriented layers, the head variance can be expressed as follows [5]:

$$\sigma_\psi^2 = (\sigma_f^2 + \sigma_\alpha^2\bar{\psi}^2)\lambda_1^2 g(\lambda/\lambda_1,\ \bar{\alpha}\lambda_1,\ \theta) \tag{2}$$

where $\sigma_\psi^2$ is the soil tension head variance; $\sigma_f^2$ is the variance of the log saturated hydraulic conductivity; $\sigma_\alpha^2$ is the variance of the pore-size distribution parameter; $\bar{\psi}$ is the mean soil tension; $\lambda_1$ and $\lambda$ are the vertical and horizontal scales of the porous media, respectively; $\bar{\alpha}$ is the mean pore-size distribution parameter; and $\theta$ is the orientation of the stratification of the media. The key finding is that the head variance is a function of the mean soil tension and the correlation scales. The head variance is directly proportional to the correlation scales. Thus, it would be inappropriate to neglect the correlation scales in the analysis of the effects of spatial variability on flows.

Another result from the analysis is the effective hydraulic conductivity of the heterogeneous media. It is found that the effective unsaturated hydraulic conductivity is a symmetric tensor of rank two. The principal components of the conductivity tensor are bounded by arithmetic and harmonic means. Furthermore, the anisotropy ratio depends on capillary pressure or moisture content.

For the special case where the soil formation is perfectly stratified, or the aspect ratio approaches infinity, the anisotropy can be written as follows [5]:

$$K_{22}/K_{11} = \exp\ [(\sigma_f^2 + \sigma_\alpha^2\bar{\psi}^2)/(1 + \bar{\alpha}\lambda_1\cos\theta)] \tag{3}$$

where the $K_{22}$ is the effective hydraulic conductivity in the direction parallel to the bedding, and $K_{11}$ is in the direction normal to the bedding. The ratio depends on the variance of saturated hydraulic conductivity, variance of α, mean capillary pressure, mean α, correlation scales, and the orientation of the stratification, θ. It is obvious that as the mean capillary pressure increases, the anisotropy will increase exponentially, since other parameters are fixed for a particular soil.

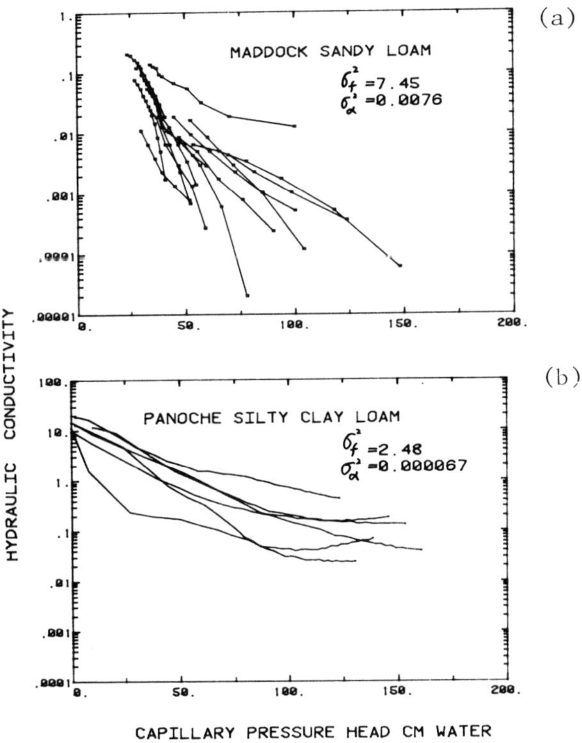


Figure 3a: Some representative K-ψ curves of Maddock sandy loam.

Figure 3b: Some representative K-ψ curves of Panoche silty clay loam.

What is the magnitude of the anisotropy in the field situation? The anisotropy ratios of two different soils; Maddock sandy loam and Panoche silty clay loam, are determined with an aspect ratio of 40.0. Figures 3a and 3b show representative K-ψ curves of these two soils along with estimated soil statistical parameters. Notice that the variation in slope of the K-ψ curves of sandy loam is relatively large in comparison to that of the silty clay loam. This is indicated by values of the variance of pore-size distribution parameter, α. The variance of α of sandy loam is about two orders of magnitude greater than that of silty clay loam.

By applying the stochastic model, Eq. (3), to these two soils, the resultant anisotropy ratios are shown in Figure 4 as a function of mean capillary pressure. The anisotropy ratio of sandy loam increases rapidly as the mean capillary pressure increases. On the other hand, the ratio of the silty clay loam remains almost constant. This is mainly attributed to the large variation in $\alpha$ of the sandy loam. This result implies that in the soils, such as the sandy loam, the horizontal conductivity could be much larger than the vertical. Thus, the water movement in this type of soil may be in the lateral direction. This lateral flow effect is intensified as the soil becomes drier.

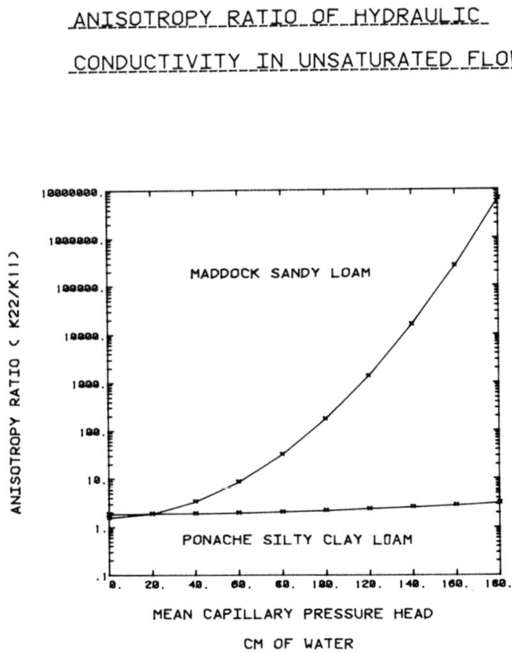


Figure 4: Anisotropy ratios of Maddock sandy loam and Panoche silty clay loam as a function of mean soil tension.

Immediate impacts of the results on classic hydrology concern the validity of the classic simulation of unsaturated flow in heterogeneous media. For example, if one wants to model the leakage beneath a disposal basin in a stratified soil such as sandy loam, the classic isotropic or anisotropic model may predict too much vertical movement of the moisture. In reality, the movement of moisture is in the lateral direction as shown by the stochastic results (see Figure 5).

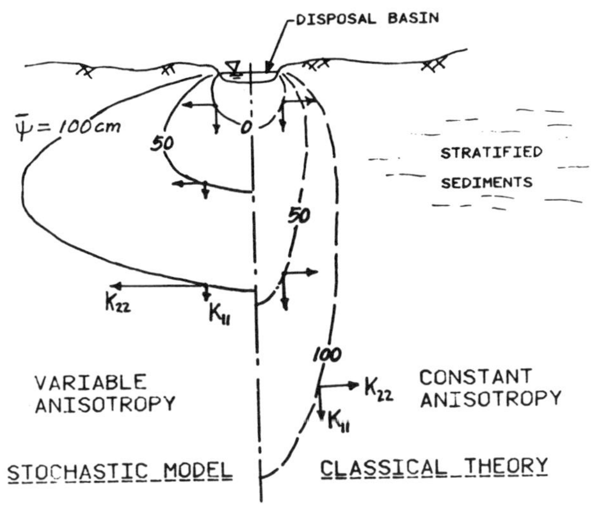


OBSERVATIONS ?

*CROSBY ET AL. (1968), WRR, 4(5)

-SPOKANE VALLEY GLACIAL OUTWASH

-MOISTURE & SOLUTE DISTRIBUTIONS BENEATH A SEPTIC TANK DRAIN FIELD

*HANFORD SITE ?

Figure 5: Soil tension distribution beneath a disposal basin.

Some field observations support this finding. Crosby et al. [6] reported the moisture and solute distribution beneath a septic tank drain field in a glacial outwash plan in Spokane, Washington are restricted to the top 25 feet of the soil formation. The soil below is relatively dry and the water table is about 90 feet below land surface. In addition, the lateral extent of the moisture and solute distribution is much larger than what they expected. Their explanation of this peculiar finding is the lateral flow effect. A similar phenomenon has also been observed at the Hanford site. Dirksen's [7] laboratory experiments indicate a moisture dependent anisotropy as well.

A moisture-dependent anisotropy may further suggest that semi-arid areas where the soil properties are similar to sandy loam may be potential sites for liquid waste disposal. Since the liquid tends to flow in the lateral direction, it is possible that the liquid waste can be confined to the upper part of the geologic formation without endangering the groundwater resources at depth.

The other direct application of the anisotropy concept is related to shallow underground waste storage facilities. Frind et al. [8] has proposed to use the unsaturated soil hydraulic conductivity properties, the "wick effect", to form a protective "umbrella" for the storage facility. Basically, the system utilizes two soils of different textures (see Figure 6). Under the unsaturated condition, fine-textured material is more conductive than gravel, moisture can be confined in the fine-textured layer, and the slope provides a lateral gradient so that water can be diverted laterally. However, one would prefer to use soil of higher anisotropy or larger horizontal hydraulic conductivity for the protective layer. The stochastic result suggests utilizing a material in which the variances of saturated hydraulic conductivity and pore-size distribution parameter are large.

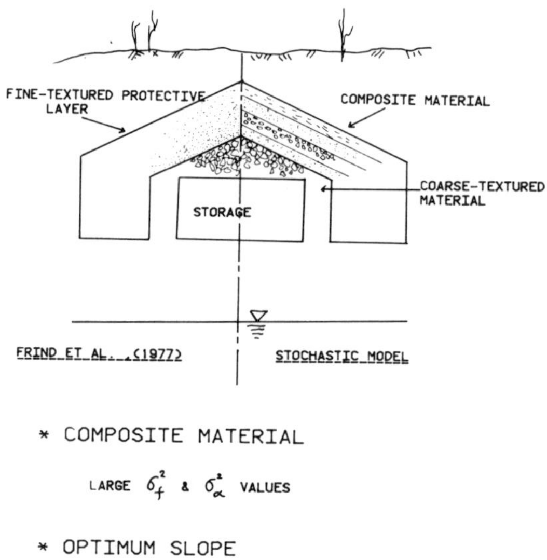


Figure 6: Configuration of the shallow underground storage facility.

This means that the protective layer should be a composite material composed of layers of different textured soils. In this way, the lateral conductivity can be much larger than the vertical, and the probability of water breaking through the interface is further reduced. In addition, the stochastic results also provides a mean to estimate the optimum slope for the design so that the maximum flux in the direction parallel to the bedding of the protective layer can be obtained.

## CONCLUSIONS

The unsaturated hydraulic conductivity anisotropy is moisture dependent. Thus, the classic approach to predict flow in unsaturated media may have to be carefully re-examined. This new finding should draw much attention to the anisotropy of unsaturated hydraulic conductivity in the future. Potential application of this anisotropy may be in liquid waste disposal and design of underground storage facilities. However, more field observations and laboratory experiments are needed to substantiate this result.

## REFERENCES


1. Nielsen, D. R., J. W. Biggar, and K. T. Erh, "Spatial Variability of Field Measured Soil-Water Properties," *Hilgardia* 42 (7), 215-260, 1973.
2. Coelho, M. A., "Spatial Variability of Water-Related Soil Physical Properties," Ph.D., Dissertation, 1974, University of Arizona, Tucson.
3. Bakr, A. A., "Stochastic Analysis of the Effect of Spatial Variations in Hydraulic Conductivity on Groundwater Flow, Ph.D. Dissertation, New Mexico Institute of Mining and Technology, Socorro, New Mexico, July, 1978.
4. Warrick, A. W., "Areal Predictions of Soil Water Flux in the Unsaturated Zone," Proceedings of National Conference, Irrigation Return Flow Quality Management, EPA and Colorado State University, May 16-19, 1977.
5. Yeh, T.-C., "Stochastic Analysis of Effects of Spatial Variability on Unsaturated Flow," Ph.D. Dissertation, New Mexico Institute of Mining and Technology, Socorro, New Mexico, July 1982.
6. Crosby, J. W., D. L. Johnstone, C. H. Drake, and R. L. Fenston, "Migration of Pollutants in a Glacial Outwash Environment, 1," *Water Resources Research*, 4(5), 1095-1113, 1968.
7. Dircksen, C., "Transient and Steady Flow from Subsurface Line Sources at Constant Hydraulic Head in Anisotropic Soil," *Transactions*, ASAE, Vol. 21 (5), 913-919, 1978.
8. Frind, E. O., R. W. Gillham, and J. F. Pickens, "Application of Unsaturated Flow Properties in the Design of Geologic Environments for Radioactive Waste Storage Facilities," *Proceedings of the first international conference on Finite Elements in Water Resources at Princeton University*, U.S.A. in July, 1976, Finite Elements in Water Resources, Pentech Press, 1977.

# ROLE OF PARTIALLY SATURATED SOIL IN LINER DESIGN FOR HAZARDOUS WASTE DISPOSAL SITES

D.B. McWhorter
Agri. Eng., Co. St. Univ., Ft. Collins, CO
J.D. Nelson
Civil Eng., Co. St. Univ., Ft. Collins, CO
T.A. Shepherd
President, Water, Waste and Land, Ft. Collins, CO
R.E. Wardwell
Principal, Water, Waste and Land, Ft. Collins, CO

## INTRODUCTION

Disposal areas containing toxic or hazardous materials are often constructed with low permeability liners to minimize the seepage losses from the impoundments. In many areas of the world, these disposal areas are located on or in natural formations which are several meters above the ground water level. In such cases, the foundation material between the impoundment and the phreatic surface of the aquifer is partially saturated. Estimation of seepage losses, required to assess potential environmental impacts and to provide input for water balance computations, should be based on the flow through partially saturated porous media since conventional saturated methods do not apply.

Previous efforts [1,2,3,4] have developed procedures which estimate seepage flows associated with those aspects of flow phenomena that have first order effects on the resulting seepage rates. The procedures developed apply to those cases where seepage is dominated by one-dimensional flow from the floor of the impoundment. The objective of the previous work was to provide engineering solutions which do not require impractical quantities of data or the use of numerical methods.

The purpose of this paper is to review the existing theory of flow from a waste disposal area and to demonstrate how the resulting seepage as estimated by the previous published procedures is influenced by the thickness and quality of the liner material. Design charts are presented which permit the selection of liner characteristics that will limit the seepage losses to desirable levels. The significance of the actual flow under a disposal area and the impacts of the liquid movement on the surrounding aquifers as predicted by the seepage model is discussed with special emphasis placed on the resulting contaminant migration.

## SEEPAGE MECHANICS

### Types of Impoundments

Hazardous waste disposal areas may be divided into two major groups: water-impoundment and non-water-impoundment areas. With the first type, the waste material is generally saturated and the disposal area is designed to retain the liquid inherent with the material upon initial disposition. Mill tailings impoundments and sludge material resulting from primary and secondary wastewater treatment facilities are examples of these types of disposal areas. With non-impoundment areas, the waste material is usually dry or at a very low water content, and is stockpiled or buried in disposal trenches. Examples of this are waste-rock material from the mining industry, stockpiles of ore for a heap leach process, and many of the low-level radioactive wastes from the uranium industry.

### Flow Mechanisms

Flow from these impoundments have been studied in previous papers. With liquid impoundment areas, flow occurs through the waste material, the liner, and the foundation soil. Eventually it may reach the ground water table. The general subsurface profile and water levels in the waste material and aquifer may be idealized as shown in Figure 1. The thickness of each layer is denoted by the symbol D with the subscripts wm, l, and f representing the thickness of the waste material, liner and foundation material to the phreatic surface, respectively. The symbol K denotes the saturated hydraulic conductivity of each particular strata, and is assumed to be uniform and isotropic.

Non-impoundment piles are usually designed so as not to retain water for long periods of time within the waste material itself. Water is introduced by precipitation or during loading. Because of this, these impoundment areas are underlain by low permeability blankets which drain the water to lateral collectors. A major concern is the potential for saturated zones to develop within the impoundment during this build-up of water and for seepage to occur through the impervious liner during this buildup.

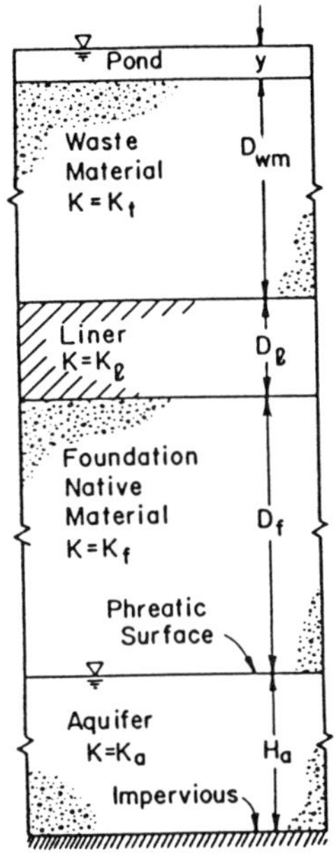


Figure 1. Schematic of water impoundment disposal area.

The techniques for analyzing the flow from both these impoundments have been presented elsewhere [1,2,3,4]. For demonstration purposes, the effects of liner design on the resulting seepage flows are illustrated using a saturated water-impoundment facility as represented in Figure 1. The same procedures used to develop design charts presented herein could also be applied to non-impoundment piles [4].

**Stages of Seepage Flow**

Seepage beneath a water impounding disposal area takes place in four distinct stages as illustrated on Figure 2. Each stage is described below.

Stage I: Wetting Front Advance - During this phase, the wetting front advances downward through the partially saturated underlying foundation material. Above the wetting front the material may or may not be saturated depending upon whether the waste material and liner

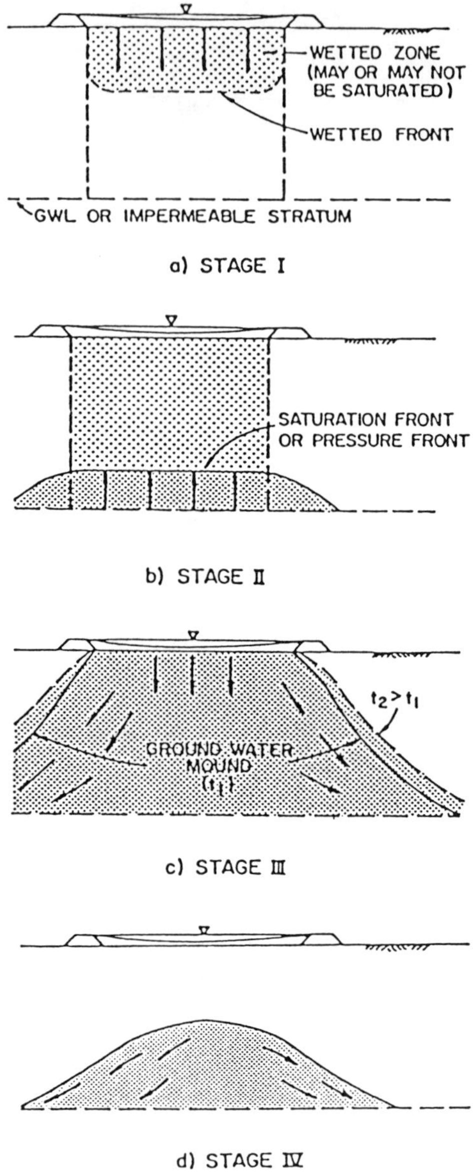


Figure 2. Stages of seepage from waste disposal area.

restricts the flow of water enough to prevent the foundation material from becoming saturated. The flow during this phase may be estimated using Darcy's Law which incorporates the effects of partially saturated flow through the foundation material.

Stage II: Rising Phreatic Surface - When the wetting front contacts either an impermeable stratum or the phreatic surface of an aquifer, a ground water mound will develop and rise toward the impoundment. The rate of rise of the phreatic surface depends upon the seepage rate, the fraction of the pore space in the foundation zone that remains available for water storage, and the tendency for ground water to spread laterally as the phreatic surface rises. If the soil behind the wetting front was saturated during Stage I, second stage activity is associated with a rising pressure front moving upward through the pore water until it contacts the bottom of the impoundment. In this case, Stage II is of short duration. Seepage rates are usually analyzed separately for the two cases [1].

Stage III: Ground Water Mound Developed - When the center of the ground water mound contacts the bottom of the impoundment, all flow takes place laterally from the saturated zone beneath the impoundment. Resistance to lateral spreading of the mound causes pore water pressures to increase at the interface between the impoundment and the foundation stratum. This causes a reduction in seepage rates. Rigorous estimations of flow in Stage III are a mathematically complex analysis. An upper bound for flow has been derived by analyzing a succession of steady state solutions [1].

Stage IV: Completion - Stage IV consists of drainage of the seepage column and dissipation of the ground water mound after the deposition of waste material is completed and after depletion of the source of water from the impoundment. The flow during Stage IV varies depending upon the seepage phase that existed when the source of water was depleted. If Stage I had not been completed, a slug of water would continue to move downward and dissipate when it reaches the ground water table. If Stage II existed, then some additional development of the ground water mound may take place before Stage IV is initiated. To analyze Stage IV requires site specific considerations of reclamation plans and decant operations.

### Contaminant Migration

The introduction of contaminants into the ground water regime is one of the most important environmental considerations when evaluating the influences of hazardous waste disposal systems on the adjacent area. Contaminants are transported by several mechanisms and are attenuated by the geochemical processes which occur in a porous media. These attenuation processes include precipitation, co-precipitation, adsorption and decay [5,6]. Because only a portion of the pore space contains liquid available for the transport, the migration of contamination in the partially saturated zone is greatly reduced. In addition, considerable storage may be available in the foundation material if the initial water content is below the residual water content. Consequently, a portion of the seepage may never reach the natural ground water. Analyses of this migration and the additional geochemical processes, which occur after the contaminants reach the ground water table and before they encounter an exit point, need to be incorporated into the design of the system.

## DESIGN ANALYSIS

### Criteria

The design of a liner for a waste disposal system incorporates a balance between the contaminant migration associated with the resulting seepage flow during the various stages of activity with the attenuation capacity of the foundation strata. It is often times advantageous to design a facility such that Stage I has not terminated when the disposal area becomes inactive. By doing this, a ground water mound does not develop during the disposal activities which, in turn, reduces the potential for any contaminants to enter an existing or new water well or other discharge points. While the seepage rates remain constant under Stage II, additional contaminants enter the ground during this phase as the ground water mound develops. Once Stage III is established, the flow rates drop, but the total volume of potentially contaminanted water becoming part of the ground water system increases as the mound spreads laterally. In addition to these reasons, limiting the seepage to Stage I conditions would provide a better situation for balancing the attenuation properties and storage capacity of the foundation materials with the seepage flows. In this manner, any contaminants which reach the critical discharge points of the ground water system would be minimal and within tolerable levels.

Stage I seepage activity will be analyzed in this paper for demonstration purposes because of the desireability of designing for this phase. The same procedures illustrated herein could be extended for the other phases of seepage flow, if desired, using previously developed flow equations [1].

Selection of Parameters

The analyses of partially saturated flow can be modeled by the application of Darcy's Law in the same manner as saturated flow as long as the effective conductivity at the resulting degree of saturation is used in place of the saturated conductivity. A discussion of the input parameters needed to analyze partially saturated flow are present elsewhere [4].

One of the most important parameters affecting the flow regime is the reduction in conductivity associated with the reduction in water content in the partially saturated zone above the ground water table. Shown in Figures 3 and 4 are typical pressure-saturation curves for

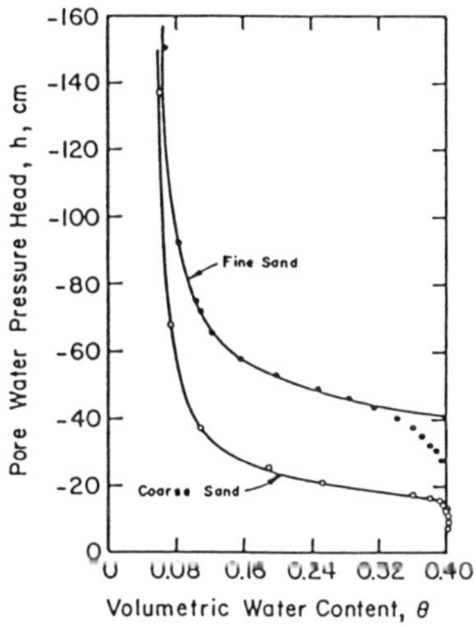


Figure 3. Typical relationship between capillary pressure head and saturation.

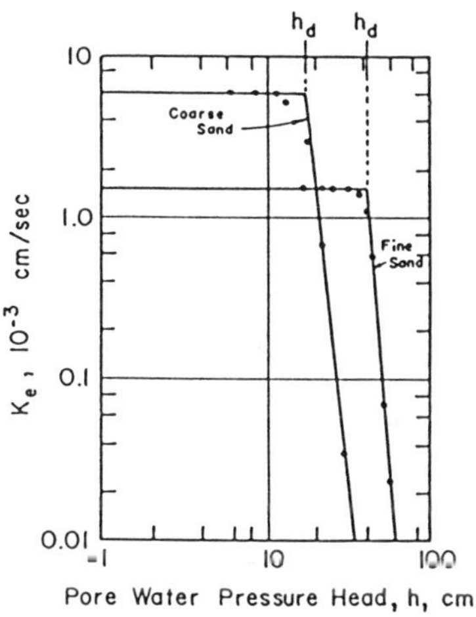


Figure 4. Typical relationship between hydraulic conductivity ($K_e$) and pore water pressure head.

two types of soils and the resulting relationships between capillary pressure and the effective hydraulic conductivity, $k_e$. In this figure, $h_d$ is the displacement pressure of the foundation material which is the negative threshold pressure when the air displacing the water becomes continuous in the strata. Very little desaturation occurs until the pore-water pressure head becomes more negative than $h_d$. The volcanic sand starts out at a higher conductivity by almost an order of magnitude than the fine sand. However, this value for the volcanic sand quickly drops since the volcanic sand desaturates at much lower capillary pressures.

Incorporation of this relationship into the Darcian flow equation permits the evaluation of partially saturated flow. Brooks and Corey [7] have shown that a functional relationship exists between capillary pore-water pressure and volumetric water content. This relationship can be expressed as a function between saturated conductivity, $K_s$, and partially saturated conductivity, $K_e$ [1]. It is recognized that the volumetric water content as a function of pore water pressure is not unique but depends upon the wetting and drying history. However the desaturation equations by Brooks and Corey have been used with success when the volumetric water content was increasing (8,9). Other curve fitting procedures relating $K_s$ to $K_e$ are available and could be used in place of the Brooks/Corey theory if desired.

Modeling Equations - Stage I

Estimation of seepage rates during Stage I flow requires the simultaneous solution of the Richard's Equation for the flow in the foundation material and the Laplace Equation for saturated flow in the waste material, both equations written for one-dimensional vertical flow [1]. The estimation of seepage rates will require different procedures depending upon whether the material behind the wetting front is saturated or partially saturated. As such, the first step in the analysis is to determine if the wetting front produces saturated or partially saturated conditions. With vertical seepage from most waste disposal areas, the hydraulic gradient is near unity and the seepage rate is close to the hydraulic conductivity of the liner and foundation strata. If the conductivity of the liner is much less than that of the foundation material, flow is controlled by the liner. In this case, the foundation area behind the wetting front will be partially saturated. If the liner conductivity is close to the foundation conductivity, then the foundation soil behind the wetting front will be saturated.

The determination of saturated or partially saturated wetting front may be expressed in terms of capillary pore water pressure by:

$$h_f < h_d \quad \text{(partially saturated)} \tag{1}$$

$$h_f > h_d \quad \text{(saturated)} \tag{2}$$

$$h_f = \left(y + D_{wm} + D_1\right) - K_f \left(\frac{D_{wm}}{K_{wm}} + \frac{D_1}{K_1}\right) \tag{3}$$

where $h_f$ is the pore water pressure at the interface between the foundation material and the liner, and $h_d$ is the displacement pressure.

Seepage for partially saturated conditions is primarily controlled by the conductivity of the waste material and the liner. Seepage rates in this case can be estimated by considering flow through a sequence of strata which yields the following equations:

$$q = \frac{\left(y + D_{wm} + D_1\right) - h_d \left(\frac{q}{K_f}\right)^{-1/(2+3\lambda)}}{\frac{D_{wm}}{K_{wm}} + \frac{D_1}{K_1}} \tag{4}$$

Seepage in this equation is the instantaneous seepage at any given time. The mean value of seepage, $q_m$, is obtained after computing q as a function of time using y, and $D_{wm}$ as functions of time as controlled by the planned placement rate of waste material.

The time required for the wetting front to travel a distance, L, may be calculated from the following:

$$t = \frac{L\left(\theta_f - \theta_i\right)}{q_m} \tag{5}$$

where $\theta_f$ and $\theta_i$ are the final and intial volumetric water contents of the foundation material.

The seepage rates under saturated wetting front are substantially influenced by the hydraulic properties of the foundation material. In addition, a time variation of q exists that cannot be ignored even if the saturated thickness of the waste material remains constant. Generalizing the Green-Ampt infiltration equation to account for the retarding effect of the waste material and liner yields the following seepage equation:

$$q = K \left( \frac{L + A - h_c}{L + K\ B_f} \right) \tag{6}$$

where:

$$h_c = h_d \left( \frac{2+3\lambda}{1+3\lambda} \right) \tag{7}$$

$$A = y + D_{wm} + D_1 \tag{8}$$

$$B = \frac{D_{wm}}{K_{wm}} + \frac{D_1}{K_1} \tag{9}$$

and L is the distance from the liner to the wetting front. The time for any wetting front penetration is given as:

$$t = \frac{n-\Theta_i}{K_f} \left[ L + \left( K_f B - A + h_c \right) \ln \left( \frac{L+A-h_c}{A - h_c} \right) \right] \tag{10}$$

where n is the porosity of the foundation strata. To solve these equations, a depth of penetration, L, is selected and the corresponding time and seepage rate calculated.

## INFLUENCES OF LINER THICKNESS AND CONDUCTIVITY

The effect of liner thickness and conductivity on the resulting seepage rates may be analyzed by solving equations (4) through (10) using varying values of $D_1$ and $K_1$. For a given situation, a set of design charts can be prepared relating $D_1$ and $K_1$ to the resulting seepage rates. The equations presented herein may be solved using hand held calculators and are easily programed on desk top computers. Codes are presently available for micro-computers which will generate design charts in a expeditious manner [2]. With these charts and a knowledge of available materials at a particular site, an economic balance can be achieved between liner quality and thickness.

As an example and to provide the basic relationship between liner characteristics, a sample waste disposal impoundment is analyzed. The properties of the various materials are presented in Table 1. Note that for simplicity of discussion the depth and saturated thickness of the waste material was taken as a constant. This is generally not the case with impoundments containing mill tailings which are constantly being placed by hydraulic means. However, with saturated sludges and other landfilled materials, the waste product is often times placed in a thick, uniform lift. In this case, the approximation presented here is reasonable for this situation. For tailings or other products which are filled at a certain rate, the equations and subsequent relationships can be modified to incorporate the time variance of thickenss and hydraulic conductivity if desired [1,2,3].

The general influence of various impoundment and substratum properties on seepage during the various phases of flow from an impoundment area have been presented elsewhere [10]. This paper uses the same basic properties to evaluate in detail the effects of liner configuration on the resulting seepage flows during the critical Stage I stage.

The first step in analyzing the Stage I activity is to determine if the area behind the wetting front is saturated or partially saturated. Partially saturated conditions exist when the capillary pore-water pressure head at the liner/foundation interface just equals the displacement pressure head, $h_d$, of the foundation material. A relationship can be derived from Eq.(3) which gives the critical liner thickness for a given conductivity which will result in partially saturated conditions to exist behind the wetting front. For the parameters given in Table 1, the equation of this liner thickness as related to conductivity is:

$$D_1 = \frac{\left(7.7 \times 10^4\right) K_1}{1 - \left(K_1 \times 10^4\right)} \tag{11}$$

Presented in Figure 5 is the relationship between the combinations of liner conductivity and thickness which will achieve a partially saturated conditions behind the

Table I. Impoundment and Strata Properties

| Parameter | Strata | | |
|---|---|---|---|
| | Waste Material, wm | Liner 1 | Foundation Material, f |
| thickness, D (meters) | 6 | varies | 20 |
| Saturated Hydraulic Conductivity, K (cm/sec) | $2x10^{-3}$ | varies | $1x10^{-4}$ |
| porosity, n | - | - | 0.4 |
| volumetric water control, $\theta$ | | | |
| initial $\theta_i$ | - | - | 0.0 |
| residual $\theta_r$ | - | - | 0.13 |
| Pore Size Distribution, $\lambda$ (Brooks-Corey) | - | - | 2 |
| Apparent Specific Yield, $S_{ya}$ | - | - | 0.2 |
| displacement pressure head, $h_d$ (meters) | - | - | -2.3 |

wetting front for the parameters presented in Table 1. For a given conductivity, a liner thickness thinner than the relationship shown on Figure 5, and conversely, for a given liner thickness, a more permeable liner than the conductivity shown on Figure 5 will result in saturated conditions existing behind the wetting front.

The relationship between the seepage rate and liner conductivity is shown on Figure 6 for various liner thickness. As may be seen on the figure, the seepage rate drops significantly as the liner conductivity reduces. The small drop in seepage rates is observed with the higher conductivity liners, especially with the thinner liner thicknesses. These rates are associated with saturated conditions existing behind the wetting front. Once the wetting front becomes partially saturated, the seepage rates reduce at a near linear fashion with conductivity when plotted on logarithmic scales. This would be expected since partially saturated seepage as calculuated by equation (4) relates exponentially to the impoundment parameters.

By comparing equations (4) and (6) with Figure 6, it can be seen that flow from a given waste material is controlled primarily by the conductivity and thickness of the liner when the area behind the wetting front is partially saturated. In this case, the liner is so impervious that it restricts the flow of water from the impoundment to such an extent that the seepage rate is basically independent of the foundation hydraulic properties. With the constant thickness of waste material, the seepage rate is constant with time for a given liner configuration since y and $D_{wm}$ are constant with time.

For saturated conditions, seepage rates are substantially influenced by the hydraulic properties of foundation material. The liner does not restrict the flow enough so that the foundation material is unable to accept the seepage flow from the impoundment. This results in an increase in pore-water pressure at the interface which in turn reduces the flow as the wetting front advances and the hydraulic gradient decreases. The seepage rates calculated in Figure 6 during saturated flow were calculated after a 10 day period. While the linear portion of the curves would be the same at different time periods, the non-linear portion of the curve would change with time and approach the straight line relationship associated with partially saturated conditions.

As may be seen on Figure 6, seepage rates reduce drastically with decreasing conductivity and increasing

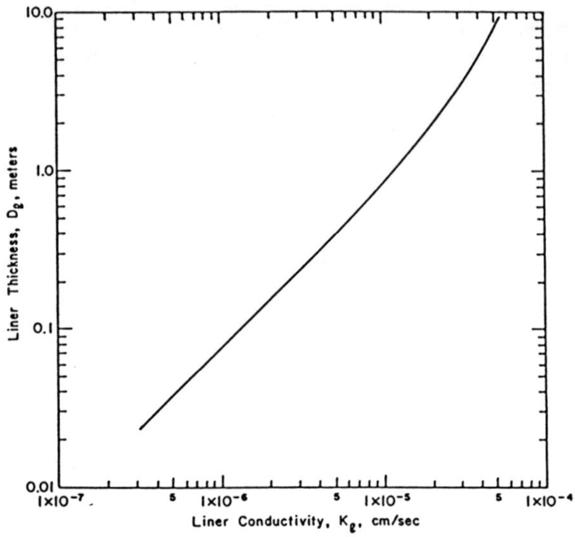


Figure 5. Liner thickness and conductivity for partially saturated wetting front.

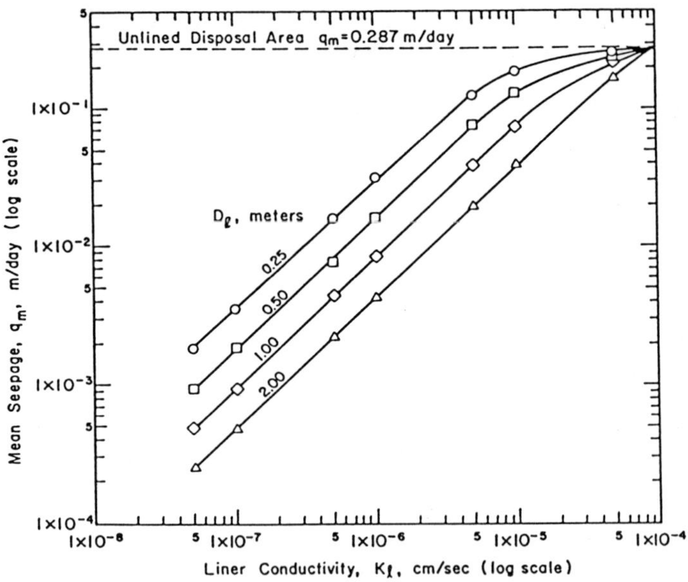


Figure 6. Stage I flow: Effects of liner conductivity on seepage flow.

liner thickness. However, it appears that reducing conductivity has a much greater influence in reducing seepage than by increasing the liner thickness. This effect is illustrated in Figure 7 when liner thickness is plotted against mean seepage for various conductivities. The large reduction of seepage rates for thin liners illustrates the transition from saturated conditions to partially saturated conditions behind the wetting front.

The curve for $K_1 = 5 \times 10^{-5}$ in/sec shown on Figure 7 shows little reduction in seepage rates as the liner thickness increases. With this conductivity, the wetting front is saturated and, as such, the seepage rate is not highly influenced by the thickness of the liner. To maximize the utilization of the liner, partially saturated conditions need to be established behind the wetting front. With this situation, the resulting seepage is highly influenced by the characteristics of the liner itself and somewhat independent of the foundation material.

These two figures illustrate that an interaction between liner thickness and conductivity influences the final seepage rates from impoundment areas. However, there is a limited range of practical liner thicknesses which can be used in an impoundment area. This, plus the fact that seepage is more influenced by changes in conductivity, indicate that the selection of the physical behavior of the liner material is a more important criteria when attempting to design a waste disposal area than by increasing the liner thickness.

## DESIGN PROCEDURES

Figures 5, 6, and 7 were plotted for the given parameters shown in Table 1. As such, they could be used as design charts for the selected situation in an effort to balance liner thickness and conductivity for the desired allowable seepage. Similar design charts could easily be established for other situations using relatively easy analytic methods and simple computer codes [1,2]. With these charts, the design of the liner maybe done on both an economic and an environmentally sound manner.

The basic steps to design a liner system involve an analyses of the potential contaminant migration which might occur with a given seepage rate from an impoundment and the potential attenuation properties of the foundation material both in the partially saturated zone above the

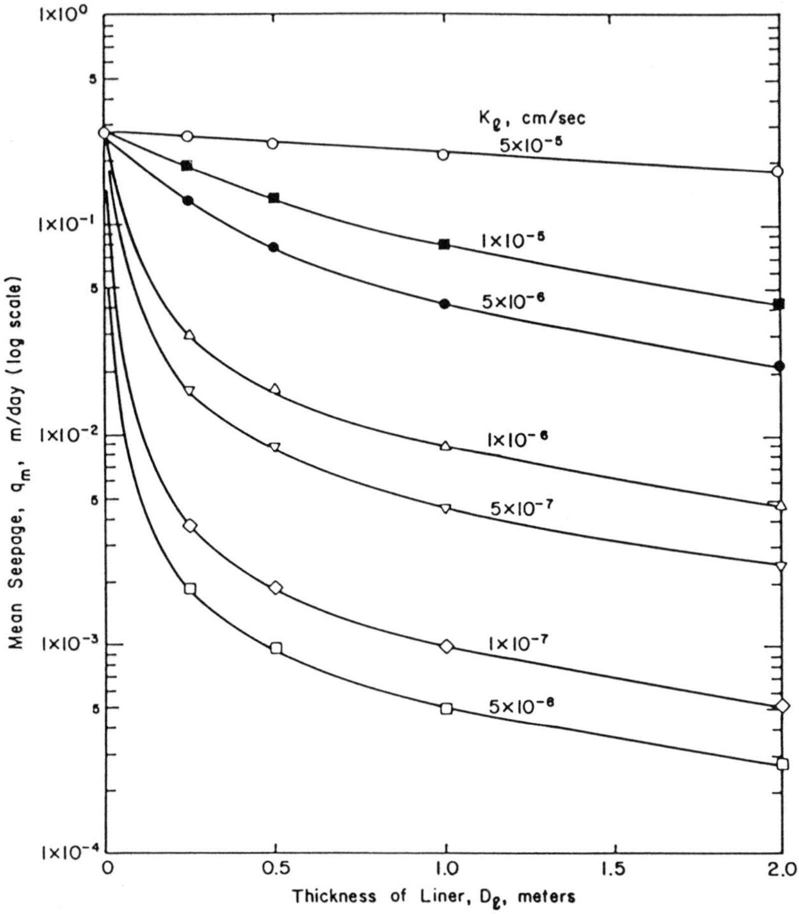


Figure 7. Stage I flow: Effects of liner thickness on seepage flow.

water table and in the aquifer itself before the nearest critical discharge point is reached. Procedures have been established which will analyze the geochemical processes occurring during this process. These will help determine the allowable seepage which will not cause detrimental impacts on the adjacent environment [5,6].

Once the allowable seepage has been determined based on an analysis of the contaminant migration, design charts such as the ones presented herein can be established for the site specific situation. The charts aid in selecting the quality and quantity of liner needed to achieve the allowable seepage. The charts presented in Figures 5, 6, and 7 were for Stage I flow from an impoundment with a constant depth of waste material and a homogenous foundation strata. The time rate of waste placement can easily be incorporated into the analysis by using a time rate of $D_{wm}$ and y increase during the operational life of the impoundment. Stratified foundation soils and stratified waste materials such as slimes and sands from mill tailings could also be analyzed by evaluating the seepage stages that would occur as the wetting front encounters other strata of different conductivities. At these points, the McWhorter/Nelson approach [1] used to analyze the flow when the ground water table or an impervious boundary was encountered could be modified for seepage through the different strata.

Several other assumptions were made in the case presented herein to simplify the analysis. These can easily be altered to better model a given field situation. The liner was assumed to be saturated at the time of waste material placement. This is a reasonable assumption considering the fact that earthen materials are usually compacted close to optimium moisture content which results in a near saturated material. If the liner is determined to be partially saturated, the flow of waste water through this strata could also be analyzed using these same procedures.

The flow in this example was associated with Stage I activity since this is generally the critical conditions which need to be evaluated. Flow during the other phases could be analyzed using previously developed equations [1,2,3]. However often times it is more advantageous to design the system primarily for Stage I activity and to create partially saturated conditions behind the wetting front. Under this situation, a minimal amount of sepage will enter the subsurface area, reducing the amount of contaminants that can potenially impact the ground water regime. By establishing partially saturated conditions, the pore area occupied by water and thereby the pore area available for contaminant transport will be limited. Any contaminants would also have close proximity to the surface of the soil matrix as the saturation of the porous media decreases. This close proximity especially at low degrees of saturation would assist in the efficiency of the geochemical processes. Through judgicial selection of the liner characteristics, a balance might be obtained between the toxins entered into the foundation area and the attenuation capacity of the strata itself.

## CONCLUSION

A method has been presented which assists in selecting an economic but environmentally sound liner system by incorporating the flows through partially saturated foundation strata and by analyzing partially saturated and saturated conditions behind the wetting front. Design charts were prepared for a specific case and procedures outlined to derive these for other conditions and to modify them for a variety of other situations. Use of this technique and the analysis of the partially saturated flow regimes has been applied to the design of the liner material and thickness but could easily be adapted to the evaluation of other parameters influencing the resulting environmental impact of a hazardous waste disposal area.

## REFERENCES


1. McWhorter, D.B., and Nelson, J.D., "Unsaturated Flow Beneath Tailings Impoundments", Journal of Geotechnical Engineering Division, ASCE, Vol. 105, GT 11, November, 1979, pp. 1317-1334.

2. Davis, Lyle A., Computer Analysis of Seepage and Groundwater Responses Beneath Tailings Impoundments, Final Report, National Science Foundation, DAR-7917109, April, 1980.

3. Nelson, J.D., and McWhorter, D.B., "Water Movement in Non-Impounding Waste Dumps", Proceedings, Fall Meeting, AIME, November, 1981.

4. McWhorter, D.B., and Nelson, J.D., "Seepage in the Partially Saturated Zone Beneath Tailings Impoundments", Mining Engineering, Vol. 32, No. 4, April, 1980, pp. 432-439.

5. Shepherd, T.A., and Cherry, J.A., "Contaminant Migration in Seepage from Uranium Tailings Impoundments-An Overview", Symposium on Uranium Mill Tailings Management, Geotechnical Engineering Program, Civil Engineering, Colorado State University, Fort Collins, November, 1980.

6. Shepherd, T.A., "Contaminant Migration in Seepage from Tailings Impoundments", Geotechnical Aspects of Tailing Impoundment Design, Short Course Notes, 1982.

7. Brooks, R.H. and Corey, A.T. "Properties of Porous Media Affecting Fluid Flow", Journal of Irrigation and Drainage Division, ASCE , Vol. 92, IR 2, June, 1966, pp. 61-80.

8. Ahuja, L.R. "A One-Step Wetting Procedure for Determining Both Water Characteristics and Hydraulic Conductivity of Soil Core", Soil Science Society of America Proceedings , Vol. 39, 1975, pp. 418-423.

9. Jeppson, R.W., et.al., "Use of Axisymetric Infiltration Model and Field Data to Determine Hydraulic Properties of Soil", Water Resources Research , Vol 11, No. 1, 1975, pp. 127-138.

10. Nelson, J.D., and McWhorter, D.B., "Influence of Impoundment and Substratum Configuration on Seepage From Impoundments", First International Conference on Uranium Mine Waste Disposal , AIME, New York, 1980, pp. 193-204.

MOISTURE CHARACTERISTICS OF COMPACTED SOILS FOR USE IN TRENCH COVERS

S. Klein, T. M. Johnson, and K. Cartwright
Illinois State Geological Survey
Champaign, Illinois 61820

## INTRODUCTION

The hydraulic behavior of unsaturated, compacted, fine-grained materials affects the performance of several types of engineering facilities, such as covers for waste disposal sites, dam cores, and highway subgrades. Unsaturated moisture movement may significantly affect infiltration through trench covers; highway subgrades are also subjected to seasonal temperature or moisture variations.

This study, sponsored by the U. S. Nuclear Regulatory Commission, investigates the hydraulic conductivity of two geologic materials that are relatively common in central Illinois—windblown silt (loess) and glacial till. The importance of compaction moisture content is illustrated for unsaturated and saturated conditions.

## SUMMARY OF PREVIOUS STUDIES

In an early study of the hydraulic conductivity of compacted, fine-grained soils Lambe [1] pointed out that the hydraulic conductivity was influenced by soil composition, chemistry of the pore fluid, void ratio (or porosity), soil structure, and degree of saturation.

The influence of soil structure on the hydraulic conductivity of fine-grained soils is one of the more interesting factors. Several investigators [1, 2, 3] have observed much lower values of saturated hydraulic conductivity for samples compacted wet of optimum moisture content than for samples compacted dry of optimum. Lambe [4] explained this behavior in terms of differences in soil structure (or fabric) for samples compacted at moisture contents wetter or drier than optimum. According to this theory, fine-grained soils compacted at above optimum water content exhibit a more flocculated structure (Figure 1) and those compacted at below optimum water content have a more dispersed structure. The more flocculated structure implies a more random orientation of the soil particles and therefore larger pores, which explains the greater hydraulic conductivity.

Direct measurement of the unsaturated hydraulic conductivity is a difficult process, but has been done [5, 6, 7]. However, the current trend is to estimate the unsaturated hydraulic conductivity from the results of simpler, less complex laboratory tests. The estimates are based on the pore size distribution of the materials (reflected by the moisture content-capillary pressure relationship during desaturation) and measurement of the saturated hydraulic conductivity. Marshall [8] first proposed this approach, and his method has been modified and evaluated in several other studies [7, 9, 10] , which have concluded that empirical methods can provide reasonable estimates of the unsaturated hydraulic conductivity.

## PURPOSE

Previous researchers have shown the importance of pore size in determining the unsaturated hydraulic properties of earth materials. The influence of compaction on the pore size and saturated hydraulic conductivity of fine-grained soils has also been documented in the literature. This study will show the influence of compaction and the resulting soil structure on the unsaturated hydraulic properties of two fine-grained soils.

## DESCRIPTION OF SOILS

A laboratory testing program was conducted on two soils that are relatively common Quaternary deposits in Illinois.

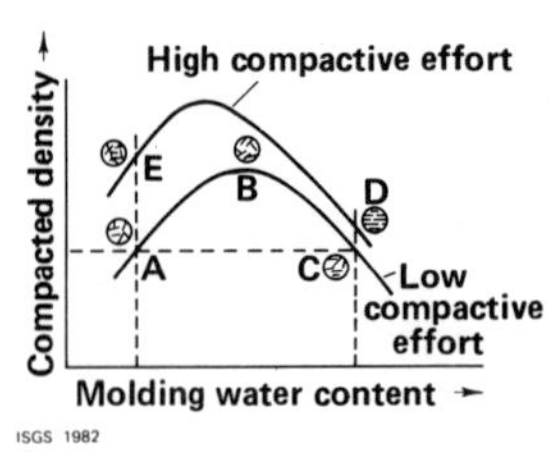


Figure 1. Effects of compaction on structure [Lambe, 1980].

Samples of a brown sandy silt (loess) and a gray sandy clay (glacial till) were obtained from St. Clair and De Witt Counties, respectively.

The silt, referred to as Peoria Loess, is a product of Wisconsinan glaciation. A significant portion of the clay-sized fraction of the loess consists of expandable clay minerals. The unit averages 1.5 m in thickness over 90 percent of Illinois, and obtains a maximum thickness of 23 m [11]. The clay is glacial till from the Fairgrange Till Member of the Wedron Formation deposited during the Wisconsinan period of glaciation. The Wedron Formation occurs over about a third of Illinois; the till consists of sand, silt and clay with some gravel-sized particles. Illite is the dominant clay mineral in the till. The results of classification tests performed on the soils are summarized in Table I.

TABLE I. Results of Classification Tests

| Soil | Atterberg Limits | | Particle Size Analysis | | | |
|---|---|---|---|---|---|---|
| | Liquid Limit | Plasticity Index | Gravel (%) (>2mm) | Sand[b] (%) (2-.0625mm) | Silt[b] (%) | Clay[b] (%) (<0.002mm) |
| Peoria Loess | 24 | NP[a] | 1 | 7 | 82 | 11 |
| Fairgrange Till | 26 | 11 | 4 | 28 | 49 | 23 |

a - Nonplastic
b - % of < 2mm fraction

## LABORATORY TEST RESULTS

The laboratory testing program consisted of standard compaction curves, moisture content-capillary pressure tests, and tests to determine the saturated hydraulic conductivity.

### Falling Head Permeability Tests

The saturated hydraulic conductivity was determined by using an air-pressure permeameter. Remolded soil samples are saturated under vacuum and then subjected to a water

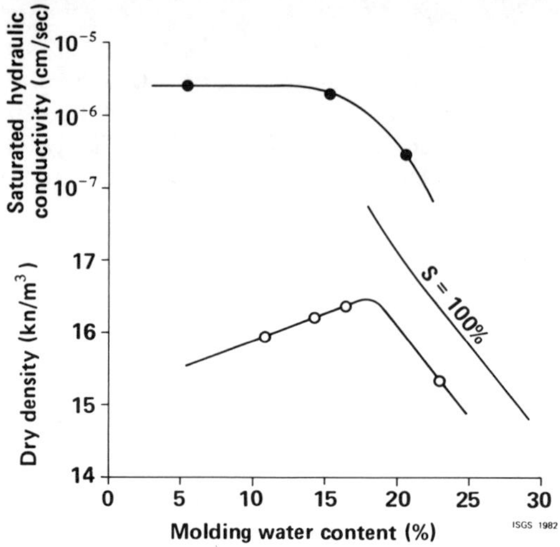


Figure 2. Results of compaction and hydraulic conductivity tests on Peoria Loess.

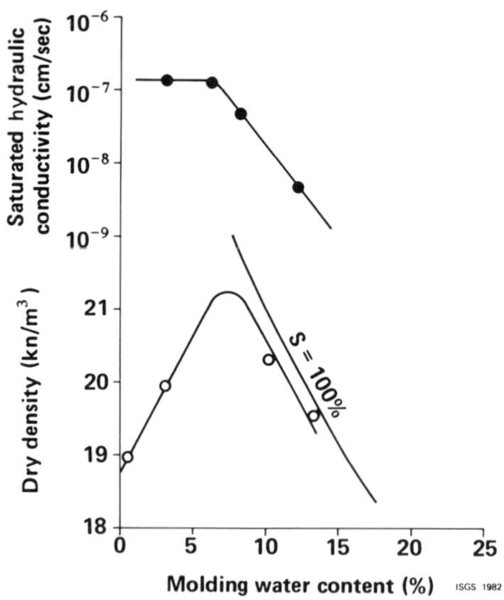


Figure 3. Results of compaction and hydraulic conductivity tests on Fairgrange Till.

pressure held constant by a storage tank and a flexible diaphragm with pressured air behind it. Water flows through the sample and rises in a standpipe, decreasing the head loss across the sample. Since the decrease in head loss across the sample is small, the test can also be interpreted as a constant head test. Hydraulic gradients varied from 11 to 21.

Samples were prepared, using the Harvard Miniature Compaction apparatus. The samples were compacted in a cylinder 33mm in diameter and 72mm high, using five layers to obtain a predetermined dry density. (This was done to prepare samples compacted at various moisture contents to the same porosity.) In this manner the influence of the soil structure can be investigated by eliminating the influence of sample porosity. Samples were prepared to densities of about 95% of the maximum density for the loess and 90% for the till as determined from compaction tests shown in Figures 2 and 3. The tamper had a 14.06 kg/cm$^2$ contact pressure and the compaction energy varied from 1415 to 2550 kg-m/m$^3$ to produce samples of the desired densities.

The results of the compaction tests and the values of the saturated hydraulic conductivity are shown in Figures 2 and 3. As should be expected, much lower values of saturated hydraulic conductivity were observed for samples compacted at moisture contents higher than optimum than for samples compacted at moisture contents below optimum. Some of the decrease in hydraulic conductivity for the till sample compacted wet of optimum (Figure 3) may be due to decreased sample porosity.

## Moisture Content-Capillary Pressure Tests

Measurements of the moisture content-capillary pressure relationship were performed, using Tempe cells obtained from Soil Moisture Equipment Corporation, Santa Barbara, CA (catalog no. 1400). (Mention of trade names does not imply endorsement by the Illinois State Geological Survey or sponsoring agency.) Undisturbed or remolded specimens are placed in the cell, saturated, and subjected to air pressure that is sequentially increased. The Tempe cell apparatus can be used up to a pressure of about 1000 cm of water (1 bar). Tests conducted at higher pressure utilize a ceramic plate or pressure membrane extractor. The moisture content-capillary pressure relationship is determined from the measured reduction in weight as the sample is desaturated.

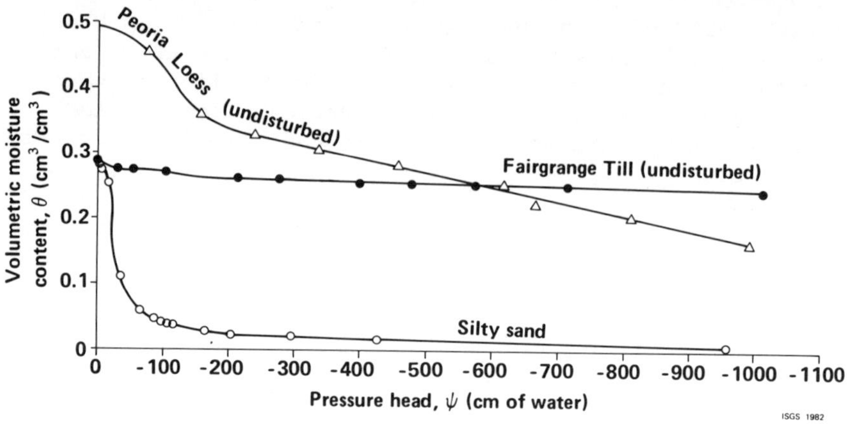


Figure 4. Soil moisture retention curves for various soils.

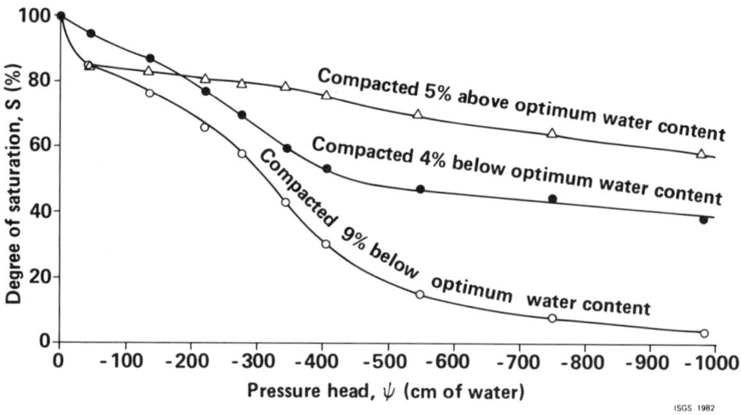


Figure 5. Desaturation curves for compacted Peoria Loess samples.

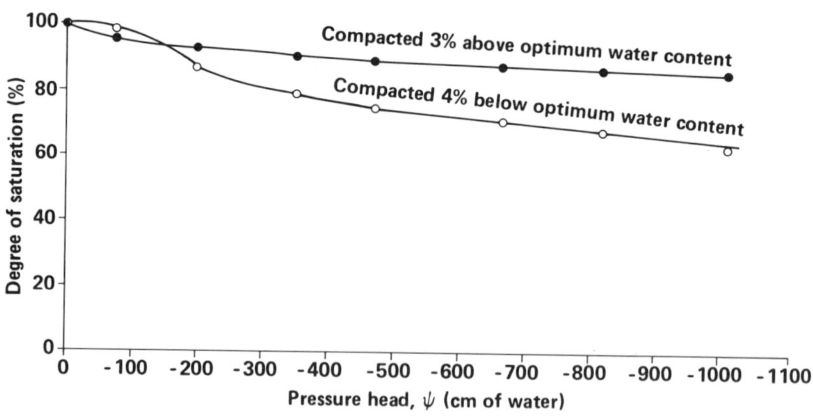


Figure 6. Desaturation curves for compacted Fairgrange Till samples.

Samples of till were compacted in brass rings approximately 54 mm in diameter and 30 mm high, using the Harvard Miniature compaction apparatus. Loess samples were obtained by pushing the same rings into soil that had been compacted into a standard compaction mold (1/30 of a cubic foot). Undisturbed samples of loess and till were also obtained from field exposures in driven brass rings.

Moisture content-retention curves of undisturbed till, loess and silty sand are shown in Figure 4. The results of desaturation tests on compacted loess and till samples are presented in Figures 5 and 6. Samples compacted at moisture contents above optimum retain more moisture at higher pressures than samples compacted dry of optimum. The smaller pores of the samples compacted on the wet side are not desaturated as easily as the larger pores of the samples compacted dry of optimum.

## CALCULATION OF UNSATURATED HYDRAULIC CONDUCTIVITY

The unsaturated hydraulic conductivity was estimated, following the procedure of Green and Corey [10]. This procedure requires the results of soil moisture-capillary pressure tests and a value for the saturated hydraulic conductivity for the same soil.

The hydraulic conductivity is estimated by dividing the soil moisture-capillary pressure curve into m-equal moisture content increments and solving the following equation at each moisture content:

$$K(\theta)_i = \frac{K_s}{K_{sc}} \cdot \frac{30\,\gamma^2}{pg\eta} \cdot \frac{\varepsilon^p}{n^2} \sum_{j=i}^{m} (2i + 1 - 2i) h_j^{-2}$$

$$i = 1, 2, \ldots, m$$

where $K(\theta)_i$ is the calculated conductivity for a specified water content or pressure (cm/min),

$\theta$ is the water content ($cm^3/cm^3$),

$i$ denotes the last water content class on the wet end, e.g., $i = 1$ identifies the pore class corresponding to the saturated water content, and $i = m$ identifies the pore class corresponding to the lowest water content for which conductivity is calculated,

$K_s/K_{sc}$ is the matching factor (measured saturated conductivity/calculated saturated conductivity),

$\gamma$ is the surface tension of water (dynes/cm),

$p$ is the density of water ($g/cm^3$),

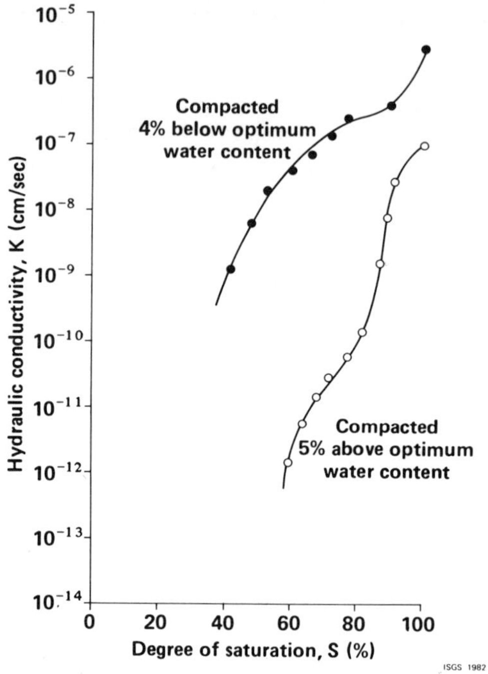


Figure 7. Hydraulic conductivity vs degree of saturation for Peoria Loess.

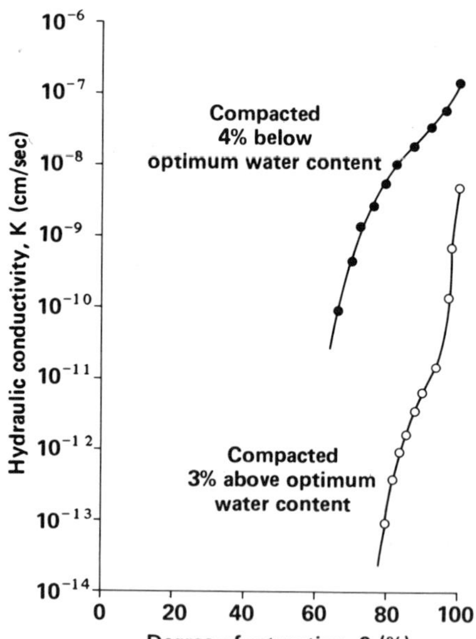


Figure 8. Hydraulic conductivity vs degree of saturation for Fairgrange Till.

$g$ is the gravitational constant ($cm/sec^2$),
$\eta$ is the viscosity of water ($g/cm\ sec^{-1}$),
$\varepsilon$ is the porosity ($cm^3/cm^3$), equal to $\theta_s$,
$p$ is a parameter that accounts for interaction of pore classes, equal to 2.0 for these calculations,
$n$ is the total number of pore classes between $\theta$ = 0 and $\theta_s$, the saturated water content:

$$n = m\left(\frac{\theta_s}{\theta_s - \theta_i}\right)$$

$h_j$ is the pressure for a given class of water-filled pores (cm of $H_2O$).

The pressure is evaluated at the midpoint of each moisture content increment. A matching factor Ks/Ksc is used to bring the saturated hydraulic conductivity into agreement with calculated value. This essential step serves to calibrate the procedure for each soil type [7].

The results of the hydraulic conductivity calculations as a function of degree of saturation are shown in Figures 7 and 8. Calculations were performed for the loess and the till samples compacted wet and dry of optimum. The figures indicate an increasing difference in the hydraulic conductivities of soils compacted wet and dry of optimum as the soils are desaturated. When the two soils are saturated the ratio of hydraulic conductivity of the soil compacted dry of optimum to that of the soil compacted wet of optimum is 30 and 26 for the loess and till, respectively. This ratio increases to 2500 and 60,000 (again for the loess and till) when the degree of saturation drops to 80%.

## DISCUSSION AND CONCLUSIONS

Previous studies have shown that hydraulic conductivity of saturated soils compacted wet of optimum moisture is much lower than that for soils compacted dry of optimum. This study has shown that this behavior is also observed for unsaturated soils and that, in fact, the difference in hydraulic conductivities increases as the soils become less saturated.

The range of capillary pressures investigated in this study was limited to between 0 and 1000 cm of water; the range of capillary pressures commonly observed in the relatively humid eastern part of the United States. It should be noted that observed capillary pressures in arid regions will be much greater. The conclusions presented here will probably also be valid in arid regions, but we have not explicitly shown this.

## ACKNOWLEDGMENTS

This research was sponsored by the U. S. Nuclear Regulatory Commission through the University of Illinois Contract No. NRC-02-80-074) as part of a study of trench covers to minimize infiltration at waste disposal sites.

In addition, the authors would like to acknowledge the contributions of Chris Stohr, Tony Devine and Becky Bianchini of the Engineering Geology Section who performed the classification and compaction tests, and Walter Morse of the Hydrogeology Section who performed some of the soil moisture retention permeability tests.

## REFERENCES

1. Lambe, T. W. "The Permeability of Fine-Grained Soils," *ASTM Special Technical Publication 163*, Philadelphia, PA., pp. 56-67 (1954).
2. Cary, A. S., B. H. Walter and H. T. Howard. "Permeability of Mud Mountian Dam Core Material," *Transactions of the ASCE*, 108:719-737 (1943).
3. Mitchell, J. K., D. R. Hooper and R. G. Campanella. "Permeability of Compacted Clay," *Journal of the Soil Mechanics and Foundations Division*, 91(SM4):41-65 (1965).
4. Lambe, T. W. "Compacted Clay: Structure," *Transactions of the ASCE*, 125(pt I):682-716 (1960).
5. Childs, E. C. and N. Collis-George. "The Permeability of Porous Materials," *Proceedings of the Royal Society of London*, A(2-1):392-405 (1950).
6. Brooks, R. H., and A. T. Corey. "Hydraulic Properties of Porous Media," *Hydrology Paper No. 3*, Colorado State University, Fort Collins, CO (1964).
7. Elzeftawy, A. and K. Cartwright. "Evaluating the Saturated and Unsaturated Hydraulic Conductivity of Soils," *ASTM Special Technical Publication 746*, Philadelphia, PA., pp. 168-181 (1979).
8. Marshall, T. J. "A Relation Between Permeability and Size Distribution of Pores," *Journal of Soil Science*, 9:1-8 (1958).
9. Millington, R. J., and J. P. Quirk. "Permeability of Porous Media," *Nature*, 183:387-388 (1959).
10. Green, R. E. and J. C. Corey. "Calculation of Hydraulic Conductivity: A Further Evaluation of Some Predictive

Methods," *Soil Science of America, proceedings*, 35:3-8 (1971).

11. Willman, H. B. and J. C. Frye, "Pleistocene Stratigraphy of Illinois," *Bulletin 94*, Illinois State Geological Survey, Urbana, Illinois (1970).

# FIELD EXPERIMENTS TO DETERMINE SATURATED HYDRAULIC CONDUCTIVITY IN THE VADOSE ZONE

Daniel B. Stephens
Assistant Professor of Hydrology
New Mexico Institute of Mining and Technology
Socorro, New Mexico 87801

Scott Tyler and Kevin Lambert
Research Assistants
New Mexico Institute of Mining and Technology
Socorro, New Mexico 87801

Scott Yates
Research Assistant
University of Arizona
Tucson, Arizona 85721

## INTRODUCTION

Borehole infiltration tests are widely used by geotechnical engineers to determine the saturated hydraulic conductivity, $K_s$, in the vadose zone. The particular borehole test under consideration is one in which a constant head of water is maintained, without pump pressure, in an open borehole until the flow rate becomes steady. Approximate analytic solutions for $K_s$ from this test have been developed by Glover [1], Nasberg [2], and Terletskata [3] using the concept of a free surface as shown in Figure 1.

$$K_s = \frac{0.159Q}{H^2} \left[\sinh^{-1}\left(\frac{H}{r}\right)-1\right] \text{, Glover [1]} \quad (1)$$

$$K_s = \frac{0.423Q}{H^2} \log_{10}\left(\frac{2H}{r}\right) \text{, Nasberg-Terletskata [2,3]} \quad (2)$$

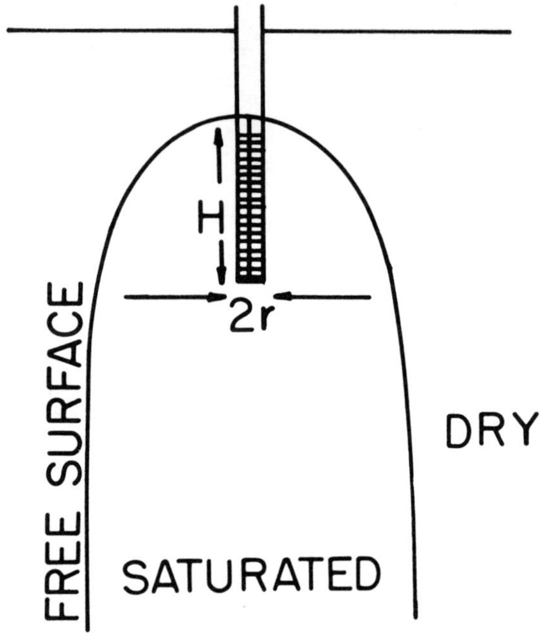


Figure 1. Classical free surface concept of flow from a borehole above a deep water table.

Stephens and Neuman [4] discuss the limiting assumptions behind these solutions, the most important of which may be the omission of flow in the partially saturated zone due to capillarity. Using numerical simulations of the constant head borehole infiltration problem which included both saturated and unsaturated flow, Stephens and Neuman [5] proposed an empirical solution for the steady state case:

$$\log_{10}C_u = 0.658 \log_{10}(H/r) - 0.238\sqrt{\alpha} - 0.398\log_{10}H + 1.343 \tag{3}$$

where $C_u = \dfrac{Q}{rHK_s}$

$Q$ = steady flow rate into the borehole

$r$ = borehole radius, meters

$H$ = height of water in borehole, meters

$\alpha$ = slope of natural logarithm of the relative hydraulic conductivity, $K_r$, versus pressure head, $\psi$, meters$^{-1}$

Their saturated-unsaturated simulations also showed that at steady state, the zone of saturation is limited to a relatively small region surrounding the borehole, outside of which unsaturated flow occurred. This is in sharp contrast to the free surface flow concept in which the entire flow region is assumed to be saturated and bounded by a zero gauge pressure surface.

None of the above mentioned analytical and empirical solutions takes into account air which may become trapped in the pore space during infiltration. Entrapped air may significantly reduce the measured hydraulic conductivity of the soil. For example, Christiansen [6] found that in laboratory experiments entrapped air reduced hyraulic conductivity by 60 to 85%. Bouwer [7] indicates that hydraulic conductivities determined in the field using an air-entry permeameter are one-half the true $K_s$ due to entrapped air.

The purpose of this paper is to report preliminary results of field experiments to analyze the reliability of methods to determine $K_s$ using borehole infiltration tests.

## SITE DESCRIPTION

The test site, located within the Sevilleta National Wildlife Refuge near Socorro, New Mexico, is on the edge of a sand dune adjacent to the flood plain of the Rio Salado. Average annual rainfall is estimated to be 23 centimeters per year. The site is nearly level and is sparsely vegetated with shrubs and grasses. Sieve analyses of 100 $cm^3$ soil core samples indicate that the site is underlain by more than three meters of unconsolidated, fine grained, poorly graded sand with a few very thin silt layers. The $d_{50}$ grain size is about 0.25 millimeters and the uniformity coefficient is approximately 2.5.

Hydraulic properties characterizing the soil at the site were also obtained from core samples. The soil moisture characteristic curve for imbibition (Figure 2a) was determined with a Büchner funnel and a hanging water column. This curve was then used to generate the relative hydraulic conductivity versus pressure head curve (Figure 2b) from the predictive procedure of Mualem [8] and a computer program by Van Genuchten [9]. The porosity of the sand ranges from 35 and 43% and the initial field water content is between 8 and 12%.

## FIELD PROCEDURES

Since August, 1981 twelve borehole infiltration tests have been conducted at three different locations within the 100 $m^2$ area comprising the site. The present discussion is limited to tests four, six, and seven, at the third borehole location within the primary site, designated as S3T4, S3T6, and S3T7, respectively. Figure 3 is a schematic of the instrumentation associated with these tests. The borehole is

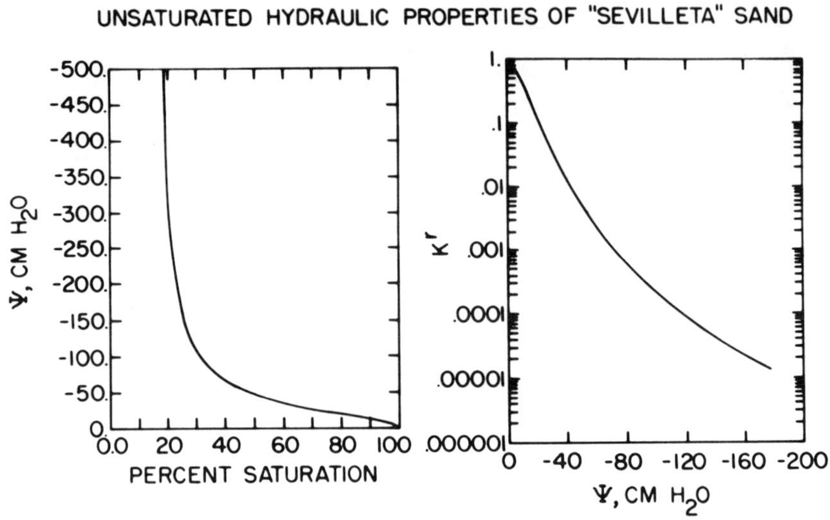


Figure 2a. Water retention curve.
Figure 2b. Relative hydraulic conductivity versus pressure.

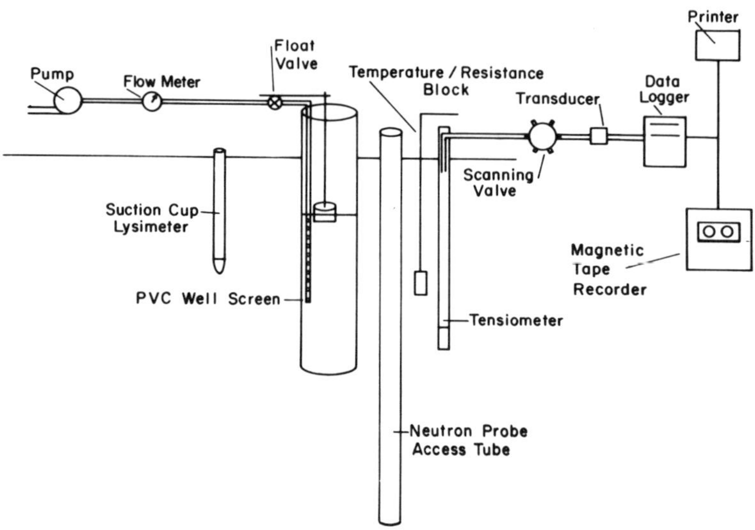


Figure 3. Schematic of borehole infiltration test equipment and instrumentation.

excavated with a post hole digger to a depth of 122 centimeters and is cased with an 18 centimeter diameter PVC pipe which is plugged at the bottom. The casing is slotted horizontally with a hacksaw and is covered with a fine screen to prevent sand from entering the casing. The water supply for the tests is pumped from well points in an alluvial aquifer 200 meters north of the site. Flow rate into the borehole is measured by an in-line, totalizing flow meter. A constant head of water is maintained by regulating the flow rate with a float attached to a stock tank valve in the same manner as prescribed by the U.S. Bureau of Reclamation [10].

To better understand the nature of flow from the borehole, pressure head, water content, temperature, and water chemistry data are required. Within a radius of 75 centimeters and to a depth of 2 meters, pressure head measurements are obtained from five tensiometers connected to a mercury manometer and 22 tensiometers connected to a scanning valve (Scanivalve Corp., San Diego, CA) and a differential pressure transducer (Model PDCR22, Druck, Inc., London, England). Data from the 22 tensiometers is collected by a data logger (Model CR 21, Campbell Scientific, Inc., Logan, UT) which controls the frequency of data recording and storage. For these tests, the data logger signals the scanning valve to switch between tensiometer ports at pre-determined time intervals. Following a switch, the pressure is allowed to equilibrate before the voltage is measured by the transducer. Time of measurement, port number, and output voltage are stored on paper tape with a thermal printer and on magnetic cassette tape for convenient data transfer to a DEC-20 computer system. To reduce instrument error due to temperature fluctuations, all tensiometers, tubing, and instruments are insulated. To further reduce errors, the transducer is calibrated during each scan by the use of two reference ports on the scanning valve; one located at the transducer level and one 200 centimeters below. Pressure heads from the transducer/scanning valve system are estimated to be accurate to within 5 centimeters of water.

Two neutron moisture probe access tubes are installed at distances of 50 and 100 centimeters from the center of the borehole for the purpose of determining water content profiles using a Troxler 3222 Depth Moisture Gauge. Soil temperature is recorded to an accuracy of 0.56°C (1°F) by thermistors (Model MC-312, Soiltest, Inc., Evanston, IL) buried at 30 centimeter intervals to a total depth of 3 meters. Two strings of temperature blocks are located at 25 and 50 centimenters from the center of the borehole. Water temperature in the borehole is measured by a soil temperature block and thermometer. At the second borehole location, the supply water and soil water extracted from the soil by suction cup lysimeters were analyzed for principal chemical constituents which could influence hydraulic conductivity.

## RESULTS

All three tests were run in April, 1982. Initial soil temperature was about 18°C and the injection water temperature was 23°C. Initial soil pressure head was in the range of -60 to -80 centimeters of water. The initial hydraulic gradient was about 1.1, indicating downward vertical flow.

For test S3T4, the height of water in the borehole was 96.5 centimeters; however, only 88.9 centimeters of the borehole was screened. Total hydraulic head from tensiometers at 15 centimeters from the borehole practically stabilized 15 to 20 minutes after the test began and did not change appreciably for the six hour duration of the test (Figure 4). A contour map of total hydraulic heads for this test is shown

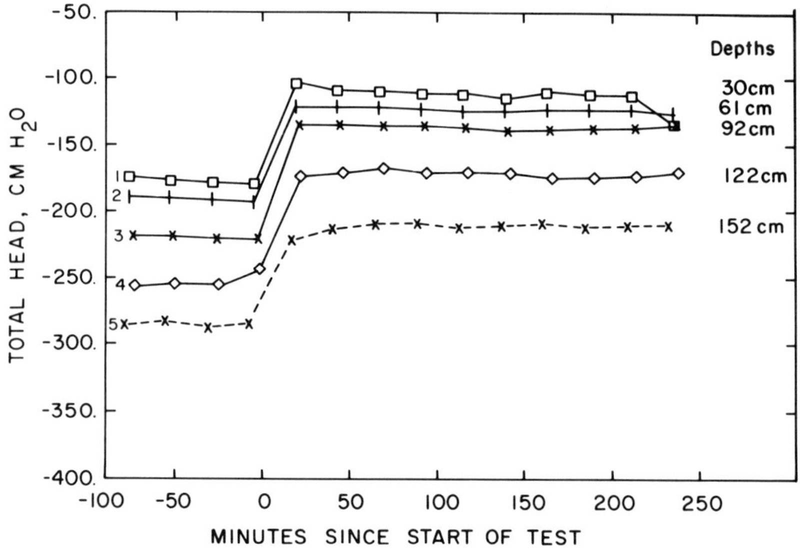


Figure 4: Total hydraulic head as measured by tensiometers during borehole infiltration test.

in Figure 5. Hydraulic gradients increase toward the base of the borehole, as would be expected if flux increases with depth along the borehole, as presumed by Glover [1] and indicated by numerical studies of Stephens [11]. Therefore in heterogeneous media, hydraulic conductivities obtained from borehole infiltration tests are more likely to represent conditions in the lower part of the borehole.

Figure 5 also shows the approximate location of the theoretical free surface (1). Observed gradients in Figure 5 indicate that the actual flow field is larger than expected by the free surface approach. Most of the lateral spread is caused by capillarity and probably a small degree of anisotropy.

At 90 minutes into S3T4, the flow rate and pressure within 75 centimeters of the borehole stabilized and the resulting pressure head distribution is shown in Figure 6. Comparing Figure 5 and 6 water flowing from the borehole along any path will pass from a saturated to an unsaturated region. This observation contradicts the free surface

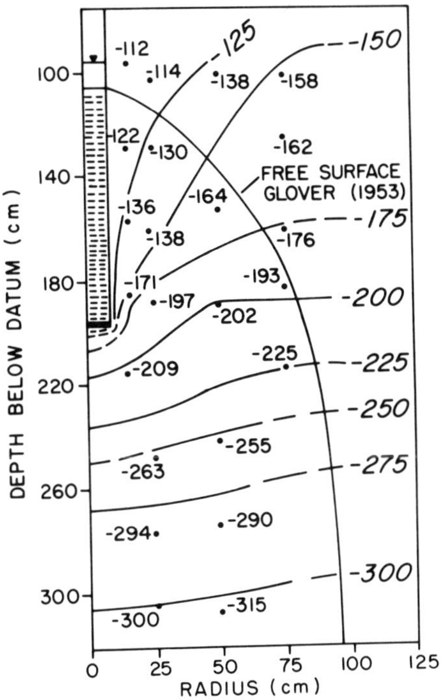


Figure 5: Total hydraulic head throughout flow field at steady state, S3T4.

approach, and supports the results of saturated-unsaturated numerical simulations.

Due to capillarity the volume of saturated soil sampled is only a small part of the flow domain. This particular sand has hydraulic properties (Figure 2) which are very similar to the coarsest soil used by Stephens [11] in his numerical studies. That study suggests that for a particular borehole construction, the region where saturation exists becomes smaller as $\alpha$ decreases (Equation 3). This implies that the volume of soil sampled may depend on the properties of the soil as well as on borehole conditions.

The preceeding discussion of the zone of saturation does not account for any gas which may be trapped in the pore space during infiltration. If air were entrapped, the hydraulic conductivity at pressure heads greater than zero would be less than $K_s$ and the saturation percentage associated with this region would be less than 100%.

Table 1 summarizes the results of the borehole infiltration tests. In general, the values of $K_s$ from all methods

TABLE I. BOREHOLE INFILTRATION RESULTS

| | R (cm) | H (cm) | $Q_s$ (1/min) | K (a) ($10^{-3}$ cm/sec) | K (b) ($10^{-3}$ cm/sec) | K (c) ($10^{-3}$ cm/sec) | COMMENTS* |
|---|---|---|---|---|---|---|---|
| S1T1 | 3.24 | 113 | 3.1 | 3.8 | 2.8 | 3.4 | 1,2 |
| S2T1 | 4.67 | 60.2 | 2.2 | 3.6 | 6.5 | 4.7 | 1,2 |
| S2T2 | 4.7 | 94 | 5.1 | 4.1 | 6.5 | 6.2 | 1,2 |
| S2T3 | 4.7 | 103 | 3.45 | 2.4 | 4.1 | 3.8 | 1,2 |
| S2T4 | 5.71 | 92.5 | 3.96 | 3.1 | 5.4 | 4.6 | 3,5 |
| S2T5 | 8.9 | 94.0 | 4.21 | 2.6 | 4.8 | 4.1 | 3,5 |
| S3T1 | 6.0 | 90 | 5.74 | 4.5 | 7.4 | 6.8 | 3,6 |
| S3T2 | 8.9 | 96.5 | 5.76 | 4.9 | 6.0 | 5.5 | 3,6 |
| S3T3 | 8.9 | 96.5 | 8.41 | 7.2 | 8.8 | 8.0 | 4,5,8 |
| S3T4 | 8.9 | 96.5 | 6.60 | 4.0 | 7.4 | 6.3 | 4,5,8 |
| S3T6 | 8.9 | 100 | 15.05 | 8.5 | 16.0 | 13.7 | 4,5,7,8 |
| S3T7 | 8.9 | 97 | 9.4 | 6.5 | 10.5 | 8.9 | 4,5,8 |
| Average of all tests | | | | 4.61 | 7.2 | 6.33 | |

*Comments: 1. Test terminated prior to steady state; 2. Gravel filled borehole; 3. Johnson PVC screen 0.006 inch slots; 4. Hacksaw cut PVC casing w/wire mesh; 5. Closed casing bottom; 6. Open casing bottom; 7. $CO_2$ injection preceded test; 8. Open interval slightly less than H.

(a) Glover (1953) (b) Nasberg-Terletskata (1954) (c) Stephens (1979); $\alpha = 9.2 m^{-1}$

fall between 2 x $10^{-3}$ and 7 x $10^{-3}$ cm/sec. According to Bouwer and Jackson [12], these values represent hydraulic conductivity primarily in the horizontal direction.

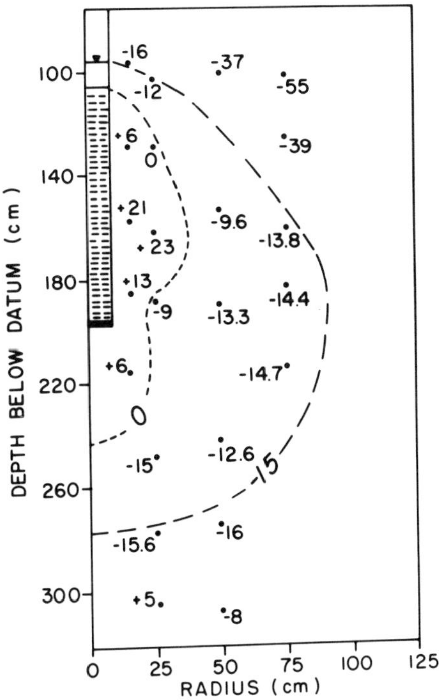


Figure 6: Pressure head throughout flow field at steady state, S3T4.

As a comparison to $K_s$ values obtained from the borehole tests, several air entry permeameter tests (Topp and Binns [13]) were conducted within the site (Table 2). This permeameter only measures hydraulic conductivity in the vertical direction. These results are not adjusted for entrapped air as suggested by Bouwer [7] and are comparable to $K_s$ values obtained from the borehole method. A preliminary conclusion may therefore be that the soil is not highly anisotropic.

Values of $K_s$ were also determined in the laboratory using 100 $cm^3$ ring samples obtained from the neutron access tube holes. The laboratory permeameter is specifically designed to test undisturbed ring samples. As indicated in Table 2, most laboratory values of $K_s$ at sites 1, 2 and 3 fall in the range of 1 x $10^{-2}$ to 3 x $10^{-2}$ cm/sec, significantly greater than all field-measured values. Some of these

TABLE II. FIELD AND LAB PERMEAMETER RESULTS

| METHOD | NUMBER OF SAMPLES | DEPTH (cm) | $K(10^{-3}$cm/sec) | COMMENTS |
|---|---|---|---|---|
| Field: | | | | |
| Air Entry Permeameter | 1 | 5 - 23 | 3.0 | 3 m east of Site 3 |
| | 1 | 5 - 21 | 3.2 | 6 m southwest of Site 2 |
| | 3 | 30 - 45 | 3.2 | Within 1.1 m of Site 2 |
| | 1 | 5 - 21 | 7.3 | 1 m north of Site 1; $CO_2$ injected |
| | 1 | 32 - 48 | 9.7 | 1.5 m west of Site 3; some $CO_2$ injected |
| Lab: | | | | |
| Constant Head; 100 cc rings | 17 | 30 - 305 | 21.6 | Site 1; Tap water |
| | 10 | 30 - 305 | 25.6 | Site 2; Tab water |
| | 5 | 30 - 305 | 14.5 | Site 2; Field water |
| | 23 | 40 - 56 | 11.5 - 15.5 | At Site 2 AEP's; Misc. filtered, unfiltered, flocculated; 8 to 36 hrs; Field and Tap water |
| | 8 | 30 - 101 | 22.6 | 3.7 m north of Site 2; Tap water |
| | 78 | 40 cm | 16.4 | Traverse through Sites 1, 2, 3; $\sigma^2 \ln K = .25$; CV = 38.6%; Tap water |
| | 48 | 30 - 305 | 37.9 | Near Dune Crest; $\sigma^2 \ln K = 0.34$; CV = 62%; Tap water |
| Constant Head; Shelby Tube | 1 | 30 - 101 | 9.0 | 3.7 m north of Site 2; Tap water |

cores were obtained from the same soil volume sampled by the air entry permeameter. These laboratory tests were carefully monitored for sources of error due to sample handling and water chemistry; however, it was apparent that these factors were relatively insignificant. This implied that if the laboratory results represent the "true" value of $K_s$, then either all the solutions for the borehole infiltration and air entry permeameter tests are in error, or that air entrappment is significant in the field.

To check the latter possibility, the top of the borehole at site 3 was packed off and approximately 5000 liters of carbon dioxide was injected into the borehole over a 16 hour period. The solubility coefficient of carbon dioxide is about 28 times greater than oxygen and 57 times greater than nitrogen at 20°C (Hillel [14]). It was presumed that the carbon dioxide would displace the soil air and then dissolve into the infiltrating water, thereby leaving the soil closer to complete saturation.

Figure 7 shows the infiltration rate versus time for tests S3T6 and S3T4. Rates have been corrected for fluid viscosity at 20°C. For about the first thirty minutes, the borehole at S3T6 was taking on water as fast as the pump could supply, and the borehole was only partially filled with water. After about 60 minutes, the water depth in the borehole stabilized at 97 centimeters and the flow rate then slowly diminished. The infiltration rate for the case where $CO_2$ was injected prior to testing was more than twice that

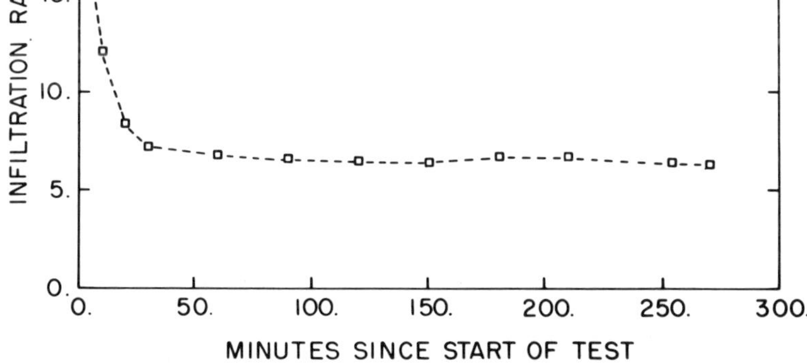


Figure 7: Infiltration rate into boreholes showing effects of $CO_2$ flooding.

of S3T4 under practically the same test conditions. The S3T6 values of $K_s$ from Equations 1 - 3 range from $7.6 \times 10^{-3}$ to $1.6 \times 10^{-2}$ cm/sec, and are considerably closer to the laboratory values (Table 2).

During the period prior to flow rate stabilization at S3T6, water content and pressure head data indicated that the soil reached a maximum wetness and then drained slightly. This could be attributed to the build up of silt on the face of the borehole or to gas coming out of solution from the injected water. Silt accumulation, air accumulation, or incomplete displacement of air prior to infiltration could all cause the field value to be less than the true value of $K_s$.

To check whether a piping channel had developed during the time interval between tests S3T4 and S3T6, the soil was allowed to drain and S3T6 was repeated without prior injection of $CO_2$. In this test, S3T7, the final infiltration rate was 9.4 liters per minute; and $K_s$ falls between the values from tests S3T4 and S3T6. Water content in the zone closest to the borehole was 24%, 31%, and 26% for tests S3T4, S3T6, and S3T7, respectively. The greatest water content is observed when $CO_2$ is flooded prior to the infiltration test, and the smallest occurred for the test run prior to $CO_2$ injection.

## DISCUSSION

A comparison of flow rate and water content data for S3T4, S3T6, and S3T7 suggests that air is entrapped during infiltration from the borehole. Flooding the soil with a highly water soluble gas, such as carbon dioxide, prior to the infiltration test seems to be an effective means of removing a significant portion of the entrapped air. Differences in infiltration rates and water contents during the three tests may be explained as follows: During S3T4 the smallest diameter pore spaces contain air entrapped behind a rapidly advancing wetting front. During S3T6 these gas bubbles are replaced with water when the carbon dioxide dissolves. During drainage after S3T6 most of the smallest pores still contain water held by capillary forces. When the water begins to infiltrate again during S3T7, most of the small pores which trapped air during S3T4 are filled with water and only a few remaining pores have structures conducive to air entrapment.

The cause of air entrappment in the field may be due in part to the large hydraulic gradients (on the order of 100) imposed on the soil during the rapid filling of the borehole. In the laboratory, the 100 $cm^3$ rings were wetted slowly from below under gradients which were much less than those found in the field. No significant change in hydraulic conductivity was observed after as many as 80 pore volumes of water passed through the laboratory samples. The laboratory values of hydraulic conductivity seem quite reasonable for this sand and do not appear to suffer to any great extent from entrapped air, although tests are in progress to further check the laboratory cores for entrapped air.

In some soils the problem of entrapped air may be greater than problems due to poor assumptions which form the basis of the analytical and empirical solutions for $K_s$. If the borehole method is used to collect data for predicting infiltrability of soil for rainfall-runoff over a waste repository for example, then perhaps the uncorrected $K_s$ value may be appropriate because infiltration from ponding which may be rapid and intermittent is likely to entrap air. On the other hand, if the borehole method is to form the basis for seepage computation where prolonged ponding is likely to eventually cause complete saturation of the soil, then air entrapment must be taken into consideration.

## ACKNOWLEDGEMENTS

Financial support for this research is generously provided by the U.S. Department of Interior, Office of Water Research and Technology and the New Mexico Water Resource Research Center under project B-073-NMEX. We are grateful to Mr. Ron Perry and staff of the U.S. Department of Interior, Bureau of Sport Fisheries and Wildlife at the Sevilleta Wildlife Refuge for their cooperation in providing access to the field area.

## REFERENCES

1. Glover, R. E. "Flow From a Test-hole Located Above Groundwater Level," in *Theory and Problems of Water Percolation*, U.S. Bureau of Reclaimation Engineering Monograph No. 8 pp. 69-71 (1953).
2. Nasberg, V. M. "The Problem of Flow in an Unsaturated Soil For Injection Under Pressure," *Izvestja Akademia Nauk*, SSSR, odt tekh Nauk, No. 9 (1951).
3. Terletskata, N. M. "Determination of Permeability of Dry Soils," *Hydroelectric Waterworks*, No. 2, Moscow (1954).
4. Stephens, D. B. and S. P. Neuman. "Vadose Zone Permeability Tests: A Review," *A.S.C.E. Hydraulics Div. J.*, Vol. 108, No. HY5, pp. 623-639 (1982).
5. Stephens, D. B. and S. P. Neuman. "Vadose Zone Permeability Tests: Steady State," *A.S.C.E. Hydraulics Div. J.*, Vol. 108, No. HY5, pp. 640-659 (1982).

6. Christiansen, J. E. "Effects of Entrapped Air Upon the Permeability of Soils," *Soil Science*, Vol. 58, pp. 355-365 (1944).

7. Bouwer, H. "Rapid Field Measurements of Air Entry Value and Hydraulic Conductivity of Soil as Significant Parameters in Flow System Analysis" *Water Resources Res.*, Vol. 2, No. 4, pp. 729-738 (1966).

8. Mualem, Y. "A New Model for Predicting the Hydraulic Conductivity of Unsaturated Porous Media," *Water Resources Res.*, Vol. 12, No. 3, pp. 513-522 (1976).

9. Van Genuchten, T. "Calculating the Unsaturated Hydraulic Conductivity With a New Closed-Form Analytical Model" Research Report No. 78-WR-08, Princeton Univ., 65 p. (1978).

10. U.S. Bureau of Reclaimation, "Earth Manual", Second Edition, U.S. Government Printing Office, Washington, D.C., 310 p. (1974).

11. Stephens, D. B. "Analysis of Constant Head Borehole Infiltration Tests in the Vadose Zone," Ph.D. Dissertation, University of Arizona, Tucson, Arizona (1979).

12. Bouwer, H. and R. D. Jackson. "Determining Soil Properties" in *Drainage for Agriculture*, J. van Schilfgaard (ed), Amer. Soc. of Agronomy Monograph 17, pp. 611-612 (1974).

13. Topp, G. C. and M. R. Binns. "Field Measurements of Hydraulic Conductivity with a Modified Air-Entry Permeameter, *Canadian Journal of Soil Science*, Vol. 56, pp. 139-147 (1976).

14. Hillel, D. "Fundementals of Soil Physics, "Academic Press, New York, 413 pp. (1980).

# COMPOSITION, DENSITY AND FABRIC EFFECTS ON BULKY WASTE CAPILLARY RETENTION CHARACTERISTICS

George E. Veyera and Joseph P. Martin
Department of Civil Engineering
Colorado State University
Fort Collins, Colorado

## INTRODUCTION

Capillary retention curves are one of the basic data sources used in evaluating hydraulic properties of unsaturated porous media. Parameters such as residual saturation, pore size distribution and displacement pressure head can be determined from such curves. These parameters are used in analyses such as determination of drainable liquid and formulation of unsaturated hydraulic conductivity functions. Retention curves are also used in an implicit manner to estimate the value of capillary pressure head at a given moisture content for a soil sample. This paper presents data and interprets it with respect to the factors influencing retention characteristics of two types of bulky wastes obtained by mineral extraction: uranium mill tailings and spent oil shale. The testing method used rapidly determines the capillary retention curves while controlling density and soil structure.

The retention curve for a given soil mass is determined by a number of factors including those related to the material composition, porosity, fabric and interaction between the pore fluid and the solid matrix. These factors are addressed in the data and analysis presented in this paper. Other retention influences such as wetting hysteresis are not addressed.

## BACKGROUND

In civil engineering applications of porous media analysis, variations in soil materials and their field conditions must be considered. This is true for both saturated and unsaturated porous media. For example, on a given earthwork project the same material may exist in both natural deposits and in remolded compacted form. One major class of engineering projects is the safe disposal of large volumes of contaminated wastes where initial drainage and long term fluid transport are of concern. At most such waste sites there are usually several materials present, each with its own fluid and mechanical properties. The properties of the natural geologic formations can often be estimated by experience. The containment system (usually liner and cover) may include both undisturbed and compacted soils. The wastes themselves are usually by-products from material processing. Depending upon the particular project, such wastes must often be considered a manufactured soil with special properties. A particular consideration is the effect of the contaminants in the liquids on the hydraulic properties of the site materials. Correlations have been made between the salt content of the soil-water and the osmotic suction [1]. Other authors have noted variations in clay soil conductivity with the pH of the permeant.

## CONTROLS ON RETENTION CHARACTERISTICS

### Material Composition

As a first approximation in predicting and interpreting actual data for waste material retention, it can be reasonably assumed that the factors which affect saturated porous media properties also influence the unsaturated properties. Certain soil composition indices have been found to be very useful in predicting many forms of soil behavior. Physical index properties include the predominant soil types (silty sand, etc.) and the grain size distribution. These parameters indicate the general pore size, porosity and hydraulic conductivity of a soil or waste deposit. To a lesser extent, particle shape and roughness have also been found to be influential. Physio-chemical influences, mostly manifest as particle surface interaction with the fluid, depend upon the nature of the fluid as well as the soil material. Index properties for such phenomena include plasticity, salt content and sodium adsorption ratio.

## Soil Mass Conditions

Quantifying the next set of retention influences, namely the conditions of the soil mass, is more difficult other than measurement of porosity. The influence of this parameter is obvious in that it sets the pore volume per unit volume of soil in which fluid activity can occur. The detailed arrangement of the soil particles and voids is referred to as the soil structure or fabric. The fabric controls other microscopic hydraulic transmission and retention influences including particle stratification and the shape, orientation, size distribution and interconnections (tortuosity) of the voids.

Fabric effects can be seen to be of major importance in unsaturated analysis since the larger voids with the most favorable transmission characteristics are occupied by the non-wetting fluid. The capillary pressure at which desaturation begins, known as the displacement pressure head, is related to both the pore size distribution and the gross porosity. As a soil desaturates a larger proportion of the liquid that is present is under significant attractive influence of the soil particle surface forces. Thus, residual retention and the flow tortuosity are also influenced by pore size distribution.

## Fabric Description

The fabric of a soil is customarily described in terms of the conditions present at the time of deposition and the deformation resulting from subsequent changes in the state of stress. For example, investigations [2] have shown that clays used in earthfills exhibit a flocculated fabric if they are compacted dry of optimum moisture content. The same material compacted wet of optimum moisture content exhibits a dispersed particle arrangement. Both the hydraulic and mechanical properties of clays are influenced by fabric. It has also been generally observed that clays sedimented under water are flocculated. When such deposits are compacted as borrow material, the tendency to dispersion is attributed to a mechanical disturbance of the original interparticle bonding.

Two general classes of soil fabric can be postulated that include most engineering applications. One results from grain-by-grain sedimentation, which would include wind and water deposits in river, marine and waste pond environments. This type of fabric will be referred to as "hydraulic deposition". The other classification is that of soil deposited and compacted in large amounts, often under controlled moisture content. Further long term fabric changes can result from post-deposition static compression superimposed on the original soil fabric [3].

## MATERIALS AND RESEARCH PROGRAMS

The data presented herein is primarily that for uranium mill tailings from two inactive mill sites, one in Grand Junction, Colorado and the other in Gunnison, Colorado. Retention data is also shown for spent oil shale derived from the Tosco process on materials obtained from Garfield County, Colorado. Data from tests on a modified (passing #10 sieve) ASTM C33 concrete fine aggregate sand is also presented to verify the test procedure.

The uranium tailings are the waste products from milling a sandstone ore using a sulfuric acid process. The waste product includes salts as well non-recovered radioactive elements. Studies of the properties of the tailings are being done under a contract with the United States Department of Energy (D.O.E.) through the Uranium Mill Tailings Remedial Actions Program (UMTRAP). A score of such inactive sites are under study at the present time.

The tailings were hydraulically discharged on the beaches of sedimentation ponds. Fine sands deposit on the beaches while finer fractions, referred to as slimes, settle in the ponds. The Grand Junction slimes are a nonplastic silt. The sand particles are angular but regular in shape. The slimes at the Gunnison site consist of both silt and clay size particles, with the sands being generally subrounded. The differences between the tailings from these two sites are the result of both the ore types and the milling processes. Data from both sites is presented for comparison.

Changes in the deposition patterns with mill operations caused material type stratifications in the impoundments such that variations exist in both the vertical and the horizontal directions. After deposition, the tailings in the lower levels were subject to overburden pressures. After mill closure, the ponds drained leaving the tailings in an unsaturated condition. Undisturbed samples of the hydraulic deposits were recovered from exploratory borings. These samples will be referred to as "intact samples".

Large quantities of disturbed material were also recovered during the field explorations and tested for various possible reclamation alternatives. The overall program is one of site and material characterization including technology development for reclamation purposes. Various reclamation alternatives can involve the disturbance and mixing of the hydraulic deposits such that the materials are redeposited in a compacted state. Experiments were conducted on these materials for the two fabrics mentioned. A detailed presentation of results is given by [4,5]. The basic material descriptions are presented in Table I with the gradation curves shown in Figures 1 and 2.

The major concern in the oil shale study, also conducted for D.O.E., is the mechanical properties of very deep deposits of spent oil shale. These properties include unsaturated material shear strength and compressibility. Retention curves were generated to estimate the suction at various compacted water contents in test specimens [6].

## RETENTION CURVES

### Uranium Tailings

#### Composition Effects With Hydraulic Fabric

Figure 3 shows retention curves for two intact samples from the Grand Junction site, a beach sand and a sandy slimes material. Both of these samples contain very thin stratifications. The curve for the beach sand is similar to the classical shape exhibited by granular materials. Approximate values for the predominant pore size, residual saturation and displacement pressure head can be obtained by inspection. This sample contains about 6% material passing the # 200 sieve. The silt-sized slimes show a higher retention at each pressure head level typical for the smaller void spaces present. However the curvature is gentler in all respects than that for the beach sand.

Figure 4 shows the retention curves for the intact samples from the Gunnison site. As would be expected the retention increases with increasing amounts of fines present in the sample. This effect is increased since the fines contain plastic material. This is evident in the retention curve inflection points for the finer materials which are very subdued. Table II shows the volumetric data for Figures 3 and 4.

#### Comparison of Fabrics

It was necessary to investigate the retention characteristics of tailings gradations for the Grand Junction site that are in between those shown on Figure 3. These

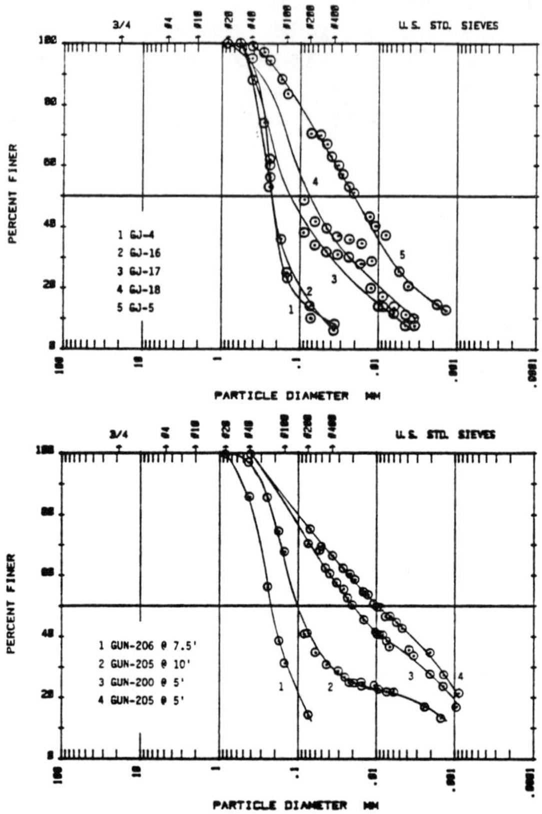


Figs. 1 & 2. Gradations of uranium tailings, Grand Junction and Gunnison sites.

Table I. Index Properties of Uranium Tailings Samples

| Sample | Description | Specific Gravity | %-#200 Mesh | Liquid Limit | Plastic Limit | USC Class |
|---|---|---|---|---|---|---|
| Grand Junction Site | | | | | | |
| GJ-4 | Fine Sand, Trace Silt. | 2.70 | 14 | - | - | SP-SM |
| GJ-5 | Silt, some fine sand | 2.76 | 70 | 37 | 3 | ML |
| GJ-16 | Fine sand, Little silt | 2.71 | 16 | - | - | SM |
| GJ-17 | Fine sand, Some silt | 2.73 | 32 | - | - | SM |
| GJ-18 | Fine sand and silt | 2.73 | 40 | 32 | NP | SM |
| Gunnison Site | | | | | | |
| GUN-200 @ 5' | Silty clay, Some F. sand | 2.75 | 70 | 34 | 22 | CL |
| GUN-205 @ 5' | Silty clay Ltl. F. sand | 2.75 | 76 | 37 | 20 | CL |
| GUN-205 @ 10' | Silty clay & Fine sand | 2.71 | 42 | 27 | 22 | SM |
| GUN-206 @ 7½' | Med-fine sand, Trace silt | 2.70 | 14 | - | - | SP-SM |

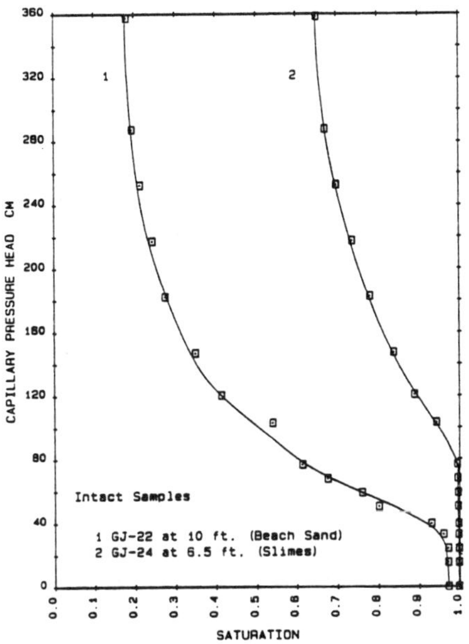



Fig. 3. Retention curves of intact samples, Grand Junction.

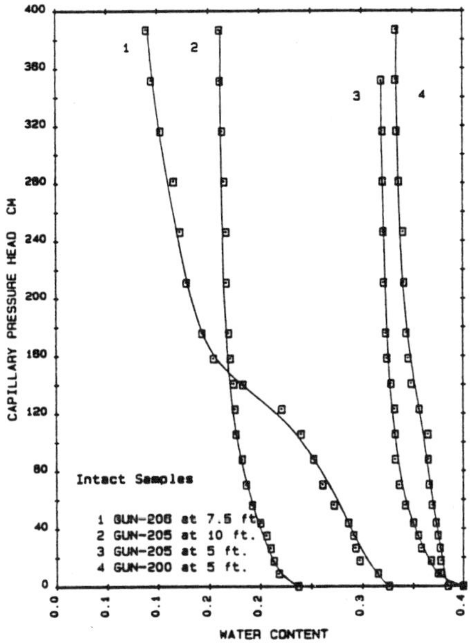



Fig. 4. Retention curves of intact samples, Gunnison.

Table II. Volumetric data for Undisturbed (Intact) Uranium Tailings Field Samples

| Sample | Dry Density grams/cm$^3$ | Void Ratio | Field Water Content (%) | Field Saturation |
|---|---|---|---|---|
| Gunnison Site: | | | | |
| Gun 200 @ 5' | 1.33 | 1.07 | 33.8 | 0.87 |
| Gun 205 @ 5' | 1.29 | 1.12 | 40.0 | 0.98 |
| Gun 205 @10' | 1.45 | 0.86 | 16.8 | 0.53 |
| Gun 206 @ 7.5' | 1.41 | 0.91 | 1.7 | 0.05 |
| Grand Junction Site: | | | | |
| GJ 22 @ 10.2' | 1.57 | 0.59 | 3.9 | .18 |
| GJ 24 @ 6.3' | 1.05 | 1.69 | 41.6 | .68 |

intermediate gradations are present at the site and would also result from reclamation mixing of the zones of interbedded sands and slimes. Both hydraulic deposition and compaction were used to prepare the samples tested. Figure 5 shows the results for tailings deposited grain-by-grain using a tube packer. The samples shown are sand-slime mixtures in various proportions. For the laboratory hydraulic deposition, the retention is roughly proportional to the amount of nonplastic fines present. Figure 6 shows the same material retention curves for a dense compaction, approximately 90% of Standard Proctor density (ASTM D-698) for each mixture. In contrast to the hydraulic fabrics, these curves show a dominant effect of fines on the material properties. This is characteristic of mechanically deposited soil structures. Once a certain amount of fines are present, the retention is not very sensitive to further variations in gradation.

Figures 7, 8 and 9 compare the effects of density and fabrics on retention for a beach sand with some slimes, a nearly evenly proportioned sand-slimes mixture and a predominantly slimes sample. The fairly uniform nonplastic slimes are not sensitive to fabric whereas the silty beach sands and the proportioned mixtures are.

Figure 10 shows fabric effects for a plastic slimes sample from the Gunnison site. The two remolded sample curves bracket the intact sample curve with respect to density, but each has a lower retenton than that of the stratified sample. This illustrates another common observation in soil mechanics practice that fabric influences can be more important than density effects for plastic soils.

## Other Effects on Retention

Figure 11 shows the same tests as those shown in Figure 10, but is presented in terms of water content by weight. This is perhaps a more useful presentation of the data if it is desired to determine the quantity of fluid that will drain from a given waste material tonnage. However, most unsaturated porous medial flow analyses are developed in terms of saturation.

The similarity of water content retention values for the compacted samples at two different porosities is most likely attributable to a dominance of retention by surface absorption in the ionic double layer. The Kelvin capillary model interface geometry (matric suction) effect is small. The retention for the intact sample is much higher, indicating that the pore size distribution and other stratified effects have a similar order of influence on the retention in addition to the material surface moisture attraction.

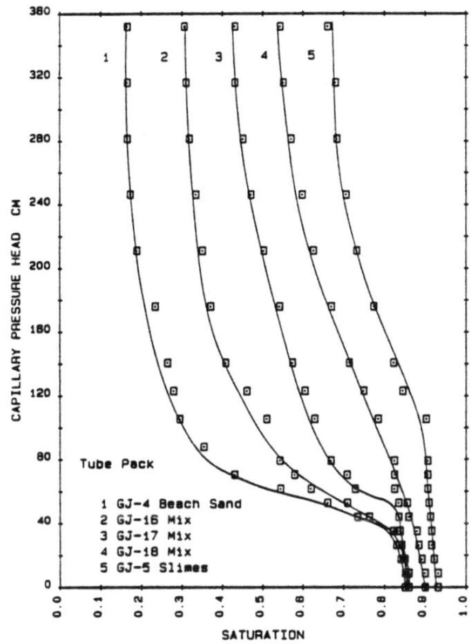


Fig. 5. Laboratory hydraulic fabric curves, Grand Junction site tailings.

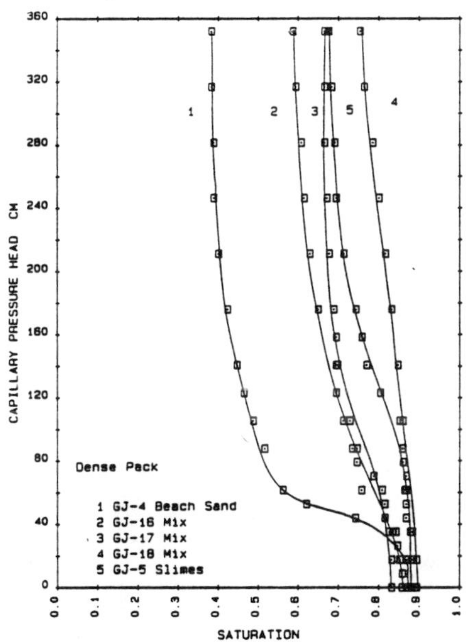


Fig. 6. Dense compaction fabric curves, Grand Junction site tailings.

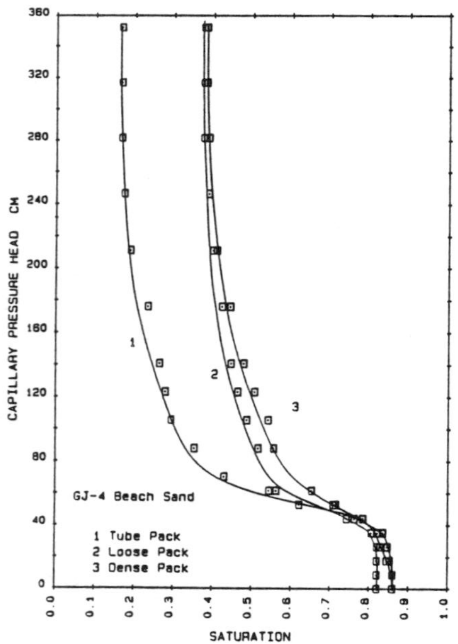


Fig. 7. Experimental fabric and density effects, Grand Junction site (beach sand).

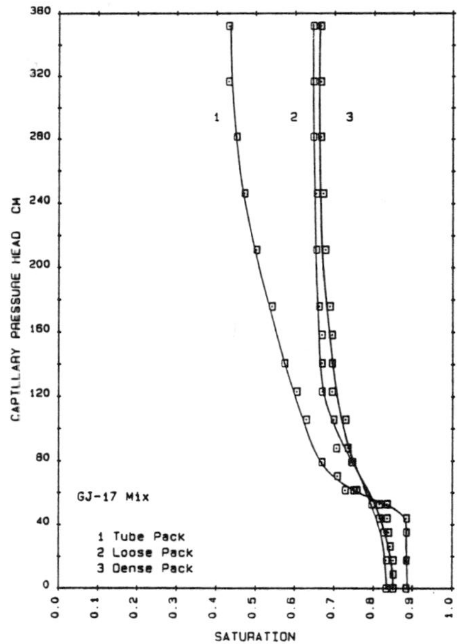


Fig. 8. Experimental fabric and density effects, Grand Junction site (mixture).

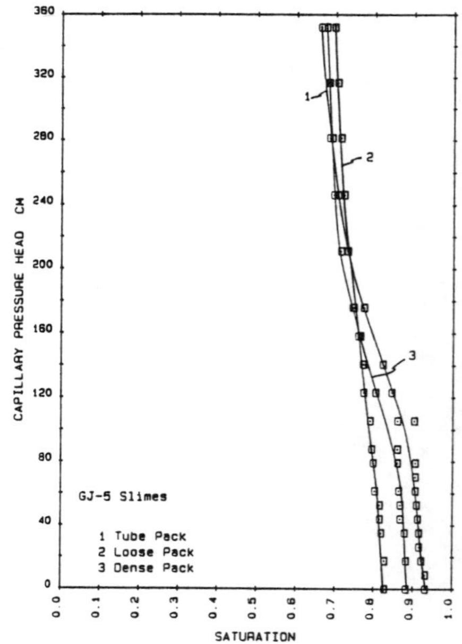


Fig. 9. Experimental fabric and density effects, Grand Junction site (slimes).

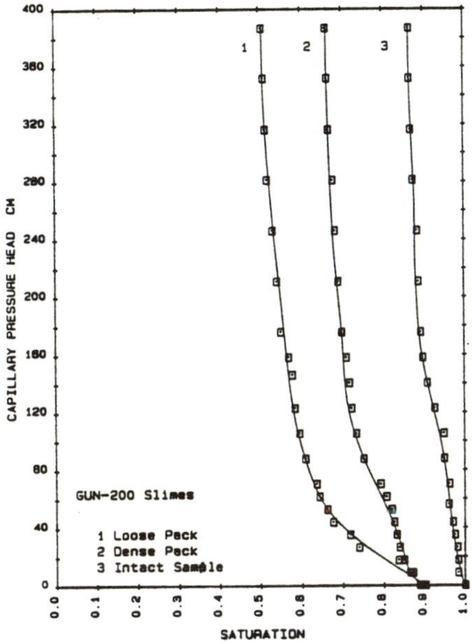


Fig. 10. Comparison of compacted and intact sample fabrics, Gunnison plastic slimes (saturation measure).

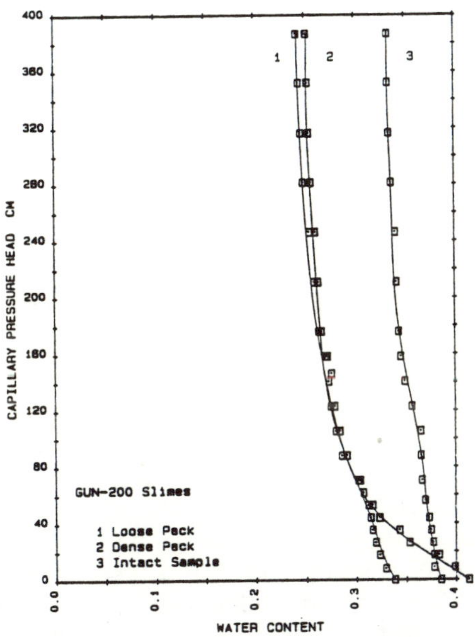


Fig. 11. Comparison of compacted and intact sample fabrics, Gunnison plastic slimes (water content measure).

Figure 12 shows a stress history effect for the two compacted Gunnison slimes samples. Both of these samples decreased in volume during the first desaturation. The cells were then resoaked with the fluid expelled during desaturation in order to restore the soil salt content as much as possible. The samples swelled to nearly their original volume and were desaturated under capillary pressure increments again. This time higher retentions were observed in both the loose and the dense samples. One possible explanation is that the fabric pore size distribution may have changed during the first volume decrease such that the compression in the larger voids was not recoverable during resoaking swell. There is insufficient data available at this time to draw a more detailed conclusion.

Studies of the mechanical properties of the material from the Grand Junction site indicated that the high salt content of the soils may have a major influence [3] on the hydraulic properties. Figure 13 shows the retention curves for the beach sand and slimes both before and after drip leaching with distilled water. These samples were prepared using the tube packer for grain-by-grain deposition with densities similar for the leached and unleached cases. In addition to a decrease in the osmotic suction, the removal of the salts covering the silica particles surfaces may have resulted in a slightly different soil structure on the second packing.

## Spent Oil Shale Data

The spent oil shale is a radically different material from the original ore, such that it is a black, slightly oily silt. Description using standard soil index properties is difficult especially with respect to evaluating gradation.

The effort in retention curve generation was in support of mechanical shear strength determinations. Figure 14 shows the effects of compaction density on the retention of waste material packed at optimum moisture content. The differences in retention between densities is not evident until fairly high suctions are imposed. It is anticipated that the waste will be deposited in embankments tens of meters high so that this is an important result with respect to initial drainage.

Tests on samples compacted at the same density but at different moisture contents showed minimal retention variations. This silty material does not appear to be sensitive to compaction conditions.

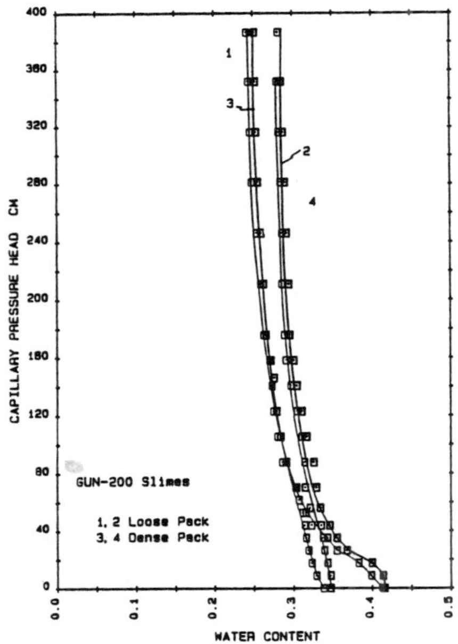


Fig. 12. **Stress history** effects on compacted slimes, Gunnison site.

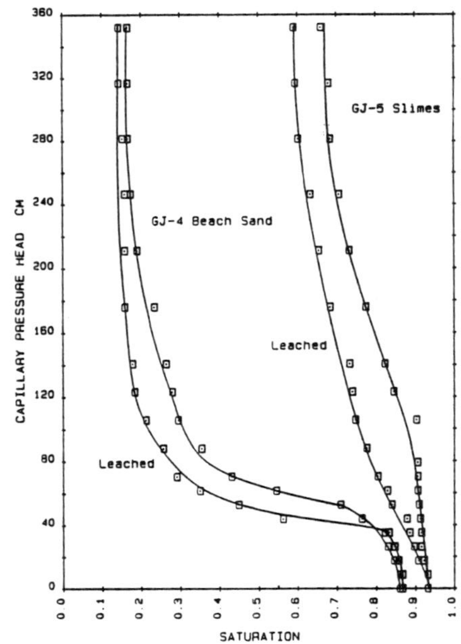


Fig. 13. **Leaching effects** on retention, Grand Junction tailings.

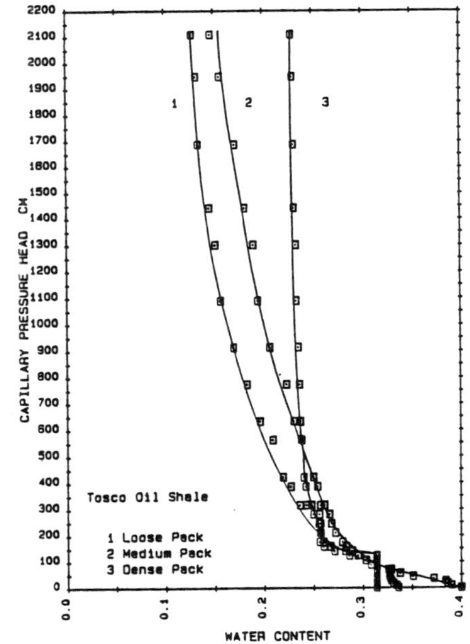


Fig. 14. **Density effects** on retention, Tosco spent oil shale.

## DATA REDUCTION

### Determination of Hydraulic Parameters

By themselves, retention curves are only general indicators of unsaturated porous media hydraulic properties. It is sometimes possible to determine the residual moisture content and displacement pressure head for clean sands graphically. For other materials, one of several curve fitting methods can be used to reduce retention data to these and other parameters. Such methods are also used to directly form unsaturated porous media conductivity functions [7,8]. Table III shows the reduction of data from the Grand Junction site tests previously obtained using the Brooks-Corey [9] method. These results have been adapted from [5]. The sensitivity of these unsaturated hydraulic parameters to fabric and fines content is shown graphically in Figures 15 and 16.

Table III

Unsaturated Material Parameters

Grand Junction Site Retention Tests

Brooks-Corey Method

| Sample | Dry Density (grams/cm$^3$) | #-200 mesh | Residual Moisture, % by weight | Residual Moisture, Saturation | Displacement Pressure Head cm. |
|---|---|---|---|---|---|
| Intact Field Samples | | | | | |
| GJ-22 | 1.57 | 6 | 3.1 | 0.14 | 73 |
| GJ-24 | 1.05 | 82 | 36.7 | 0.60 | 120 |
| Bulk Samples, Hydraulic Model Fabric | | | | | |
| GJ-4 | 1.66 | 14 | 5.8 | 0.25 | 46 |
| GJ-16 | 1.63 | 16 | 9.7 | 0.29 | 57 |
| GJ-17 | 1.32 | 32 | 14.7 | 0.38 | 80 |
| GJ-18 | 1.24 | 40 | 26.5 | 0.50 | 97 |
| GJ-5 | 1.12 | 70 | 31.9 | 0.60 | 108 |
| Bulk Samples, Dense Compacted | | | | | |
| GJ-4 | 1.62 | 14 | 9.0 | 0.37 | 42 |
| GJ-16 | 1.71 | 16 | 10.8 | 0.55 | 58 |
| GJ-17 | 1.63 | 32 | 16.5 | 0.67 | 85 |
| GJ-18 | 1.62 | 40 | 17.9 | 0.70 | 102 |
| GJ-5 | 1.25 | 70 | 33.3 | 0.66 | 108 |

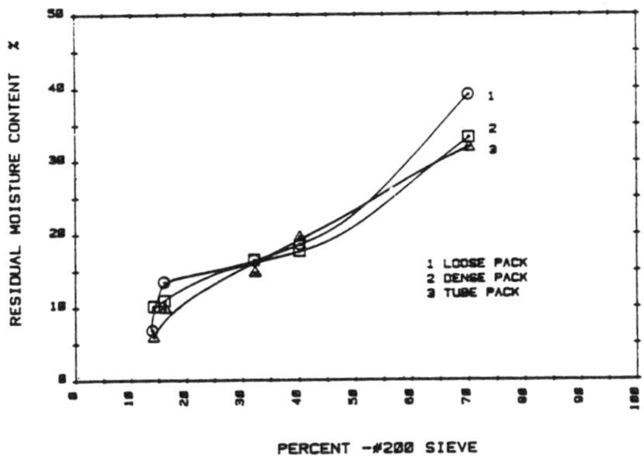


Fig. 15. Effects of fabric and fines content on residual moisture content, Grand Junction site tailings.

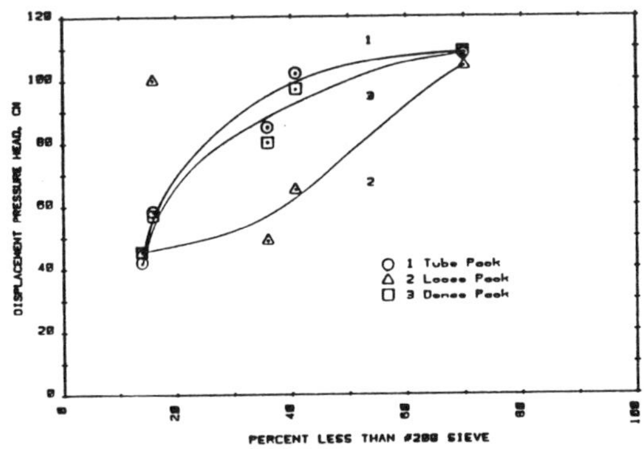


Fig. 16. Effects of fabric and fines content on displacement pressure head, Grand Junction site tailings.

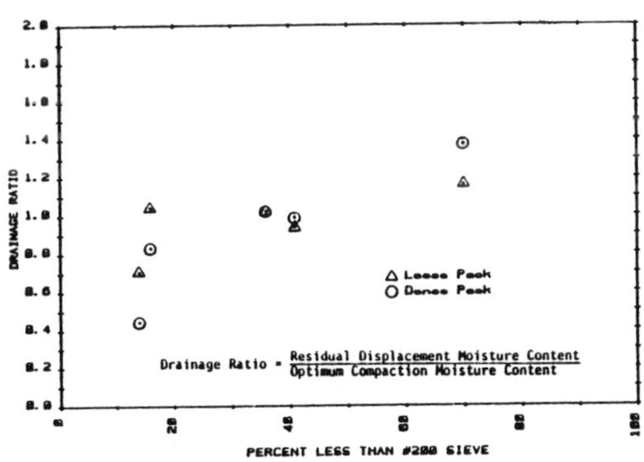


Fig. 17. Comparison of optimum and residual moisture content, Grand Junction compacted tailings.

## OTHER USES OF RETENTION CURVES

Unsaturated porous media properties are related to other engineering properties. For tailings from the Grand Junction site, it was found that blending clean sands and slime deposits to a more well graded material yielded several favorable results. Among these were better material handling characteristics, excellent compactibility and reasonable shear strength. Combining the dry sands and wet slimes in varying proportions at field moisture contents also yielded a final moisture content near that of Proctor optimum. A linear relationship was observed between fines content and optimum moisture content. A comparison of the optimum moisture contents and the residual moisture contents for these materials at compacted densities is shown in Figure 17. It appears that little contaminated fluid drainage would result from mixtures of field materials. This is a site-specific result, however, but it does show that capillary retention data can be related to other material properties to give more insight into material behavior for waste disposal.

## EXPERIMENTAL METHODS

The necessity of developing hundreds of retention curves for different materials and fabrics as part of waste material research projects, led to the development of a method to generate the curves rapidly and simply. The laboratory procedure is discussed in detail by [5].

Among the requirements for the system are the use of samples large enough that density and fabric can be accurately measured and controlled. A low sample aspect ratio (depth to width) is also desirable to minimize hydrostatic pressure and edge effects. It is also easier to use increasing air pressure rather than negative water pressure to generate capillary pressure head increments. This practice reduces the external volume of water coming in contact with the test sample, by which diffusion of salts can occur. The last requirement is that both remolded samples and undisturbed field samples be tested in a similar manner so that reasonable comparisons can be made between them.

Figure 18 shows the pressure cylinder used for testing remolded samples. In testing undisturbed field samples, end caps are attached to the sample as shown in Figure 19. A porous, plastic, reuseable membrane is usually employed, but when it is necessary to carry the retention curves to higher

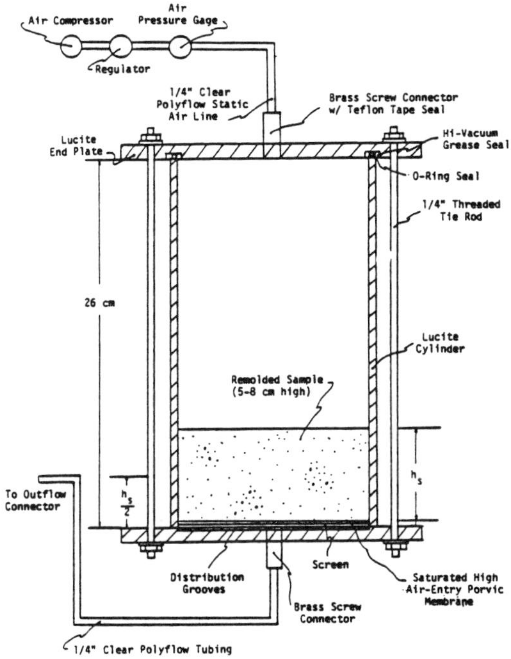


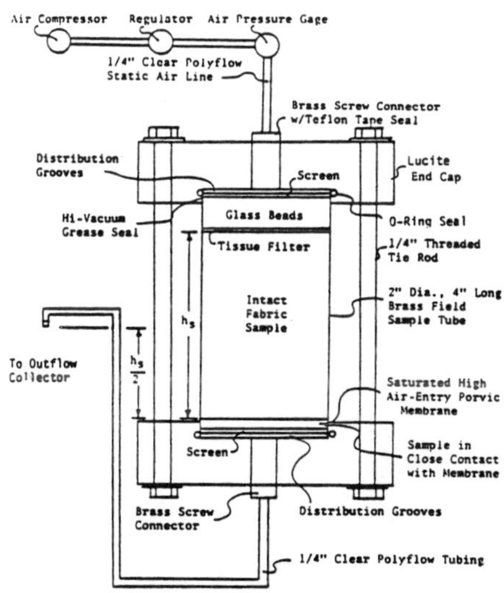


Figs. 18 & 19. Schematic of apparatus for testing remolded and intact samples.

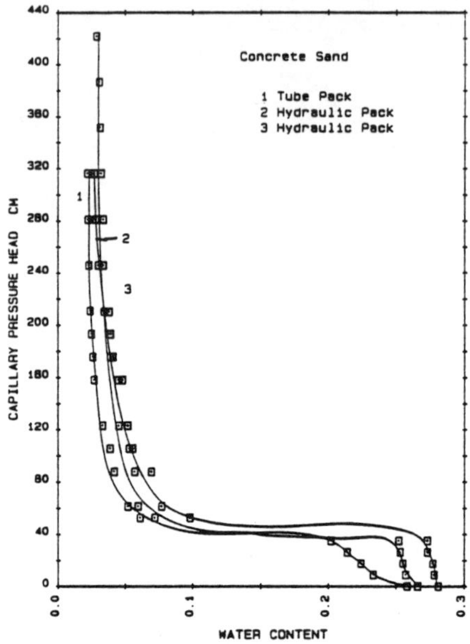


Fig. 20. Comparison of laboratory hydraulic deposition models, screened concrete sand.

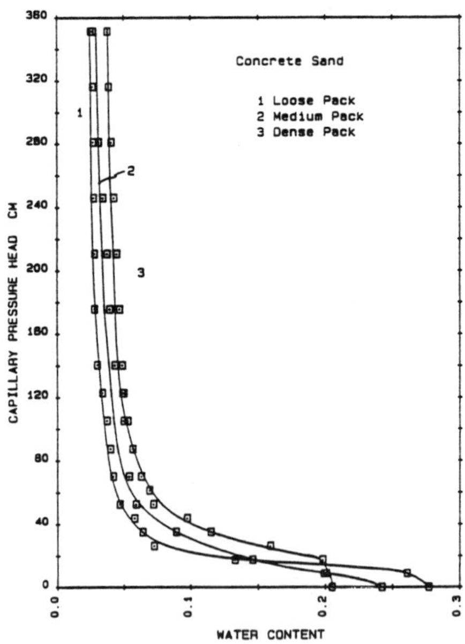


Fig. 21. Compacted density effects on retention for screened concrete sand.

capillary pressures, other types of membranes were required. Retention is monitored by periodic weighing of the sample at equilibrium points throughout the test.

There are a limited number of undisturbed samples recovered from a field investigation that are available for these tests. Therefore, it is often necessary to recreate a hydraulic deposition fabric in packing the pressure cells. A tube packer developed in the Agricultural Engineering laboratories at Colorado State University is used to replicate sedimentation. This method is discussed in detail by [10]. Retention curves for a screened concrete sand packed both by tube and subaqueous hydraulic deposition is shown in Figure 20.

The ability to control density with the large sample size can be seen in the compacted concrete sand results shown in Figure 21. The curves shown in Figure 20 and 21 are similar to those obtained by other authors [9] for the same type of material.

## SUMMARY

A laboratory method has been presented which rapidly and easily yields capillary retention data for bulky waste materials. The data was generated for remolded samples by controlling composition, density and fabric. Intact field samples were also tested.

Test results indicate that uniform clean sand properties do not seem to be particularly sensitive to fabric . Soils having plastic behavior have been shown to be very sensitive to fabric. Most waste materials and natural soils would appear to fall in between these two extremes in regards to fabric sensitivity.

Capillary retention data generated from these tests provides basic data which is useful in evaluating the hydraulic properties of unsaturated porous media.

## REFERENCES

1. Krahn, J. and Fredlund, D. G., "On Total, Matric and Osmotic Suction," Soil Science, Vol. 114, No. &, Pages 339-348 (1972).
2. Lambe, T. W., "The Permeability of Compacted Fine Grained Soils," Special Technical Publication No. 163, ASTM (1955).

3. Martin, J. P., "Unsaturated Embankment Consolidation With Two Fluid Continuity and Finite Strain," Doctoral Dissertation, Colorado State University, Fort Collins, Colorado (1982).
4. Martin, J. P., Veyera, G. E., Nasiatka, D. M., Osorno, L. F., "Characterization of the Inactive Tailings Sites," Proceedings of the Third Symposium on Uranium Mill Tailings Management, Colorado State University, Fort Collins, Colorado (1980).
5. Veyera, G. E., "Unsaturated Flow Parameters of Uranium Mill Tailings," M.S. Thesis, Colorado State University, Fort Collins, Colorado (1980).
6. Colorado State University, Department of Civil Engineering, "Long Term Stability Aspects of Spent Oil Shale Waste," Report Prepared for Oil Shale Task Force, U.S. Department of Energy (1982).
7. Brustaert, W., "Some Methods of Calculating Unsaturated Permeability," Transactions, ASAE, Vol. 10, Pages 400-404 (1967).
8. Klute, A., "The Determination of the Hydraulic Conductivity and Diffusivity of Unsaturated Soils," Soil Science, Vol. 113, Pages 264-276 (1972).
9. Brooks, R. H. and Corey, A. T., "Hydraulic Properties of Porous Media," Hydrology Paper No. 3, Colorado State University, Fort Collins, Colorado (March, 1964).
10. Shoostari, M. M., "One Dimensional Transient Drainage,22M.S. Thesis, Department of Agricultural Engineering, Colorado State University, Fort Collins, Colorado (1975).

# LABORATORY ANALYSIS OF WATER RETENTION IN UNSATURATED ZONE MATERIALS AT HIGH TEMPERATURE

Jim Constantz
Water Resources Division
U.S. Geological Survey
Menlo Park, California

## INTRODUCTION

The disposal of high-level radioactive wastes in the unsaturated zone may cause large temperature gradients in the immediate vicinity of the disposal site. Predicting the fate of high temperature wastes necessitates an adequate understanding of heat and mass transfer through the surrounding material. A quantification of the influence of temperature upon water retention characteristics in unsaturated materials is essential for prediction of heat and mass transfer. Since high-level radioactive wastes generate sufficient heat to create 100°C temperatures in the surrounding unsaturated zone, the water holding characteristics of these materials should be examined up to 100°C.

The influence of temperature on the water characteristics in soil has been observed for over half a century [1,2]. Gardner [3] and Taylor [4] found a consistent trend for soils to lose water with increasing temperature. Wilkinson and Klute [5] and Jury and Miller [6] found this trend to be inadequately predicted by a simple capillary pressure versus temperature relationship. Peck [7] pointed out that entrapped air bubbles can influence water retention, resulting in a change in water content with temperature which is greater than predicted by a simple surface tension theory.

These investigations indicate that the influence of temperature upon water holding characteristics is significant and that additional work is necessary for a more complete understanding of the magnitude of the effect. No measurements of water retention between 50°C and 100°C have been reported, leaving theoretical work completely untested in this range. The purpose of this study is to extend the temperature range investigated and to extend the experimental conditions to incorporate both draining and wetting processes. In the present investigation, the water retention is measured up to 95°C using a new experimental technique. This technique provides a means of examining the usefulness of applying simple capillary retention theory to pore water retention above 50°C.

The influence of temperature on pore water retention can be estimated from the Laplace equation applied to nondeformable porous media. The pore water matric potential, $\psi$ , is expressed as

$$\psi = \frac{-2\gamma\cos\phi}{r} \qquad [1]$$

where $\gamma$ is the air-water surface tension, $\phi$ is the water-solid contact angle, and r is the mean radius of the largest undrained pore size fraction. There is evidence [8,9] that temperature has a negligible influence on the value of cos $\phi$ for both receding and advancing air-water interfaces. Furthermore, at a given bubble-free volumetric water content, the value of r can also be considered independent of temperature. Assuming $\phi$ and r are constants, if $\psi$ is known at one temperature, then the value of $\psi$ at a second temperature is calculated by a simple ratio of the surface tension of water at the two temperatures:

$$\psi_1 = \psi_1 \frac{\gamma_2}{\gamma_1} \qquad [2]$$

where the subscripts 1 and 2 represent the value of the parameters at temperatures $T_1$ and $T_2$. In the present investigation the experimental values of $\psi$ measured at 25°, 75°, and 95°C are compared to those values calculated from Eq. [2] between 100% and 20% volume saturation.

## EXPERIMENTAL EQUIPMENT AND PROCEDURE

### I. Equipment

The investigation was accomplished using a submersible pressure outflow cell (SPOC) situated in a temperature regulated circulating water bath (Fig. 1 and 2). The SPOC was developed to eliminate two problems which have made high temperature water retention measurements difficult. At high temperature conventional pressure outflow cells fail to function properly due to desorption of the porous plate and bubble formation below the plate. De-aired water flushing systems, designed to operate below the plate, have proven ineffective above 50°C. The SPOC's porous bottom plate is in direct contact with the circulating water in the bath. This permits continuous flushing beneath the porous plate, resulting in no observable bubble formation or plate desorption up to 100°C.

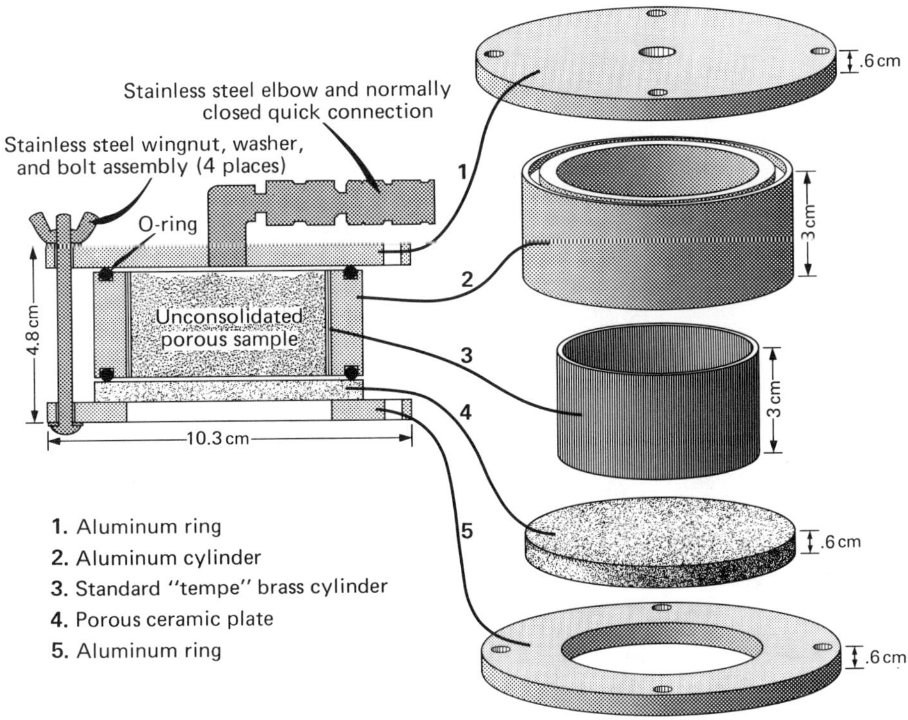


Figure 1. Cross-sectional and exploded views of a submergible pressure outflow cell (SPOC). Four bolts hold the SPOC firmly together and two o-rings form top and bottom gland seals to contain the applied air pressure.

SCHEMATIC OF THE EXPERIMENTAL CONFIGURATION

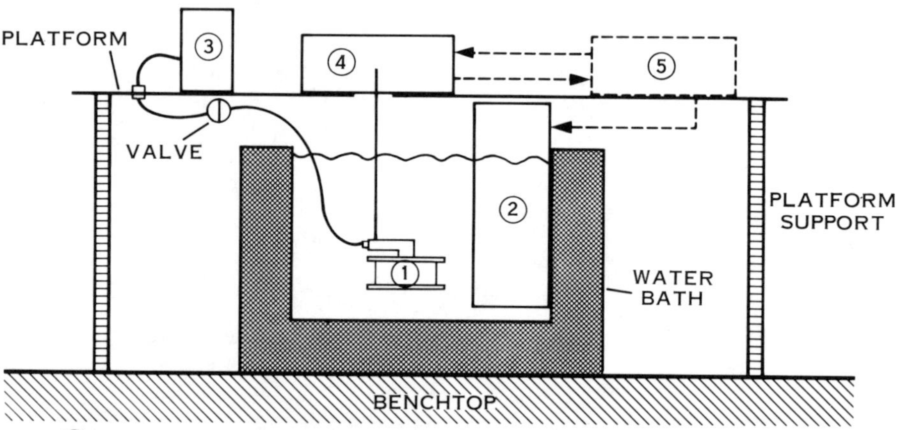


1. S.P.O.C. IN RUNNING POSITION
2. WATER CIRCULATION-TEMPERATURE CONTROL UNIT
3. AIR PRESSURE REGULATION SYSTEM
4. MASS TRANSDUCER, i.e., ELECTRONIC BALANCE OR STRAIN GAUGE SYSTEM
5. DATA ACQUISITION-EQUIPMENT CONTROL UNIT (OPTIONAL)

Figure 2. The complete experimental assembly shown during the weighing of a submergible pressure outflow cell (SPOC). The data acquisition and equipment control unit is shown as an option which was not used for this experiment.

As shown in Fig. 1, the SPOC is based on a cylinder with o-ring gland fittings on each end. A sample is place in the cylinder, and then held firmly in place between a porous ceramic plate and a top plate fitted with an air pressure port. In this study, the SPOC was submerged in a regulated temperature circulating bath with a uniformity of $\pm.01^{\circ}C$ and a stability of $\pm.05^{\circ}C$. A 5 mm brass air pressure line was attached to the top plate of the SPOC, via a quick-release coupling. Pressure was regulated to $\pm.05$ kPa. Throughout the experiments the SPOC remained visibly free of brubbles below the plate regardless of the temperature, pressure, or the duration of the experiment.

For any given change in air pressure or water temperature, the resultant change in the weight of the pore water in the SPOC was measured with an electronic balance (resolution .01 g). Figure 2 shows the complete system with a wire suspending the SPOC from the hang-down level below the balance. A detailed discussion of the SPOC system's temperature correction and tare procedure is included in Constantz and Herkelrath [10].

## II. Procedure

A Tipperary Sand (Mixed Typic Torripsamment) was collected northeast of Soda Lake, Fallon, Nevada. Soil materials were air-dried and passed through a 0.5 mm stainless steel mesh screen. Soil samples were rotary-tap packed using the method described by Ripple and others [11] to construct a packed column composed of standard Tempe cylinders (each 3 cm high x 5.2 cm I.D.). Each SPOC containing a single packed cylinder was run through two wetting and drying cycles at $25^{\circ}C$ before being thermally equilibrated at zero pressure in the water bath.

An initial experiment was run to compare the amount of drainage at $25^{\circ}$ and $95^{\circ}C$ at pressures up to 94 kPa of applied air pressure. The sample was initially drained at $25^{\circ}(\pm.1^{\circ}C)$ and rewet, then drained at $95^{\circ}(\pm.2^{\circ}C)$ and rewet, and again drained at $25^{\circ}(\pm.1^{\circ}C)$. The sample was then removed for determination of its oven dry water content.

A second experiment was performed to examine the relative effect of temperature on desorption versus sorption. This experiment compared both draining and wetting curves at $25^{\circ}$ and $75^{\circ}C$ up to 94 kPa of applied air pressure. The sample was run through a draining and wetting cycle at $25^{\circ}C$, then at $75^{\circ}C$, and again at $25^{\circ}C$ (all with $\pm1^{\circ}C$ stability). The sample was then removed for determination of its oven dry water content.

## RESULTS

The results of the first experiment are presented in Table I, where the value of $\psi$ is expressed as the difference between the water pressure and the air pressure; and the value of S is expressed as a percent volume saturation. Figure 3 shows the retention curves for the first experiment with S on the ordinate and $\psi$ on the abscissa, plotted on a semi-log scale. A comparison of the two 25°C curves in Fig. 3 demonstrates a measurable decrease in water retention above the inflection point for the second 25°C run relative to the first 25 °C run. Below the inflection point the two 25°C curves overlap, and the separation between the 25°C curves and the 95°C curve progressively widens with decreasing values of $\psi$.

---

Table I

Comparison of the Matric Potential, $\psi$ , Versus the Percent Volume Saturation, S, in Tipperary Sand for Experiment One

| Run #1, 25°(±.1°C) | | Run#2, 95°(±.1°C) | | Run#3, 25°(±.1°C) | |
|---|---|---|---|---|---|
| (kPa) (±.05) | S(%) (±.01) | (kPa) (±.05) | S(%) (±.01) | (kPa) (±.05) | S(%) (±.01) |
| -1.2 | 96.0 | -1.0 | 93.5 | -1.1 | 97.6 |
| -2.1 | 93.6 | -2.0 | 87.6 | -2.0 | 91.5 |
| -3.1 | 91.0 | -3.0 | 83.8 | -3.0 | 89.2 |
| -4.0 | 88.4 | -4.0 | 82.3 | -4.0 | 87.0 |
| -5.0 | 82.1 | -5.0 | 65.3 | -5.0 | 73.1 |
| -6.0 | 65.3 | -6.0 | 53.7 | -6.0 | 62.8 |
| -7.1 | 54.6 | -7.0 | 38.1 | -7.0 | 53.7 |
| -7.9 | 48.2 | -8.0 | 31.1 | -8.1 | 46.9 |
| -12.7 | 30.9 | -12.5 | 20.6 | -12.5 | 31.9 |
| -19.6 | 26.2 | -19.3 | 17.4 | -19.4 | 26.4 |
| -39.6 | 23.3 | - | - | - | - |
| -66.7 | 21.5 | - | - | - | - |
| -94.0 | 19.7 | - | - | -93.8 | 20.8 |

---

The results of the second experiment are presented in Table II with the draining and wetting values of S listed continuously from zero pressure down to -94 kPa and back up to zero pressure for each run. In Figures 4 and 5 the results are separated into draining and wetting curves, respectively, to isolate the two processes for a clearer comparison of the 75° and 25°C curves. For both the draining and wetting curves, the results show a measurable

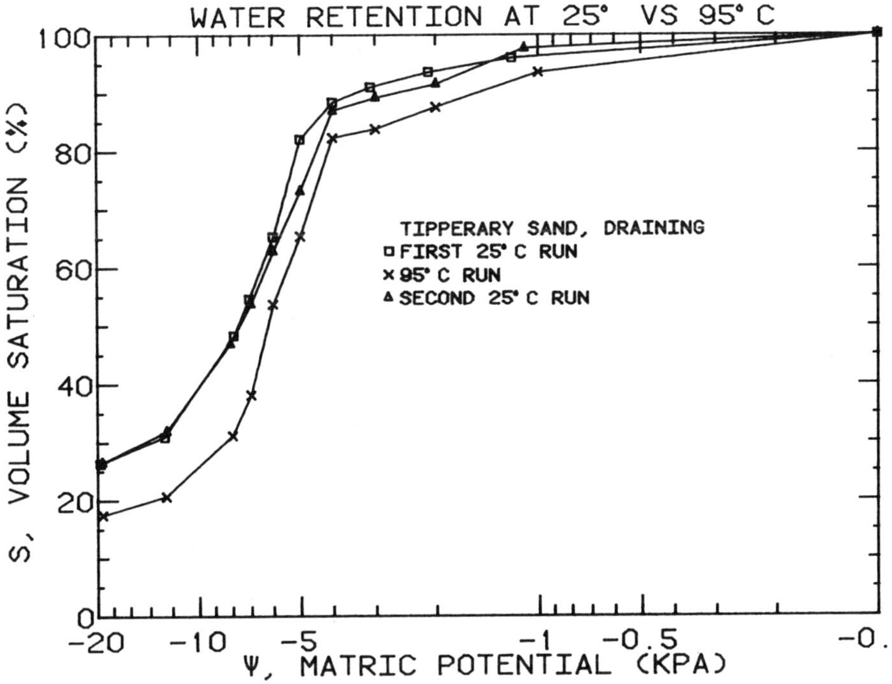


Figure 3. A comparison of S vs. $\psi$ at 25°, 95°, and 25°C in Tipperary Sand for Experiment One.

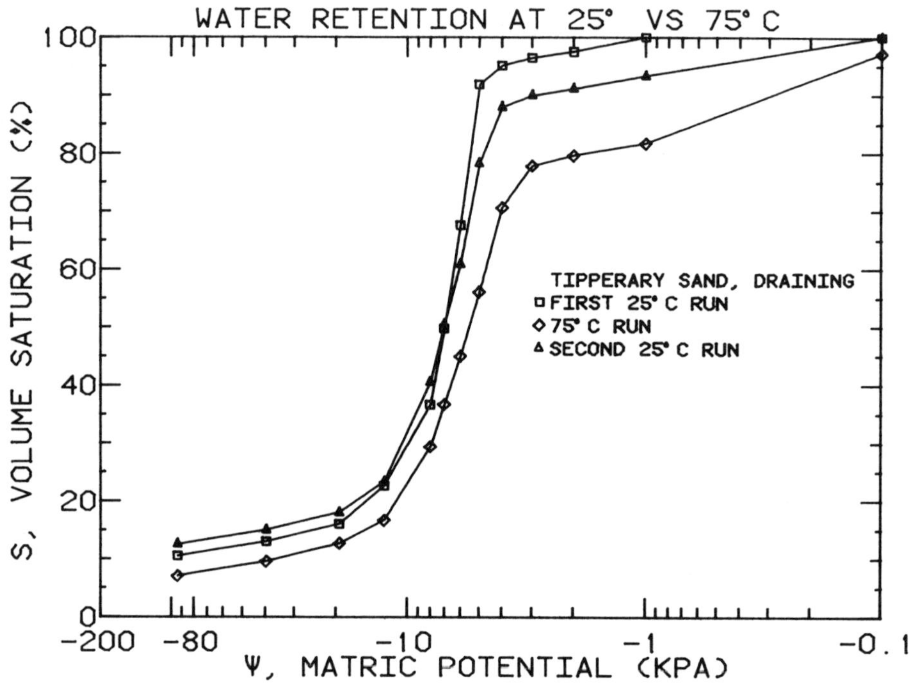


Figure 4. A comparison of S vs. $\psi$ at 25°, 75°, and 25° C during draining in Tipperary Sand for Experiment Two.

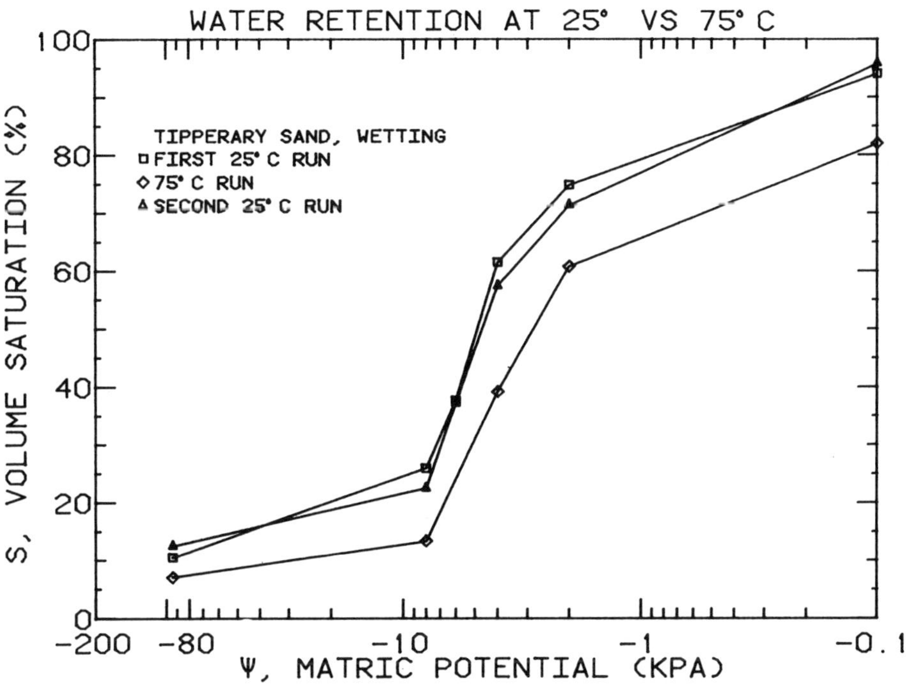


Figure 5. A comparison of S vs. ψ at 25°, 75°, and 25° C during wetting in Tipperary Sand for Experiment Two.

decrease in water retention above the inflection point for the second 25°C run relative to the first run. However, at the lowest water contents, the second 25°C run retains more water than the first run. Both 25°C runs retain measurably more water than the 75°C run at all values of $\psi$ for both the draining and wetting processes.

Table II

Comparison of the Matric Potential, $\psi$ , Versus the Percent Volume Saturation, S, in Tipperary Sand for Experiment Two

| Matric Potential S(%) | Run#1, 25°(+.1°C) S(%) | Run#2, 75°(+.1°C) S(%) | Run#3, 25°(+.1°C) S(%) |
|---|---|---|---|
| 0.0(±.05) | 100.0(±.01) | 97.14(±.01) | 100.0(±.01) |
| -1.0 | 100.0 | 81.76 | 93.32 |
| -2.0 | 97.48 | 79.65 | 91.11 |
| -3.0 | 96.41 | 77.86 | 89.95 |
| -4.0 | 95.09 | 70.62 | 87.92 |
| -5.0 | 91.80 | 56.11 | 78.27 |
| -6.0 | 67.70 | 45.03 | 60.89 |
| -7.0 | 49.75 | 36.73 | 50.54 |
| -8.0 | 36.70 | 29.38 | 40.57 |
| -12.5 | 22.60 | 16.68 | 23.30 |
| -19.3 | 15.97 | 12.70 | 17.97 |
| -39.6 | 13.04 | 9.59 | 14.97 |
| -93.8 | 10.53 | 7.08 | 12.51 |
| -8.0 | 25.96 | 13.35 | 22.46 |
| -6.0 | 37.65 | - | 37.19 |
| -4.0 | 61.51 | 39.19 | 57.41 |
| -2.0 | 74.80 | 60.84 | 71.35 |
| 0 | 93.98 | 81.95 | 95.76 |

DISCUSSION

The decrease in water retention above the inflection point in both experiments for the second 25°C run compared to the first 25°C run, may be an indication of compaction of the samples during the course of the experiments. In addition, for the lowest values of $\psi$ in Tables I and II, the second 25°C run retains more water than the first run, which is a further indication of compaction. Since the samples were stabilized by two preliminary draining and wetting cycles prior to the first 25°C run, the apparent compaction measured during the second 25°C run may have occurred during the initial draining in the high temperature intermediate run. This hypothesis is supported by the similarity in the shapes of the high temperature curve and the second 25°C curve in both experiments. Possible causes of temperature induced compaction include thermal expansion and contraction of the soil matrix or interparticle frictional reduction due to increased temperature. Irrespective of the cause of the apparent compaction, a comparison of the results of only the second 25°C run with the high temperature run in each experiment, appears more representative of the influence of temperature on water retention in these samples.

Table III

Comparison of Experimental and Theoretical Values of the Matric Potential for a given Water Content at 95°C for Experiment One.

| Percent Volume Saturation, S | $\psi$, Matric Potential (kPa) Experimental | Theoretical | Experimental / Theoretical |
|---|---|---|---|
| 100.0 (±.01) | 0.0 (±.05) | 0.0 | - |
| 93.5 | -1.0 | -1.4 | .71 |
| 87.6 | -2.0 | -3.1 | .65 |
| 83.8 | -3.0 | -3.5 | .85 |
| 82.3 | -4.0 | -3.6 | 1.1 |
| 65.3 | -5.0 | -4.8 | 1.0 |
| 53.7 | -6.0 | -5.8 | 1.0 |
| 38.1 | -7.0 | -8.9 | .79 |
| 31.1 | -8.0 | -11.2 | .71 |
| 20.6 | -12.5 | -78.0* | .16 |
| 17.4 | -19.3 | -107.0* | .18 |

*By linear extrapolation below available percent volume saturation values measured at 25°C.

Theoretical and experimental values of $\psi$ at 95°C are compared at a given water content in Table III. Theoretical values were obtained by substituting 25°C values of $\psi$ from the second 25°C run into the right side of Eq. [2]. Linear interpolation of the measured 25°C values were used to obtain 25°C values at identical water contents as those measured experimentally at 95°C. Values of $\gamma$ for pure water were used. Table III also lists the ratio of experimental to theoretical values of $\psi$. A ratio <1.0 indicates that Eq. [2] underestimates the effect of an increase in temperature upon $\psi$, and a ratio >1.0 indicates an overestimate of the influence of temperature. For this experiment, Eq. [2] best predicts the influence of temperature on $\psi$ at 95°C at intermediate values of S near the inflection point of the draining curve. A comparison of the draining curves at 95°C, based on Table III is shown in Fig. 6. As indicated by the ratios in Table III the agreement between the curves becomes increasingly worse below 50% saturation.

---

Table IV

Comparison of Experimental and Theoretical Values of the Matric Potential for a given Water Content at 75°C for experiment two.

| Percent Volume Saturation, S | $\psi$, Matric Potential (kPA) Experimental | Theoretical | Experimental / Theoretical |
|---|---|---|---|
| 97.14 (±.01) | 0.0 (±.05) | -0.4 | - |
| 81.76 | -1.0 | -4.1 | .24 |
| 79.65 | -2.0 | -4.3 | .47 |
| 77.86 | -3.0 | -4.4 | .68 |
| 70.62 | -4.0 | -4.8 | .83 |
| 56.11 | -5.0 | -5.7 | .88 |
| 45.03 | -6.0 | -6.7 | .90 |
| 36.73 | -7.0 | -7.9 | .87 |
| 29.38 | -8.0 | -10.9 | .73 |
| 16.68 | -12.5 | -18.5 | .68 |
| 12.70 | -19.3 | -79.1 | .24 |
| 9.59 | -39.6 | -139.5* | .28 |
| 7.08 | -93.8 | -188.3* | .50 |
| 13.35 | -8.0 | -86.6 | .09 |
| 39.19 | -4.0 | -5.8 | .69 |
| 60.84 | -2.0 | -3.5 | .57 |
| 81.95 | 0.0 | -1.1 | - |

*By linear extrapolation below available percent volume saturation values measured at 25°C

---

Theoretical and experimental values of $\psi$ at 75°C are compared at a given water content in Table IV. The theoretical values in Table IV were obtained using the same

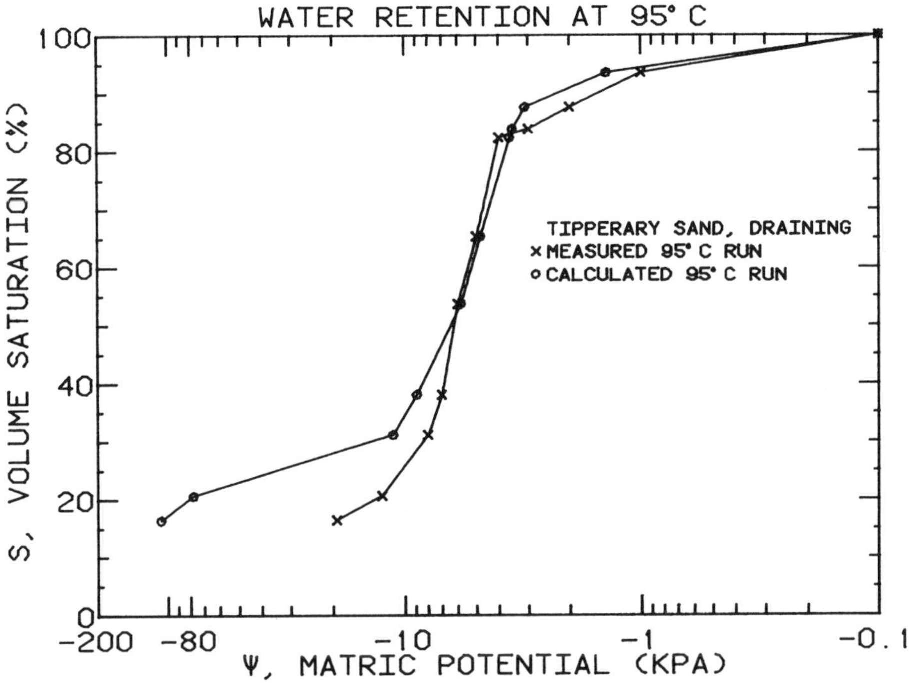


Figure 6. A comparison of S vs. $\psi$ at 95° C for experimentally measured values and theoretically calculated values from Eq.[2] in Tipperary Sand for Experiment One.

technique used to generate Table III. For the second experiment, Eq. [2] best predicts the influence of temperature on $\psi$ at 75°C at intermediate values of S for both the draining and wetting cases. Figures 7 and 8 show a comparison of the 75°C curves for the draining cases and wetting case, respectively, based on the values in Table IV. As indicated by the ratios in the table, the agreement between the curves becomes increasely worse below about 50% saturation for both cases.

Results indicate that when estimating values of $\psi$ at high temperature from values obtained at 25°C, Eq. [2] produces estimates of $\psi$ which diverge from the measured high temperature values by as much as ten fold. As shown in Figures 6, 7, and 8, the trend for increasing divergence of the calculated versus measured curves, may indicate that Eq.[2] is inadequate in its present form to describe all the temperature sensitive forces important in water retention below about 50% saturation. The failure of Eq. [2] at lower moistures may indicate that: 1) the values of $\cos\theta$ and r at a given value of S are not independent of temperature, or 2) the temperature dependence of $\gamma$ for pore water diverges from that of free water at high temperature, or 3) the Laplace equation may not be applicable at lower moistures.

For the draining cases at 95°C and 75°C, the rapid decrease in water retention for values of $\psi$ near zero may be caused by entrapped gas bubbles or by the formation of bubbles in the pore water. Chahal [12] has shown that entrapped gas bubbles can displace significant amounts of pore water near saturation. Since the propensity for air bubble nucleation increases with increasing temperature and decreasing values of $\psi$, pore water displacement by bubbles may have occurred at 95°C. The relative importance of entrapped bubbles versus nucleated bubbles in influencing water retention cannot be determined from the present results. However, it is clear that a change in $\psi$ with temperature is likely to be larger than predicted by any simple surface tension-temperature correction, for any pore water system containing bubbles.

In the case of the wetting curves at 75°C, the volume of bubbles in the pore water was negligible initially; yet the curves in Fig. 8 show a large difference between the experimental and calculated results even at high levels of saturation. Here the bubbles in the system near saturation were derived almost entirely from bubble formation during the wetting process. Therefore, increased bubble formation at 75°C most likely contributed to a lower water holding capacity during wetting than predicted by a simple surface tension-temperature correction.

In conclusion, three results of these experiments on Tipperary Sand are notable in connection with high

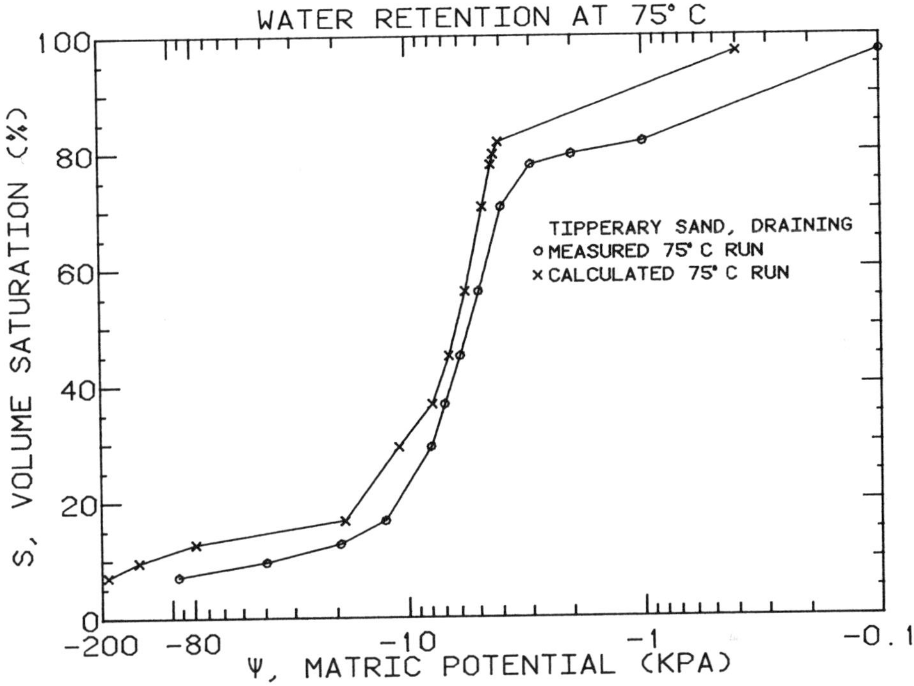


Figure 7. A comparison of S vs. $\psi$ at 75$^{\circ}$ C for experimentally measured values and theoretically calculated values from Eq.[2] during draining in Tipperary Sand for Experiment Two.

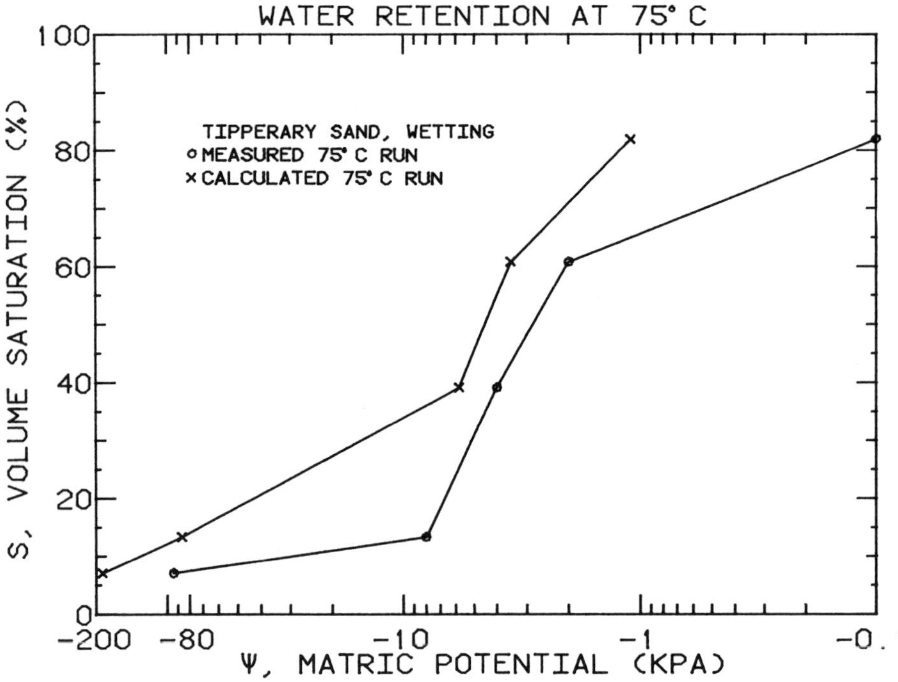


Figure 8. A comparison of S vs. $\psi$ at 75°C for experimentally measured values and theoretically calculated values from Eq.[2] during wetting in Tipperary Sand for Experiment Two.

temperature water retention characteristics. First, apparent compaction of the sand samples may have been caused by the high temperature conditions. This suggests that predictions of the physical behavior of unsaturated zone materials in the immediate vicinity of high temperature wastes should consider the effect of compaction of the material in response to thermal loading. Secondly, the results from these experiments show that measurably less water is held at high temperature below about 50% saturation than is predicted by a simple capillary retention model. Finally, results indicate that the influence of temperature or gas bubble formation within the pore water system may measurably decrease the water holding capacity of sand at high temperature above 50% saturation. Therefore, failure of the Laplace equation to predict the effect of temperature above 50°C on water retention may be explained by the inadequacies of a simple surface tension-temperature correction to properly describe capillary retention forces existing at lower values of $\psi$ , and the inability of the correction to handle bubble formation at the highest values of $\psi$. The usefulness of a simple surface tension-temperature correction is apparently limited to intermediate values of S and $\psi$ for Tipperary Sand above 50°C.

## REFERENCES


1. Bouyoucos, G.J., "Effects of Temperature on Some of the Most Important Physical Processes in the Soil", Mich. Agr. Exp. Sta. Tech. Bul. 22 (1915).

2. Moore, R.E., "The Relation of Soil Temperature to Soil Moisture: pressure potential, retention, and infiltration rate", Soil Sci. Soc. Am. Proc. 4:61-64 (1940).

3. Gardner, R., "Relation of Temperature to Moisture Tension of Soil:. Soil Sci. 79:257-265 (1955).

4. Taylor, S.A., "The Activity of Water in Soil", Soil Sci. 86:83-90 (1958).

5. Wilkinson, G.E., and A. Klute. "The Effect of Temperature on the Equilibrium Energy Status of Water Held by Porous Media", Soil Sci. Soc. Am. Proc. 26:326-329 (1962).

6. Jury, W.A. and E.E. Miller, "Measurement of the Transport Coefficients for Coupled Flow of Heat and Moisture in a Medium Sand", Soil Sci. Soc. Am. Proc. 38:551-557 (1974).

7. Peck, A.J., "Change of Moisture Tension with Temperature and Air Pressure: Theoretical. Soil Science, 89:303-310 (1960).

8. Phillips, M.C. and A.C. Riddiford, "Temperature Dependence of Contact Angles", Nature 205:1005-1006 (1965).

9. Ellison, A.H., R. B. Klemm, A.M. Schwartz, L.S. Grubb, and D.A. Petrash, "Contact Angles of Mercury on Various Surfaces and the Effect of Temperature", J. Chem. Eng. Data 12:607-609 (1967).

10. Constantz, J., and Herkelrath, W.N., "An Improved Pressure Outflow Cell for Measurement of Pore Water Properties from $0^o$ to $100^oC$", Soil Sci. Soc. Am. J. (In preparation).

11. Ripple, C.D., R.V. James, and J. Rubin, "Radial Particle-Size Segregation During Packing of Particulates into Cylinderical Containers", Powder Tech. 8:165-175 (1973).

12. Chahal, R.S., "Effect of Temperature and Trapped Air on the Energy Status of Water in Porous Media", Soil Sci. 98:107-112 (1964).

# ROLE OF DESATURATION ON TRANSPORT THROUGH FRACTURED ROCK*

Daniel D. Evans & Chi-hua Huang
Department of Hydrology and Water Resources
University of Arizona
Tucson, Arizona

## INTRODUCTION

The safe, long-term storage of high level radioactive wastes is a critical issue facing governments of different levels and private enterprises. The production of such wastes is increasing rapidly, which necessitates an early solution to the waste disposal problem. The most promising solution at present is the construction of repositories at several hundred meters below surface within a geological formation where the wastes will be isolated from the biosphere until the radioactivity decays to an acceptable level. Geologic formations of most interest are those of igneous rocks, such as granite, basalt and tuff, and salt [1].

A conceptual design of a repository is shown in Figure 1. Vertical shafts would be constructed to the repository

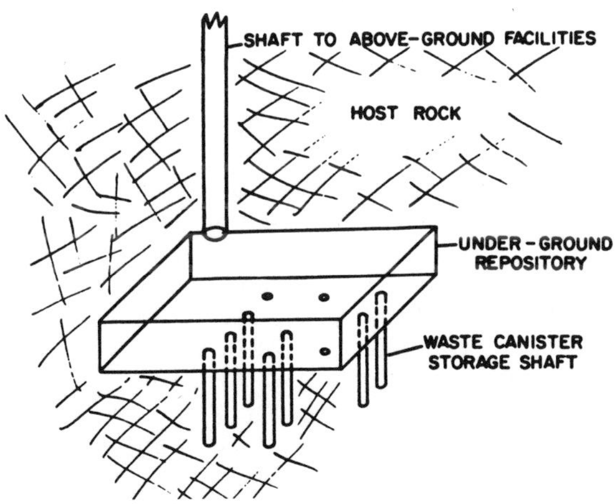


Figure 1. A conceptual drawing of a repository.

* This work was supported by the U.S. Nuclear Regulatory Commission under Contract NRC-04-81-224.

zone. Horizontal workrooms would be excavated from which vertical wells would be drilled. Waste enclosed in specially designed canisters would be placed in the wells, and individual wells would be sealed when filled to capacity. A repository complex may be operated for 50 years or longer after which all excavations would be filled with a special sealant. The location and design should minimize the probability of any released contaminants reaching the biosphere or being disturbed by man over several thousand years.

In characterizing a potential repository site, it is necessary to account for the rare chance of leakage of contaminants and their entering the regional hydrologic system and being transported to other locations. Therefore, predictions of water movement over long time spans is crucial during the site selection and licensing processes.

The potential repository zone may be initially water saturated or unsaturated. Even if the zone is initially saturated, unsaturated conditions may develop during repository construction and operation by artificial dewatering, or at some later time due to climatic change or regional water management modifications. This presentation deals with flow through unsaturated igneous rock systems. Since rock formations are invariably fractured to some degree, the fractures may be the principal conduits for water flow and contaminant transport. Hence, the focus is on the fracture system rather than on the rock matrix per se.

## FLOW THROUGH THE UNSATURATED FRACTURED ROCK

Water movement through unsaturated fractured rock is more complex than that for saturated systems. The apparent unsaturated hydraulic conductivity is not a constant but a function of the water potential. As the potential decreases (becomes more negative), smaller and smaller fractures are partially drained, changing the water conductance properties of the fracture system. Upon desaturation of a fracture, flow changes from saturated contained flow to film flow along the fracture surfaces. Also, the partially drained fracture becomes a conduit for vapor flow, leading to possible two-phase flow--gas and liquid. Because elevated temperatures are anticipated in a high level waste repository, vapor transport may be significant and even exceed liquid transport. In fact, there is some experimental evidence that liquid flow may be toward the higher temperature repository in all directions. In any case, drying of the rock surrounding the repository should be expected and the drying zone should increase in distance from the repository with time. Such drying will reduce the possibility of released contaminants entering the regional hydrologic system.

Previous research on water flow and contaminant transport through unsaturated fractured rock has been extremely limited, especially for geologic formations suitable for repositories. However, research interests are expanding because of the concern for possible contaminant transport through such media.

Several basic aspects are considered in the following sections. Capillary theory is applied to approximate the air -entry potential for different aperture size fractures. The thickness of the remaining water film upon desaturation and the change in water conducting properties as a function of water potential are considered. Following this, parallel joint sets with lognormal distributions of aperture size are examined to show the influence of the arithmetic mean of aperture size, the associated coefficient of variation and the water potential on the relative conductivity of assumed joint sets. Finally, two-phase flow through an unsaturated fractured rock system is considered.

## Air Entry Potential

Capillary theory may be applied to approximate the water potential at which a given size fracture will partially drain, commonly refered to as the air - entry potential. The well known height-of-rise equation for two parallel plates separated by a distance b is

$$h = 2\,\eta\cos\theta/b\gamma \qquad (1)$$

where h = height of rise from a free water surface, or the air entry potential,

$\eta$ = surface tension,

$\theta$ = angle of contact of the air-water interface and the plate wall, and

$\gamma$ = specific weight of water.

In appling the equation to a rock fracture, it has been assumed that the fracture is open, has smooth surfaces, has uniform aperture size and that $\theta = 0$.

Calculated results are shown in Figure 2 which show, for example, that a 1 mm fracture will drain at -1.5 millibars (mb) while a 1 $\mu$m fracture will drain at $-1.5 \times 10^3$ mb. These results demonstrate the variation of air entry potential with aperture size for an idealized fracture.

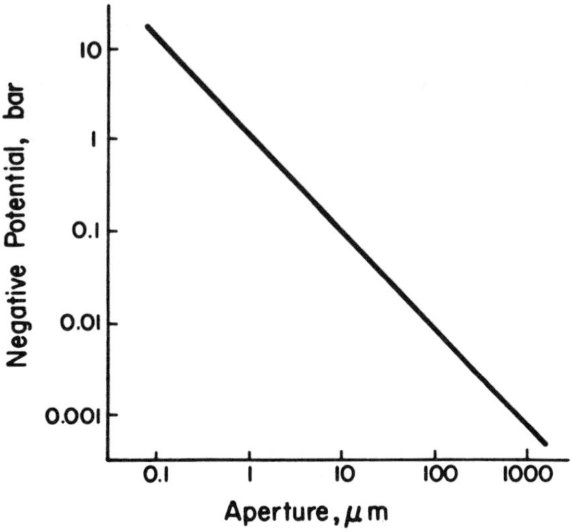


Figure 2. Air entry potential versus aperture size.

Film Flow

The remaining water film on the fracture surfaces after partial drainage provides a much reduced cross-sectional area for flow than when the fracture was fully saturated. In addition, film thickness is a function of the existing water potential and this relationship must be known to characterize film flow. Adequate theory or experimental data are not readily available to determine the relationship for natural rock surfaces. However, some experimental data exist for soil materials and sands (Figure 3) [2]. Film thicknesses for these materials are of the order of tens of angstroms in the potential range of -0.1 to -2 bars.

Assuming the data of Figure 3 approximate that for rock surfaces, the relative cross-sectional area for flow and the relative flow can be calculated. For example, an aperture of 1.5 μm will drain at -1 bar potential. Once it is drained, the film thickness on the fracture surface is approximately 30Å. In this case, the relative cross-sectional area for liquid transport in this particular fracture after drainage is 0.004 and the relative flux is $6.4 \times 10^{-8}$ of that when saturated. It is apparent that film flow may be insignificant when compared to flow through saturated fractures even for very small apertures. Note that in this simplified example, we do not consider the effects of the fracture surface on the physical and

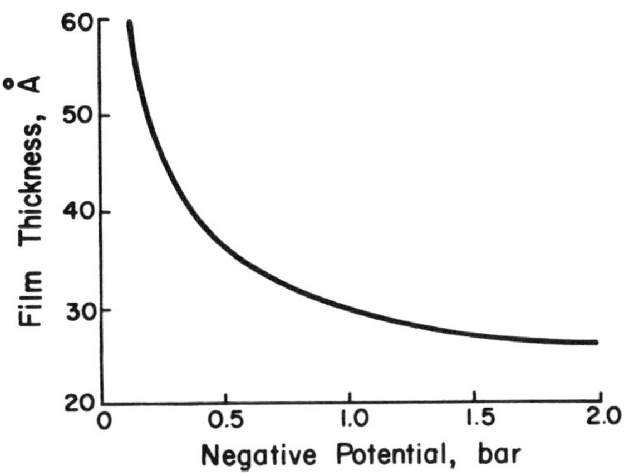


Figure 3. Film thickness versus water potential.

thermodynamical properties of the water. Results from studies of water near clay surfaces show that the properties of water are greatly altered by the charged surfaces [3]. For example, the viscosity of water is several times higher than the bulk free water at 10Å from the montmorillorite surface. Thus, the thin film liquid flow on a fracture surface could possibly be considerably smaller than that predicted based on an ideally smooth and inert surface.

## Parallel-plate Analogy and the Poiseuille Equation

Flow through fractures is often idealized as flow between two parallel plates, which can be described by the Poiseuille equation:

$$Q = \frac{\gamma}{12\mu} L\ J\ b^3 \qquad (2)$$

where $Q$ = volume flux in $L^3/T$,
$\gamma$ = specific weight of water in $M/L^2T^2$,
$\mu$ = dynamic viscosity of water in $M/LT$,
$L$ = segment length normal to the flow in $L$,
$J$ = potential gradient in $L/L$, and
$b$ = aperture in $L$.

In this parallel plate analogy, the only parameter which characterizes the fracture is the size of the fracture or the aperture. For a single fracture with smooth walls, the aperture is simply the distance between the fracture surfaces. However, fractures in rock have uneven walls and variable apertures. One way to model these fractures is to use a set of parallel plate openings with different apertures. This brings the need of a statistical description of fractures in the rock. Thus a probability density function, or frequency distribution, will uniquely describe the statistics of an aperture distribution. In fact, we can use this set of parallel plates to model a joint set, and also, by combining several sets with different orientation, to model a fracture system in three-dimensional space with several intersecting joint sets.

In this paper, we will limit our discussion to a set of parallel plates with different apertures. In this simplified case, the flow is one-dimensional, and the Poiseuille equation applies for every individual aperture.

## Lognormal Distribution for Apertures

Selection of an appropriate probability density function to describe the apertures is empirical and based on observed data. The exact form of the aperture frequency distribution is difficult to determine. The frequency distribution depends on many factors, such as the rock fabric, the mode of fracturing, the stress history of the fracture, and the size scale being considered. Existing data suggest that the lognormal distribution is appropriate for the description of aperture distribution [4].

The probability density function, or the frequency distribution, f(x), of a lognormal distribution is defined by:

$$f(x) = \frac{1}{x\sigma\sqrt{2\pi}} \exp\left[-1/2 \left(\frac{\ln x-\xi}{\sigma}\right)^2\right] \qquad (3)$$

where $\xi$ and $\sigma^2$ are the mean and variance of a normal distrubution which is the result of logarithmic transformation of the lognormally distributed variates [5]. Thus the aperture frequency function f(b) is determined by two parameters, $\xi$ and $\sigma$. The mean aperture, $\tilde{b}$, can be expressed as

$$\tilde{b} = \int_0^\infty b\, f(b)db = \ell \exp(\xi + 1/2\, \sigma^2) \qquad (4)$$

where $\ell$ is unit length. The variance, or the second moment about the mean, $\tilde{M}^2$, of a lognormal distribution is defined as

$$M^2 = \int_0^\infty (b-b_1)^2 \; f(b)db$$

$$= (\tilde{b})^2 \; [\exp(\sigma^2)-1] \; . \qquad (5)$$

If we let $\tilde{k}^2 = \exp(\sigma^2) - 1$ , then (6)

$$M^2 = \tilde{b}^2 \; \tilde{k}^2 \qquad (7)$$

and here $\tilde{k}$ is called the coefficient of variation. From Equations (4) and (7), we see that $\tilde{b}$ and $\tilde{k}$ uniquely define $\xi$ and $\sigma$, and subsequently, we can use this alternative pair of parameters, $\tilde{b}$ and $\tilde{k}$, to describe the lognormal distribution. The use of $\tilde{b}$ and $\tilde{k}$ in our discussion is preferred because they bear physical meanings.

Some examples are shown in Figures 4 and 5. Figure 4 shows several distributions with specified means and a constant coefficient of variation of 0.5, while Figure 5 shows the distributions for a range of variances with a constant arithematic mean of 4 μm.

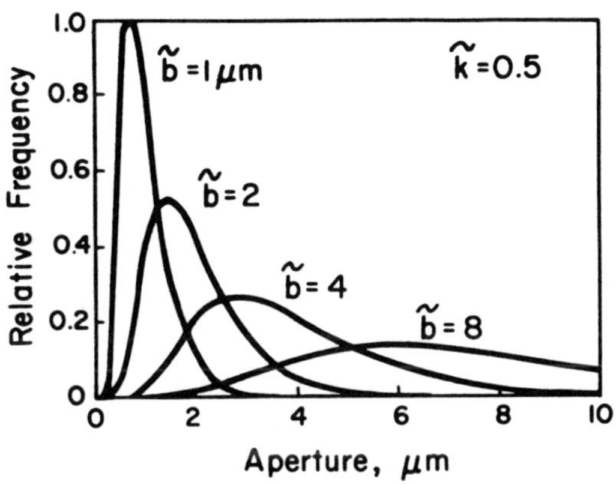


Figure 4. Lognormal distribution of aperture size with a variation $\tilde{k}$=0.5, and varying mean aperture.

## The Modified Poiseuille Equation

Combining the lognormal distribution for the fracture size and the Poiseuille equation for each individual opening, Neuzil and Tracy [6] proposed a modified Poiseuille equation, which is

$$Q = \frac{\gamma}{12\mu} \; L \; J \int_0^\infty \; b^3 \; f(b)db \; . \qquad (8)$$

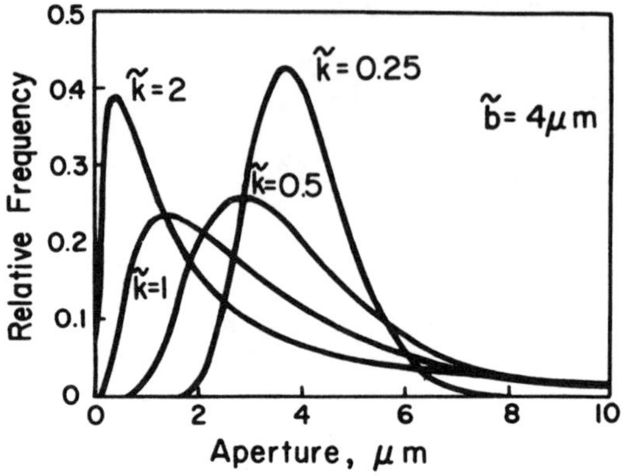


Figure 5. Lognormal distribution of aperture size with a mean aperture of 4μm and varying coefficients of variation.

An analytical function can be used to evaluate this specific integration, that is:

$$Q = \frac{\gamma}{12\mu} \; L \; J \; \ell^3 \; \exp\left(3\xi + \frac{9}{2}\sigma^2\right) \quad . \tag{9}$$

Note that this expression gives the flux for the set of parallel fractures under saturation.

Using capillary theory, certain fracture will be drained under certain potentials. When a fracture is drained, the flow is changed from a fully enclosed flow between parallel plates to flow of thin films on the surfaces. The film flow phenomenon on a flat surface can also be described by the Poiseuille's cubic law (refer to fluid dynamics textbooks, e.g., Basic Fluid Mechanics by Peerless, 1967). As we have shown on the previous section titled "Film Flow", liquid conductance from a desaturated fracture is negligible and the flux will be contributed mostly from those undrained fractures.

Combining the capillary theory and the modified Poiseuille equation (Eq.(7)), we derive an expression to describe the water flow in the set of parallel fractures under partially saturated condition. This equation is given as:

$$Q = \frac{\gamma}{12\mu} \; L \; J \int_0^{b_x} b^3 \; f(b)db \tag{10}$$

where $b_x$ is the maximum aperture that is saturated. Any opening with size greater than $b_x$ will be drained and does not contribute to liquid flow. The relative flux and cross-sectional area for liquid flow with respect to the saturated fracture set can be expressed as:

$$\frac{\int_0^{b_x} b^j \; f(b)db}{\int_0^{\infty} \; b^j \; f(b)db} \qquad (11)$$

where j=1 for cross-sectional area and j=3 for water flux. Equation (11) can be used to examine the effects of desaturation on water conductance for fracture sets with different distribution functions. The denominator of Eq.(11) can be evaluated by analytical functions (Eqs.(4) and (9)), while the numerator must be approximated by numerical integration techniques.

Results from two cases are shown in Figures 6 and 7. Figure 6 is for a mean aperture of 4 μm and a coefficient of variation of 1.0, while Figure 7 is for the same mean aperture but a different $\tilde{k}$, 0.25. In these figures, the air entry potentials, or the potentials at which certain size of fracture will be drained, are also shown. It needs to be pointed out that the horizontal scales are different by an order of magnitude in the two figures.

These figures show a very strong dependency of the flux on large saturated fractures. This phenomenon is especially significant for distributions with large variances. To

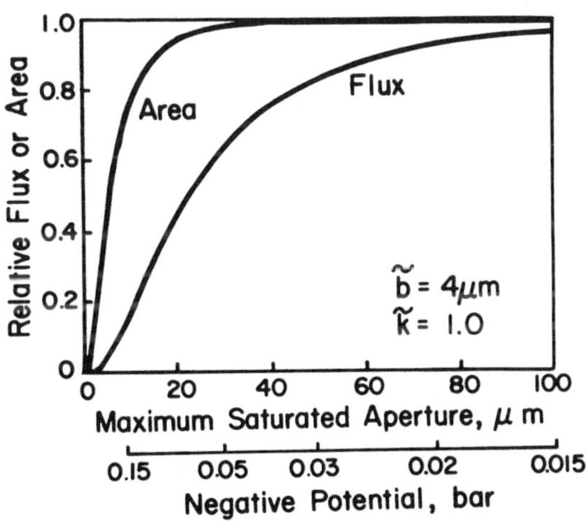


Figure 6. The effect of fracture drainage on flux or area of the flow for a set of parallel fractures with mean aperture 4 μm and a coefficient of variance 1.0.

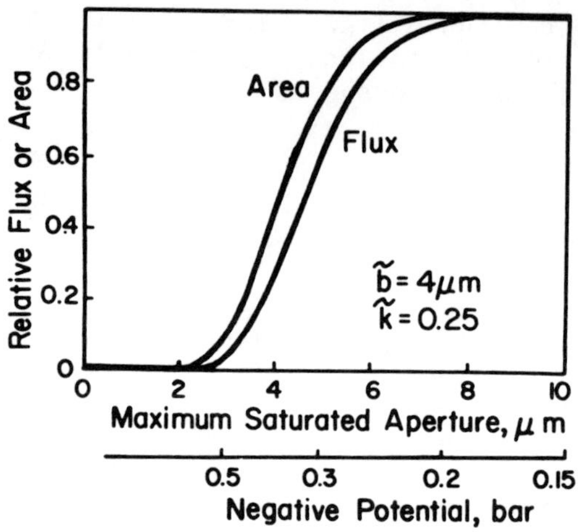


Figure 7. The effect of fracture drainage on flux or area of the flow for a set of parallel fractures with mean aperture 4 μm and a coefficient of variance 0.25.

emphasize this, some data have been tabulated from Figures 6 and 7, and these are shown in Tables I and II. For $\tilde{k}$ equal to 1.0, a 5% reduction in cross-sectional area will result in a 50% reduction of flux, while for a smaller $\tilde{k}$, e.g. 0.25 the same reduction in the available flow area will result in a 12% reduction in flow. Data in Tables I and II also show that for large $\tilde{k}$ values, more fractures are in the larger size range. For example, when $\tilde{k}=1$ and $\tilde{b}=4\mu m$, 90% of the fracture porosity is made up of fractures having apertures less than 16 μm, for $\tilde{k}=0.25$, the same volume is made up of apertures less than 5.6 μm. These data show the role of aperture distribution on the flow of partially drained fracture systems.

## NON-ISOTHERMAL TWO-PHASE FLOW

Simultaneous transport of water and energy in non-isothermal porous media has been studied extensively and to a lesser extent for saturated fractured rock. However, such transport in unsaturated fractured rock has only been examined in a qualitative sense except in the petroleum field for formations that are likely to be unsuitable for a repository. Based on basic physical laws, some understanding of water and energy transport through unsaturated fracture rock can be gained.

Table I. Numerical Example of Data Extracted From Figure 6 for a Mean Aperture of 4μm and a Coefficient of Variation of 1.0

| Rel. area | Area/ length (m) | Max. sat. aperture (μm) | Air entry potential (bar) | Rel. flux | Flux/ℓJ (m²/sec) |
|---|---|---|---|---|---|
| 1 | $4.0\times10^{-6}$ | -- | 0.0 | 1 | $4.2\times10^{-10}$ |
| 0.95 | $3.8\times10^{-6}$ | 23.2 | -0.065 | 0.51 | $2.1\times10^{-10}$ |
| 0.90 | $3.6\times10^{-6}$ | 16.0 | -0.094 | 0.34 | $1.4\times10^{-10}$ |
| 0.50 | $2.0\times10^{-6}$ | 5.6 | -0.27 | 0.05 | $2.1\times10^{-11}$ |

Table II. Numerical Example of Data Extracted From Figure 7 for a Mean Aperture of 4μm and a Coefficient of Variation of 0.25

| Rel. area | Area/ length (m) | Max. sat. aperture (μm) | Air entry potential (bar) | Rel. flux | Flux/ℓJ (m²/sec) |
|---|---|---|---|---|---|
| 1 | $4.0\times10^{-6}$ | -- | 0.0 | 1 | $6.3\times10^{-11}$ |
| 0.95 | $3.8\times10^{-6}$ | 6.2 | -0.24 | 0.88 | $5.5\times10^{-11}$ |
| 0.90 | $3.6\times10^{-6}$ | 5.6 | -0.27 | 0.77 | $4.8\times10^{-11}$ |
| 0.50 | $2.0\times10^{-6}$ | 4.0 | -0.38 | 0.31 | $1.9\times10^{-11}$ |

In an unsaturated media subject to non-isothermal conditions, water is transported via both liquid and vapor phases under the combined influence of hydraulic and thermal gradients. Experimental evidence for porous materials show that the net water transport is from warm to cool regions [7, 8, 9]. These results conform to the theoretical considerations under which several possible mechanisms have been proposed [10]. The thermally induced vapor transport is mainly due to the elevated vapor density gradient. The liquid phase transport as a response to the temperature gradient can be explained by 1) the higher surface tension at the cool region; 2) the higher matric potential at the warm region; 3) the high random kinetic energy in the water molecules associated with high temperatures, and 4) the thermally induced osmotic gradients or the Soret effect. The direction of liquid phase transport under thermal gradients is not always from hot to cold as the above mentioned theory predicts. Column studies have shown that despite the net accumulation of water in the cold region, there were increased salt content near the hot end [7, 9]. The results of Gurr et al. [7] are depicted in Figure 8. Since the salt will only be transported in the liquid phase, it indicates that the liquid water moved from the cold to the hot region. Thus a rational explanation is postulated below. As water (both liquid and vapor) moves from hot to cold under the influence of a temperature gradient, the matric potential gradually increases on the cool side, thus there is an increasing tendency for water to flow back to equalize the matric potential. At some time later, when a quasi-steady state is reached, water evaporates at the hot

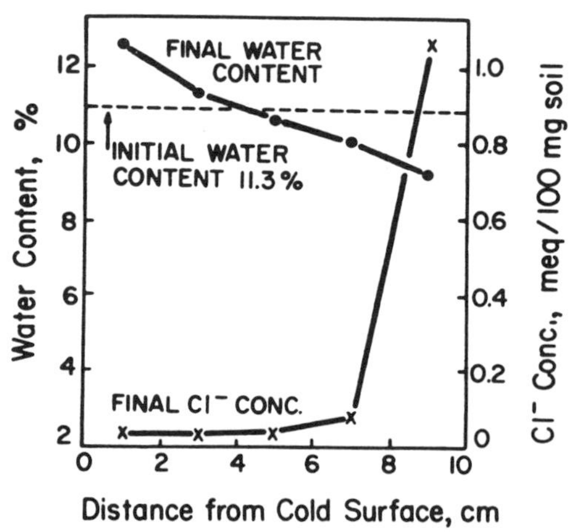


Figure 8. Moisture and salt movement under thermal gradient (results of Gurr et al. [7]).

end, moves to the cold region as vapor, condenses at the cold end, and then returns to the hot region in the liquid phase. Thus a circulatory evaporation-vapor transport-condensation-liquid transport system is established in a finite column with an imposed temperature gradient (Figure 9).

The relative contribution of transport via liquid and vapor phases and their direction of movement with respect to a repository bear important meanings in the containment of contaminants. If water is moving toward the repository in the liquid phase and away from the repository in the vapor phase, soluble contaminants will not be transported away from the repository. For a repository setting, the rock medium may be considered infinite in extent and there will be a net water movement away from the repository even with liquid movement toward the repository.

The net flux of water away from the elevated temperature region can also provide additional protection for the safeguard of the repository by the formation of a dry mantel around the repository. As mentioned in the introduction, vapor transport may be an important mechanism in relation to contaminant movement from a repository, particuarly when a temperature gradient is imposed. Connecting partially drained fractures will provide a pathway for vapor transfer, and consequently, transport away from the repository will cause a drying of the surrounding rocks. The drying zone will extend to an ever increasing distance from the repository and over long time spans the drying zone may extend hundreds of meters from the repository, thus creating a significant barrier to the movement of soluble contaminants (Figure 10). In contrast, the movement of gaseous contaminants may be enhanced by the increased cross-sectional area for gaseous diffusion as well as by the imposed temperature gradient. Thus an unsaturated environment may be less suitable for wastes with gaseous compounds.

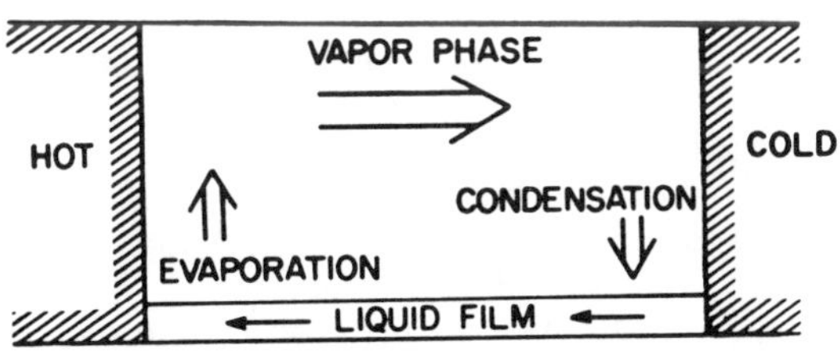


Figure 9. Mechanism of moisture transport in an unsaturated, finite medium under thermal gradients.

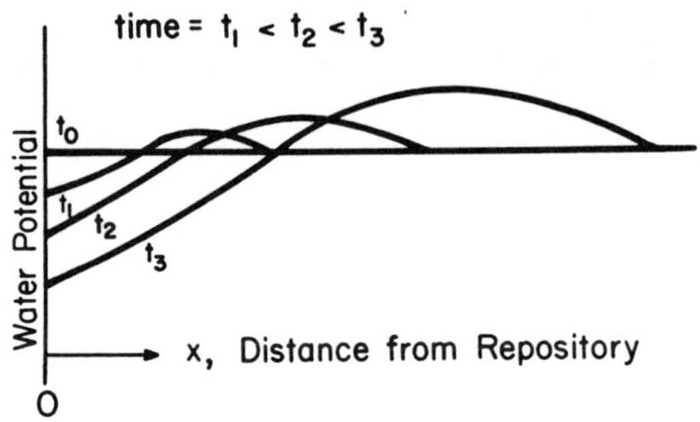


Figure 10. Hypothetical water potential vs. distance from a repository at three times.

REFERENCES


1. Gonzales, S. "Host Rocks for Radioactive Waste Disposal" Amer. Scientist 7:191-200 (1982).
2. Taylor, S. A. and G. L. Ashcroft. (ed). "Physical Edaphology" W. H. Freeman and Co. San Francisco (1972).
3. Low, P. F. "Nature and Properties of Water in Montmorillonite-Water systems". Soil Sci. Soc. Am. J. 43:651-658 (1979).
4. Snow, D. T. "The Frequency and Apertures of Fractures in Rock". Int. J. Rock Mech. Min. Sci. 7:23-40 (1970).
5. Aitchison, J. and J. A. C. Brown. (ed). "The Lognormal Distribution". University press, Cambridge. (1957).
6. Neuzil, C. E. and J. V. Tracy. "Flow Through Fractures". Water Resour. Res. 18:191-199 (1981).
7. Gurr, C. G., T. J. Marshall and J. T. Hutton. "Movement of Water in Soil Due to Temperature Gradient". Soil Sci. 74:335-345 (1952).
8. Taylor, S. A. And L. Gavazza. "The Movement of Soil Moisture in Response to Temperature Gradient". Soil Sci. Soc. Am. Proc. 18:351-358 (1954).
9. Weeks, L. V., S. J. Richards and J. Letey. "Water and Salt Transfer in Soil Resulting from Thermal Gradients". Soil Sci. Soc. Am. Proc. 32:193-197 (1968).
10. Cary, J. W. "Soil Moisture Transport due to Thermal Gradients: Practical Aspects". Soil Sci. Soc. Am. Proc. 30:428-433 (1966).

# POTENTIAL USE OF GPR IN ASSESSING GROUNDWATER POLLUTION IN PARTIALLY AND FULLY SATURATED SOILS

J.J. Bowders, Jr., R.M. Koerner and A.E. Lord, Jr.
Departments of Civil Engineering and Physics
Drexel University
Philadelphia, Pennsylvania 19104

## OVERVIEW OF EM METHODS

Electromagnetic (EM) methods have been used to probe subsurface soil and rock materials since the 1920's. Most of this activity has been by mining geologists and engineers and has centered about prospecting for minerals and other natural resources. These efforts are generally associated with searches over large distances and to great depths (on the order of hundreds of meters). Geotechnical and environmental engineers, on the other hand, have need to investigate subsurface details over limited distances and at shallow depths (on the order of a few meters). In this latter regard, typical investigative areas include the following:

- soil strata changes
- rock surface profiling
- water table definition
- seepage plume tracing
- location of subsurface voids
- detection of buried pipelines and containers

The various EM techniques can be classified according to the approximate frequency of the transmitted wave or field. They include the following:

1. electrical resistivity - essentially dc
2. extremely low frequency - 1 to 100 Hz
3. very low frequency - 100 to 10 KHz
4. pulsed radio frequency - 1 MHz to 900 MHz
5. pulsed microwave - 1 GHz to 8 GHz
6. continuous wave radio frequency - 0.1 MHz to 600 MHz

7. continuous wave microwave - 0.2 GHz to 8 GHz
8. infrared - about $10^{14}$ Hz
9. optical - about $10^{15}$ Hz

This particular paper concentrates on the technique utilizing pulsed radio frequency waves, commonly called Ground Penetrating Radar or Ground Probing Radar (GPR).

## BACKGROUND OF GPR TECHNIQUE

A major portion of subsurface probing at shallow depths has been based on the transmitting of pulsed RF waves in the frequency range from about 1 MHz to 900 MHz. The transmitted pulse travels through the soil until it infringes upon an object or material with dissimilar electrical characteristics. Part of it is then reflected back to the ground surface where it is received and the time of travel is measured. The depth "d" to the interface is then calculated from $d = (vt)/2$ where v is the wave velocity (which is equal to $c/\sqrt{\varepsilon_r}$, where c is the velocity of light and $\varepsilon_r$ is the relative dielectric constant of the material in which the wave is propagating) and t is the pulse travel time. The relative dielectric constant of numerous soils at different water contents has been evaluated by many investigators, including Okrasinski, et al. [1]. As shown in Table I, the pulsed RF method has been studied by many workers and, as a result, some systems are commercially available.

The authors have been active in adapting this particular method to problems of interest in geotechnical and environmental engineering areas. Using a system which transmits RF pulses in the 80 MHz to 900 MHz frequency range and presenting the received signal in a real time printout, a visual subsurface profile is developed. See, for example, Figure 1 which is the GPR trace of a stationary water table, located in a uniform well graded sandy soil, using a 120 MHz frequency signal. The antenna (actually a transceiver) was traversed across the ground surface from a small creek at the left side to a wooded area 100 m to the right. Although the ground surface appears level on the trace it actually rose about 2 m in this distance. The water table (seen as the distinction between light upper and dark lower areas) was quite level within the soil although it appears to be dropping off to the right. The figure has superimposed upon it the actual water table elevations which were obtained from standard water observation pipes. The GPR predicted accuracy was on the order of $\pm$ 10% of the actual measured values.

The method is extremely sensitive to small changes in dielectric constant of the soil. A laboratory study was

Table I. Details of Pulsed Radio Frequency Methods, after Lord and Koerner [2]

| Investigator(s) | Freq. (MHz) | Depth (m) | Major Application Area |
|---|---|---|---|
| Cook [3,4] | 1-100 | 225 | locating faults, walls, holes |
| Rosetta [5] | 100-200 | 15 | locating faults, caverns, water, utilities |
| Morey [6] | 100-200 | 15 | locating faults, caverns, water, utilities |
| Dolphin, et al. (7) | 15-50 | 40 | locating rock cavities |
| Unterberger [8] | 230 | 500 | salt thickness measurement |
| Harrison [9] | 35 | 2000 | determining ice thickness |
| Rubin, et al. [10] | 100-200 | 10 | detecting subway tunnels |
| Rubin and Fowler [11] | 100-200 | 15 | drilling guidance, subway tunnel monitoring, coal thickness |
| Benson, et al. [12,13] | 100-200 | 10 | general subsurface probing, locating and follow pollutants, buried containers |
| Sandness, et al. [14] | 100-200 | 10 | general subsurface probing as described above |
| Drexel Univ. [15,16] | 80-900 | 4 | located containers, water table, soil stratigraphy |

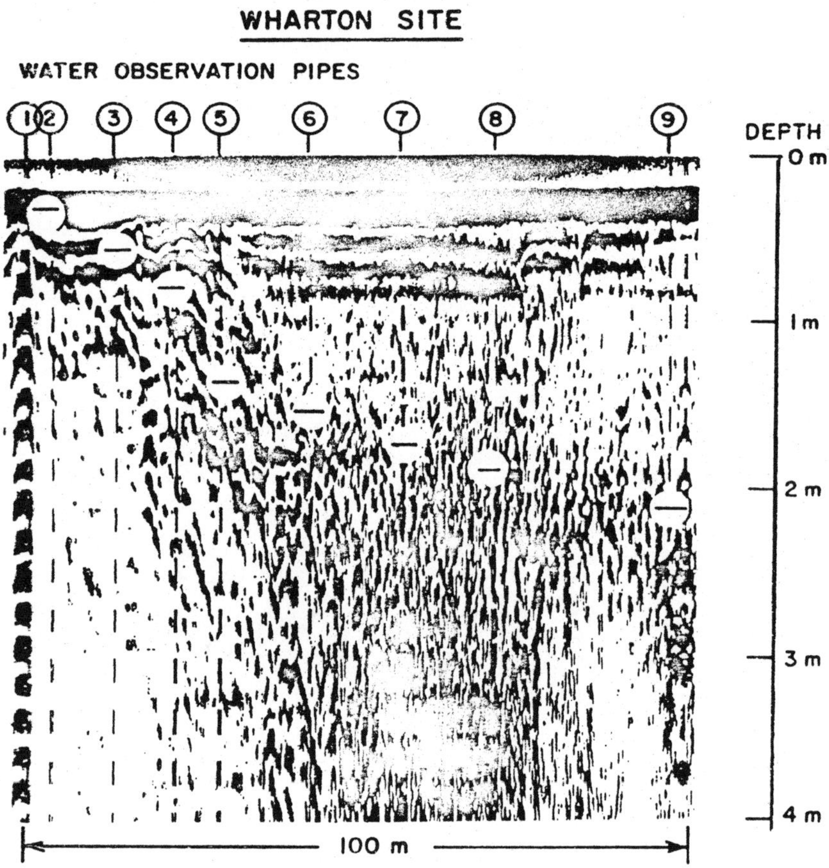


Fig. 1. GPR trace of a stationary water table surface located within a uniform sandy soil with measured water table locations superimposed on the trace.

conducted under quite "ideal" conditions where readouts were obtained from a sandy soil under uniform conditions and compared to the same soil but with an excavated trench backfilled with the soil at about 10% lower density. The trench sides were easily discernable [17]. This density change corresponds to a dielectric constant difference of less than one. This small change will cause less than 1% of the signal power to be reflected from the sides of the trench, but obviously 1% reflection causes a significant blackening of the readout. Thus the method shows great promise for determining very minute changes in subsurface structure or other features or properties of the material being penetrated, but at the same time this great sensitivity has the deleterious effect at times of masking sought-after details.

## BASIC PREMISE FOR THIS STUDY

Owing to the great sensitivity of the GPR technique, and coupled with the well established fact that high ion content pore fluids severely attenuate EM waves, it is hypothesized that the method can be used to distinguish polluted soils containing hazardous wastes from soils containing non-polluted (at least with regard to conductivity) pore water. Using these features of the technique in an inverse manner, it is felt that GPR systems might identify subsurface liquid pollutants by virtue of the strength and nature of the return signals. The confirmation of this premise is the goal of this laboratory test program.

## LABORATORY TEST SETUP AND RESULTS

A wooden box, measuring approximately 1 m x 1 m x 1 m was filled with a well graded, uniform density sand in the dry state as shown in the photographs of Figure 2. A 900 MHz transceiver was placed on the surface of the sand in the center of the box and a GPR trace was made. The system used was a GSSI-SIR Model 7 which is commercially available from Geophysical Survey Systems Inc., Hudson, New Hampshire.

The next test was conducted in the same manner but now the soil was placed in a partially saturated state, i.e., at 33% saturation and a GPR trace was made. Succeeding tests were performed at 67% and 100% saturation, thus making four separate tests in each test series. So as to have a well defined reflecting surface, the bottom of the box in this first fresh water test series had a steel plate beneath it. The results of these four tests are shown in Figure 3. Note that these traces are time traces and that the time for the signal to travel from the soil surface to the bottom of the

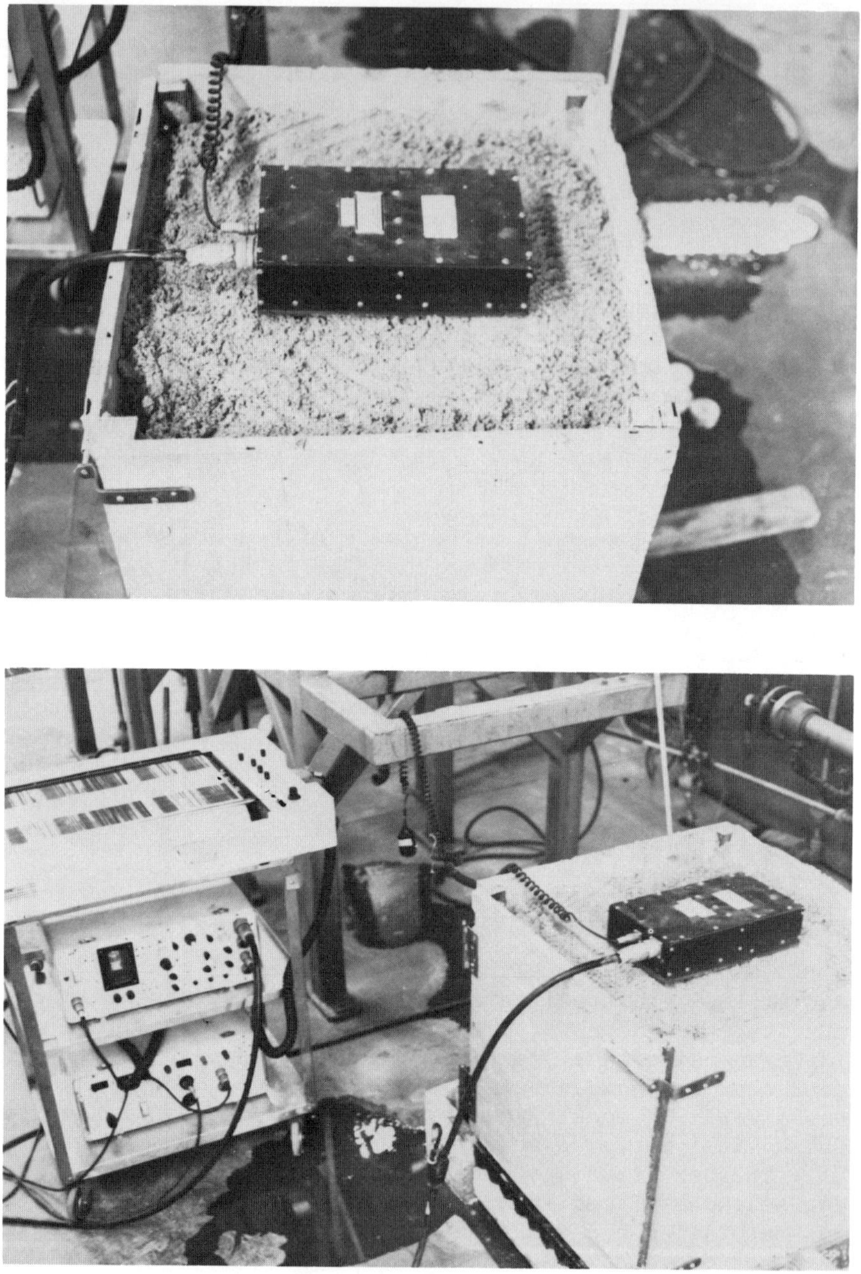

Fig. 2. Photographs of laboratory test setup showing filled container and container with GPR test equipment.

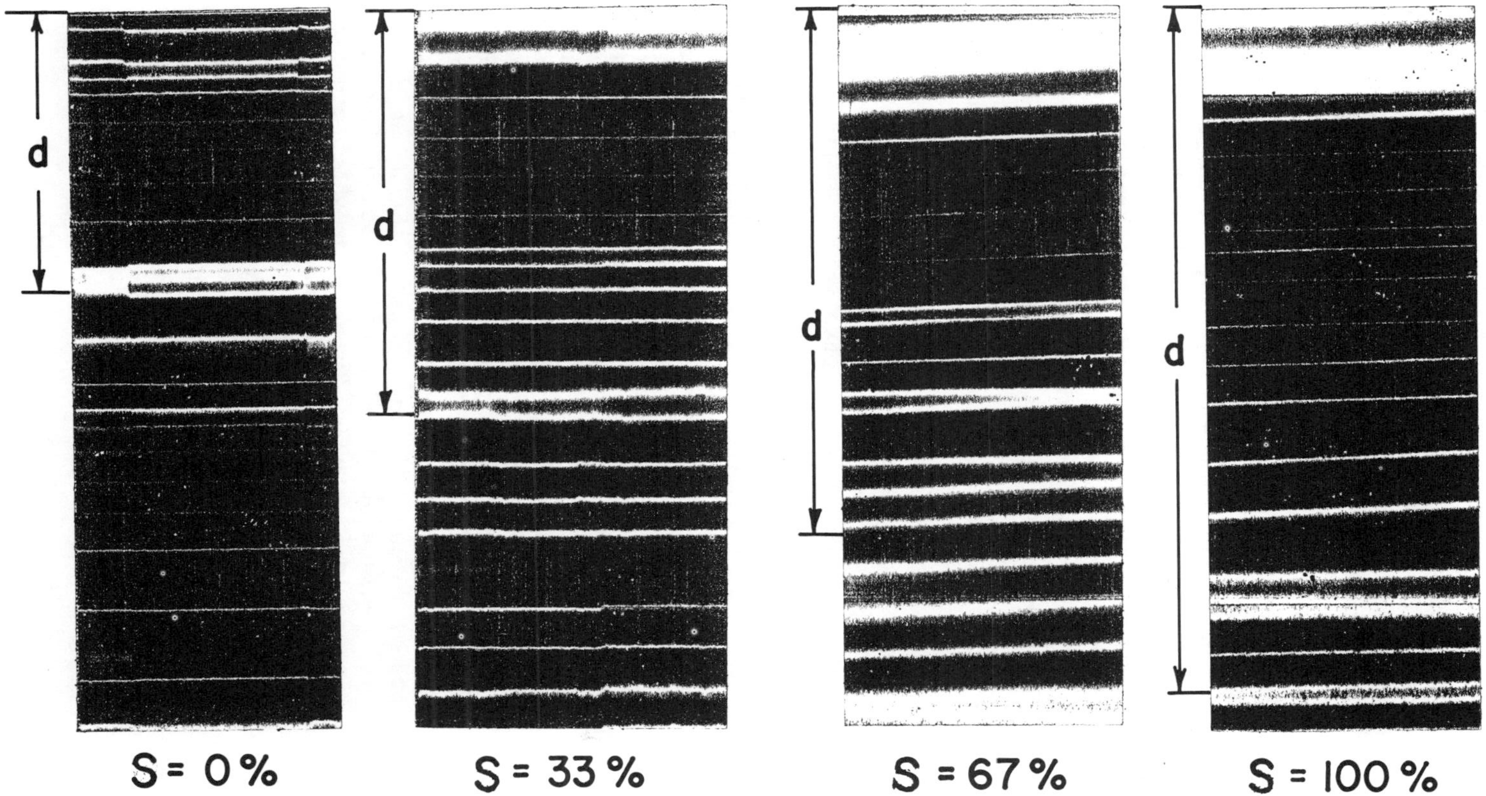


Fig. 3. GPR traces of well graded sand at varying water contents using fresh water with steel plate beneath box.

box increases as the percent saturation increases. Thus, to the signal, it appears as though the depth "d" increases as percent saturation increases. The ringing pattern on the traces is due to the individual pulses which were transmitted and their subsequent reflections.

This first test series was then repeated in an identical manner except that the pore fluid was now a highly concentrated salt solution, i.e. a 6.14 M molarity sodium chloride solution which contains 36 gm NaCl per 100 gm water. The results for the four tests at 0%, 33%, 67% and 100% saturations are shown in Figure 4. In contrasting this behavior to the first test series, the severe attenuating nature of the high ion content pore fluid is easily seen. Furthermore, the number of bands present between the soils' surface and the steel plate has been significantly decreased. Thus both qualitative and quantitative distinctions can easily be seen.

To more reasonably simulate field conditions, the steel reflecting plate beneath the box was removed and replaced with "Ecosorb", a microwave attenuating material. Both of the previous test series were repeated (with fresh water and with the highly concentrated salt solution) with the Ecosorb material beneath the box containing the soil. Results are given in Figures 5 and 6. Similar patterns to those with the steel plate are seen with the major features being that the high salt concentration pore fluid significantly lightens the GPR trace, i.e., it attenuates the signal, and the number of pulse/reflection rings are again reduced.

## SUMMARY AND CONCLUSIONS

On a qualitative basis it is easily seen that high salt concentrations in soil pore fluid significantly attenuates GPR signals. Furthermore, this attenuation, as evidenced by the lightness of the GPR traces, increases with increasing degrees of saturation.

In an attempt to quantify the process, the bands between the soils' surface and the bottom of the box have been counted and plotted in Figure 7. These bands which are due to the transmitted pulses and their reflections, might be a tool which can be used to determine the presence of high ion content pollutants in soil pore water at various degrees of saturation including the fully saturated case. Such a sweeping statement, however, must be further explored beyond the limits of this feasibility study. Our current ideas are listed in the following section.

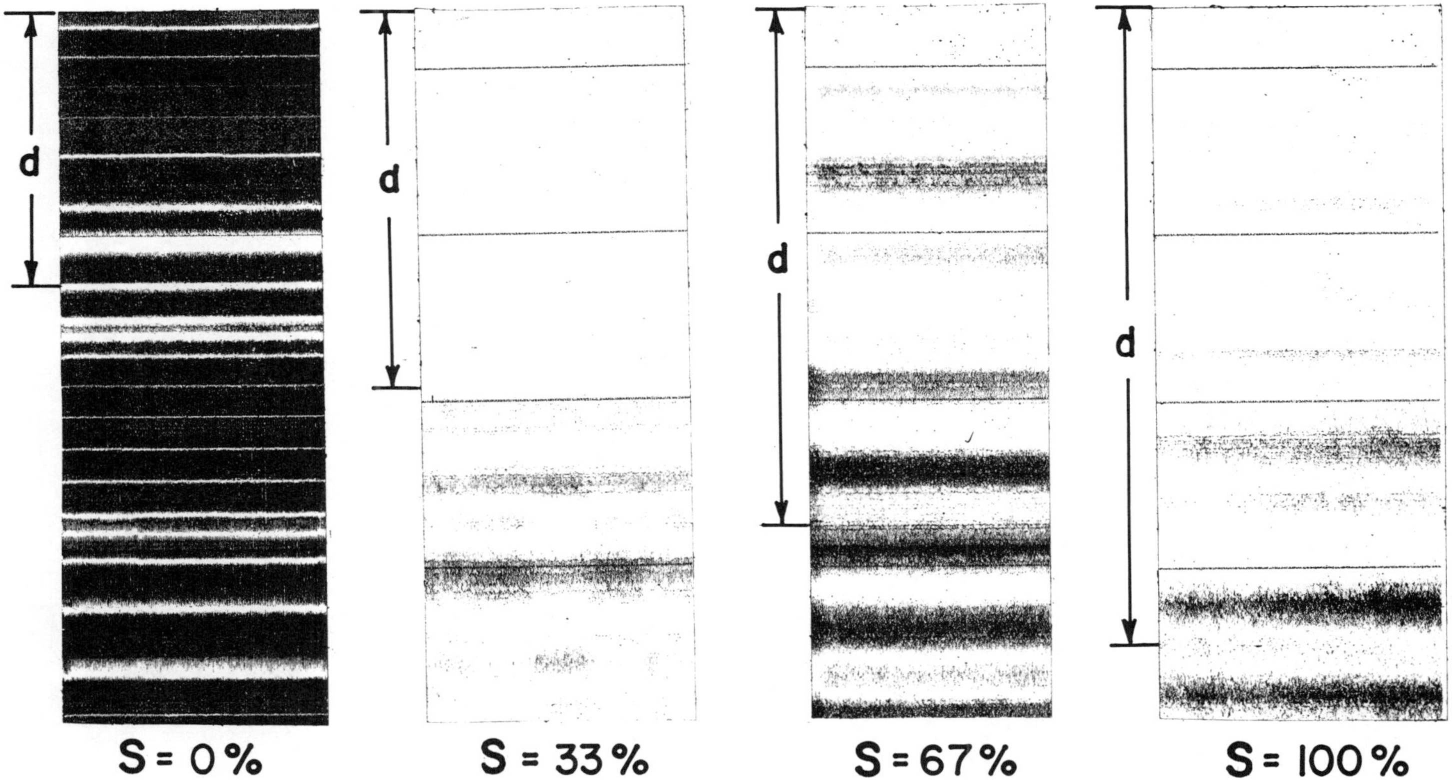


Fig. 4. GPR traces of well graded sand at varying water contents using salt saturated water with steel plate beneath box.

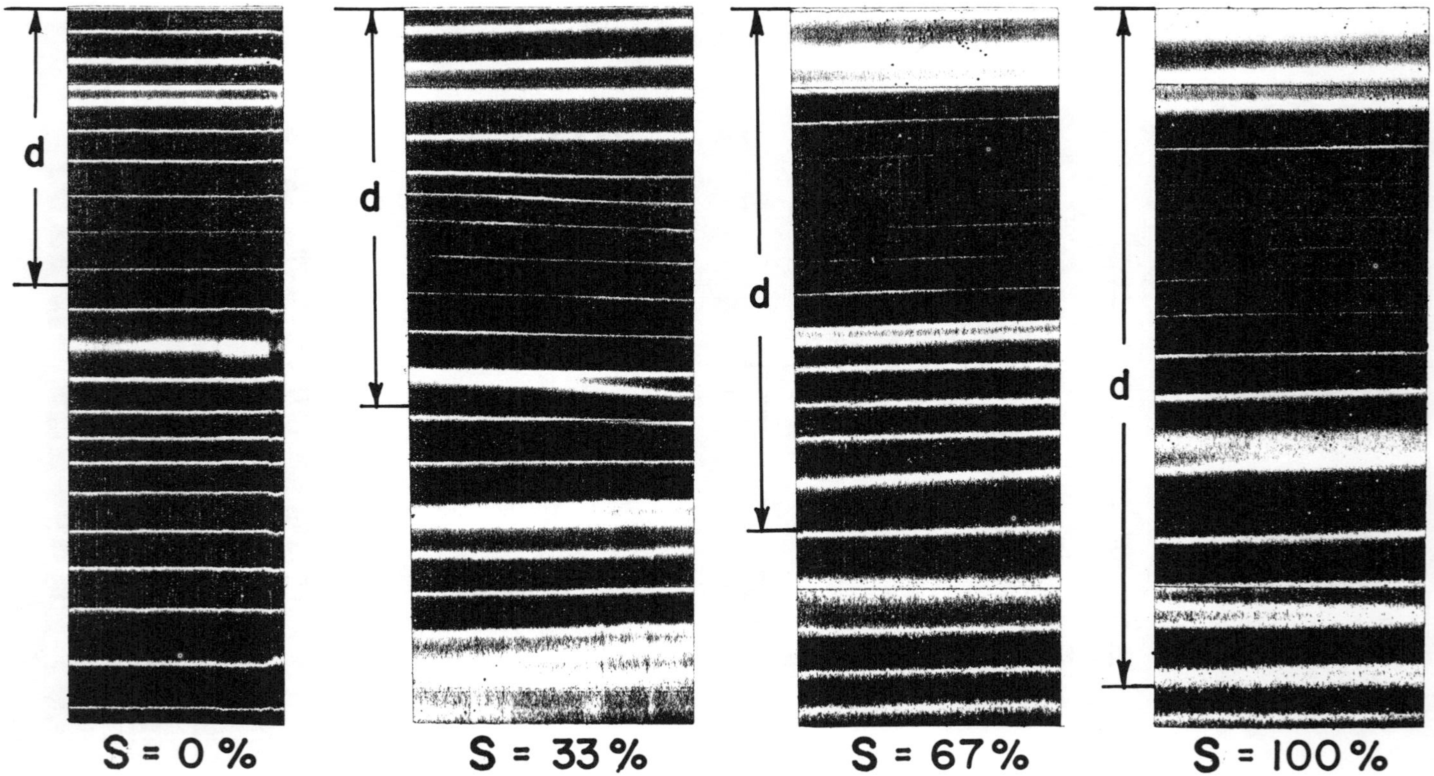


Fig. 5. GPR traces of well graded sand at varying water contents using fresh water with "Ecosorb" beneath box.

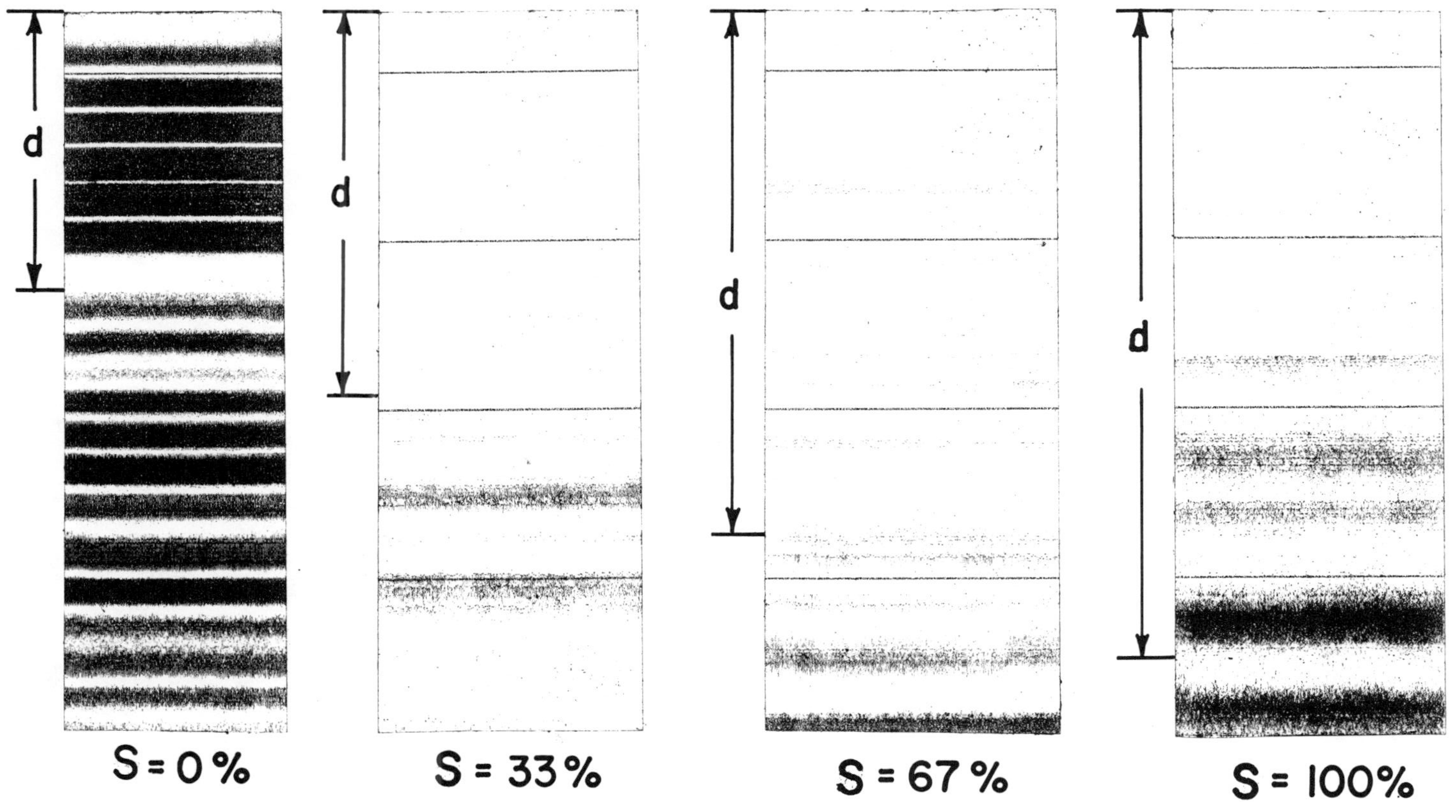


Fig. 6. GPR traces of well graded sand at varying water contents using salt saturated water with "Ecosorb" beneath box.

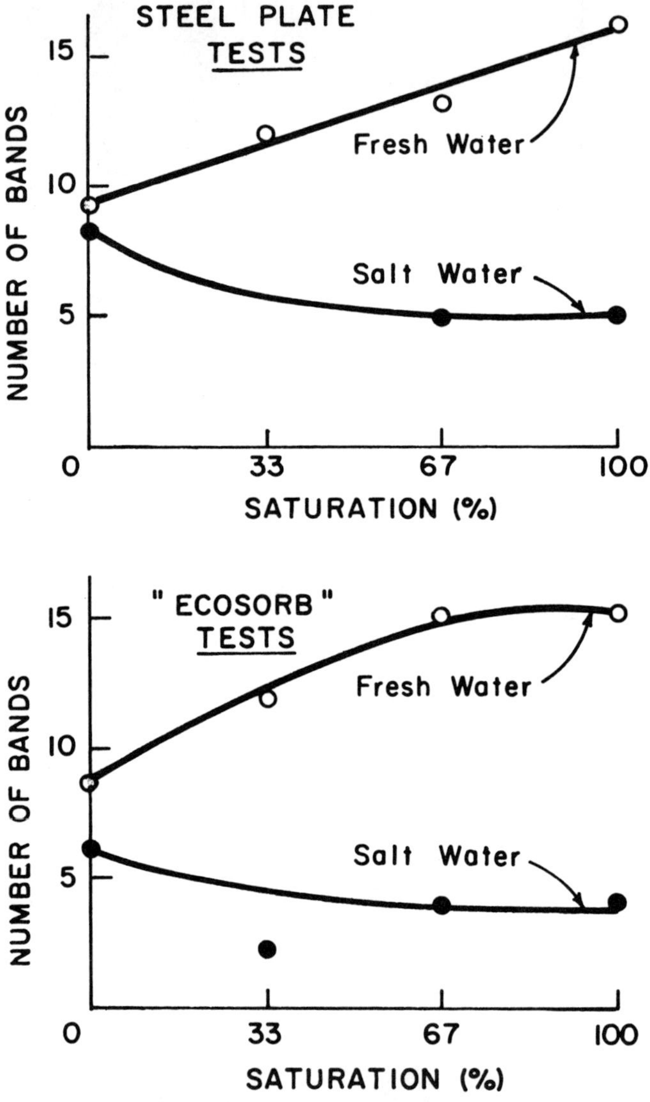


Fig. 7. Pulse/reflection bands showing differences between fresh water and salt water tests at all saturation states from dry to saturated.

## RECOMMENDATIONS FOR FURTHER STUDY

Use of a pulsed RF method, as in the GPR system used in this study, to assess ground water pollution in partially and fully saturated soils needs additional work in the following areas:

- Pore fluids with less than fully concentrated salt solutions need evaluation. Hopefully, the responses will fall within the limits shown on Figure 7.
- The detection of a salt water layer beneath a fresh water layer is important and an attempt was made at its detection with the result shown in Figure 8(a). The interface can be distinguished but it is not easily interpretable. It is therefore a limiting factor.
- Other pore fluid pollutants, than the sodium chloride used herein, should be evaluated and their response compared to the results of this study. In this regard acetone was used as a pore fluid contaminant with results shown in Figure 8(b) for the 100% saturation case. Only minor attenuation of the signal is observed and the technique is not at all responsive in this case.
- The effect of nonhomogeneous soil conditions must be evaluated. For example, a dry soil layer overlying the contaminated stratum must be evaluated to see if the stratigraphy masks out the response from the lower layer which is the layer of interest.
- Other soil types, e.g., silts and clays, need investigation.
- Penetration depths must be determined before the practical utility of the method can be suggested. The maximum limit of the system used in this study is about 10 m in dry soil and 2 to 3 m in saturated soil. When the pore fluid is more attenuating than fresh water the combination could be disasterous. Perhaps a lower frequency transmitted pulse would be helpful in this regard.
- As with all exploratory research projects of this type, where instrumentation is involved, "ground truth" measurements are essential. Actual sites of polluted groundwater are attractive, but to set up a test situation, under carefully controlled conditions is ideal to determine the advantages and limitations of the procedure.

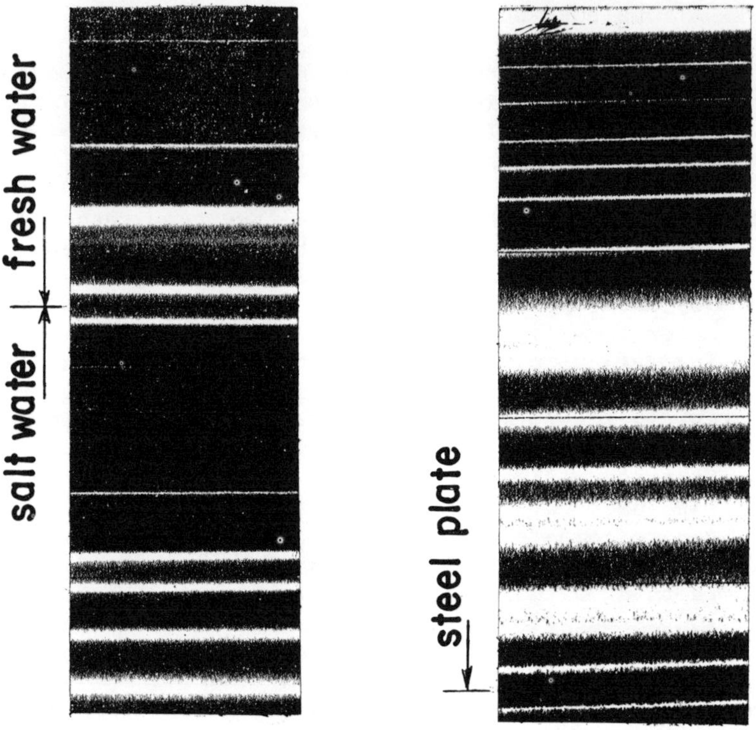


(a) 30 cm sand saturated with fresh water above 30 cm sand saturated with salt water.

(b) Uniform sand saturated with acetone as a pore fluid contaminant.

Fig. 8. GPR traces of uniform well graded sand in other possible situations.

ACKNOWLEDGEMENTS

This project has been supported through a U. S. Environmental Protection Agency Cooperative Agreement (No. CR-804763) administered through the Municipal/Environmental Research Laboratory in Edison, New Jersey. Our thanks to Dr. John E. Brugger, Mr. Frank Freestone and Mr. Ira Wilder for their constant help and encouragement.

REFERENCES

1. Okrasinski, T. A., Koerner, R. M. and Lord, A. E., Jr., "Dielectric Constant Determination of Soils at L-Band Microwave Frequencies," Geotechnical Testing Journal, ASTM, Vol. 1, No. 3, 1979, pp. 134-140.
2. Lord, A. E. Jr. and Koerner, R. M., "Electromagnetic Methods in Subsurface Investigations," to appear in Proc. of Engineering Foundation Conference on Subsurface Probing and Investigations, Anaheim, CA, January 1982.
3. Cook, J. C., "Seeing Through Rock With Radar," in Proceedings of the North American Rapid Excavation and Tunneling Conference, Port City Press, Baltimore, 1972, pp. 89-101.
4. Cook, J. C., "Ground Probing Radar," Proceedings, Subsurface Exploration for Underground Excavation and Heavy Construction," American Society of Civil Engineers, New York, 1974, pp. 172-194.
5. Rosetta, J. V., Jr., "Detection of Subsurface Cavities by Ground Probing Radar," in Symposium on Detection of Subsurface Cavities by Surface Remote Sensing Techniques, U. S. Army Waterways Experiment Station, Vicksburg, MS, 1977, pp. 120-127.
6. Morey, R. M., "Continuous Subsurface Profiling by Impulse Radar," in Proceedings of Subsurface Exploration for Underground Excavation and Heavy Construction, American Society of Civil Engineers, New York, 1974, pp. 213-232.
7. Dolphin, L. T., Bollen, R. L., and Oetzel, G. N., "An Underground Electromagnetic Sounder Experiment," Geophysics, Vol. 39, No. 1, Feb. 1974, pp. 49-55.
8. Unterberger, R. R., "Overview of Electromagnetic Methods in Geophysics and Application of Radar to Detection of Cavities in Salt," in Symposium on Detection of Subsurface Cavities by Surface Remote Sensing Techniques, U. S. Army Waterways Experiment Station, Vicksburg, MS, 1977, p. 119.
9. Harrison, C. H., "Reconstruction of Subglacial Relief from Radio Echo Sounding Records," Geophysics, Vol. 35,

No. 6, 1970, pp. 1099-1115.

10. Rubin, L. A., Griffin, J. N., and Still, W. L., "Subsurface Site Investigation by Electromagnetic Radar, Phase I Report: Feasibility," Report NSF/RA-76-0187, ENSCO, Inc., Springfield, VA, Mar. 1976.
11. Rubin, L. A. and Fowler, J. C., "Ground-Probing Radar for Delineation of Rock Features," Engineering Geology, 12, 1978, pp. 163-170.
12. Benson, R. C. and Glaccum, R. A., "Remote Assessment of Pollutants in Soil and Groundwater," Proceedings of the 1979 Conference on Hazardous Material Risk Assessment, Disposal and Management, Information Transfer, Inc., Miami Beach, April 25-27, 1979, pp. 188-194.
13. Benson, R. C. and Glaccum, R. A., "Radar Survey for Geotechnical Site Assessment," Proc. of Conf. on Geophysical Methods in Geotechnical Engineering, Atlanta, Oct. 23-25, 1979, Preprint #3794.
14. Sandness, G. A., Dawson, G. W., Mathieu, T. J. and Rising, J. L., "The Application of Geophysical Survey Techniques to Mapping of Wastes in Abandoned Landfills," Proceedings of the 1979 Conference on Hazardous Materials Risk Assessment, Disposal and Management, Information Transfer, Inc., Miami Beach, April 25-27, 1979.
15. Bowders, J. J. Jr., Koerner, R. M. and Lord, A. E. Jr., "Buried Container Detection Using Ground Probing Radar," accepted for publ. by Jour. of Hazardous Materials, Elsvier Publ. Co., Amsterdam, Holland.
16. Koerner, R. M., Lord, A. E. Jr. and Bowders, J. J. Jr., "Utilization and Assessment of a Pulsed RF System to Monitor Subsurface Liquids," Proc. Natl. Conf. on Management of Uncontrolled Hazardous Waste Sites, Oct. 28-30, 1981, Wash., DC, HMCRI Publ., pp. 165-170.
17. Bowders, J. J. Jr., Lord, A. E. Jr. and Koerner, R. M., "Sensitivity Study of a Ground Probing Radar System," accepted by Geotechnical Testing Journal, ASTM, Nov. 1981.

# NUCLEAR WASTE ISOLATION IN THE UNSATURATED ZONE OF ARID REGIONS

H. A. Wollenberg and J. S. Y. Wang
Earth Sciences Division
Lawrence Berkeley Laboratory
University of California
Berkeley, California 94720

G. Korbin
Geotechnical Consultant
46 California Avenue
Orinda, California 94563

## INTRODUCTION

The vadose zone or that subsurface region lying between the surface and the water table, is well developed in arid and semi-arid regions. In arid regions the vadose zone may be several hundred meters thick while in areas of higher precipitation it may be nonexistant or only a few meters thick. In the vadose zone, where the pore spaces between grains and where fractures in hard rock are not saturated with ground water, there would be a long hydraulic flow path between the nuclear waste repository and the saturated regime beneath the water table and thus a long, time-consuming pathway for radionuclides to be transported, ultimately to the biosphere. Primarily in this respect, the vadose zone in arid regions is considered as a possible environment for geologic isolation of nuclear waste.

There are several topographic and lithologic combinations in the vadose zone of arid regions that may lend themselves to waste isolation considerations. In some cases, topographic highs such as mesas and interbasin ranges--comprised of several rock types, may contain essentially dry or partially saturated conditions favorable for isolation. The adjacent basins, especially in the far western and southwestern U. S., may have no surface or subsurface hydrologic connections with systems ultimately leading to the ocean. Some rock types may have the favorable characteristics of very low permeability and contain appropriate minerals for the strong chemical retardation of radionuclides. Environments exhibiting these hydrologic and geochemical attributes are the areas underlain by tuffaceous rocks, relatively common in the Basin and Range geomorphic province. Adjacent valley areas, where tuffaceous debris makes up a significant component of valley fill alluvium, may also contain thick zones of unsaturated material, and as such also lend themselves to strong consideration as repository environments. The distribution of tuffaceous highlands and adjacent alluvial valleys in the northwestern Basin and Range province is shown in Figure 1.

An appraisal of waste isolation in the Sedan Crater of the Nevada Test Site [1] brought to the attention of the scientific and engineering communities the attributes of alluvial environments. Partly for this reason and also because of the present focus of waste repository investigations at the Nevada Test Site, this paper summarizes the aspects of nuclear waste isolation in unsaturated regimes in alluvial-filled valleys and tuffaceous rocks of the Basin and Range province. (A more detailed report which furnishes the basis for this paper is in preparation [2].) Highland and alluvial areas in other arid regions underlain by other rock types may serve equally well or better than those described here. Such areas include ranges and hills of basaltic and other volcanic rock in the Columbia Plateau, and large mesas underlain by sandstone in the Colorado Plateau and Great Basin.

## REQUIREMENTS FOR A WASTE REPOSITORY SITE

The principal requirement of a high-level waste repository site is that it provide long-term ($10^5$ years or more) isolation of radionuclides. The engineering aspects of the site must also be suitable for waste repository construction and for maintenance of structural integrity during repository operations.

The primary mechanism for radionuclide escape will be migration of contaminated groundwater. The requirement, at present, is that the groundwater travel time from the repository to the accessible environment must substantially exceed 1,000 years [3]. In order to meet this requirement, the proposed site must exhibit a number of favorable conditions:

- Groundwater access to and transport from the repository must be minimized; therefore, the site should have low groundwater content and its lithology should inhibit circulation within the host rock and between hydrogeologic units. Consideration must be given to potential changes in regional groundwater flow due to human activity or climatic changes.

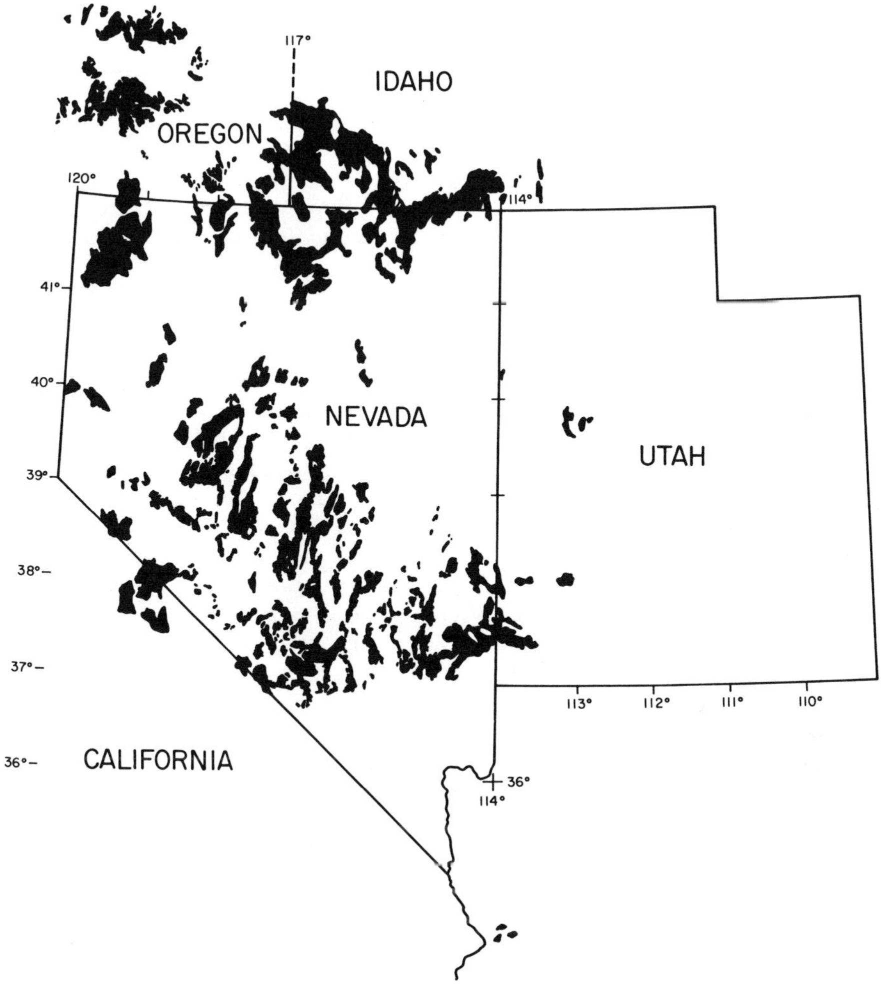


Figure 1. The northwestern portion of the Basin and Range province, showing highland areas that contain tuffaceous rocks. Valleys with appreciable components of tuffaceous alluvium lie between these highlands. [XBL 8112-4898]

- The host rock must be of sufficient thickness and extent to maximize groundwater transit time to other geologic units.
- The mineralogical and geochemical characteristics should be such that they promote precipitation and sorption of radionuclides and do not increase but possibly decrease, groundwater mobility. In response to thermal loading, these properties should remain unaltered or be enhanced.
- The response of the rock encompassing the repository to its excavation and to the heating caused by the waste should have minimal effect on the hydrological integrity of the repository.
- Any major geologic discontinuity should contribute to isolation (such as a fault zone acting as an aquitard).

Another possible source of radionuclide escape is exhumation by natural processes, tectonism, volcanism and erosion, or by man. Human intrusion could be deliberate or inadvertent. Therefore the site should offer little resource value, be geologically stable, and be at sufficient depth to make exhumation by man difficult and by erosion highly improbable.

A given site may be determined to meet the requirements as that site exists today but, aside from the changes induced by construction and waste storage, the site is subject to long-term natural processes. It will be necessary to predict those processes for tens and possibly hundreds of thousands of years. For this purpose the assumption will be made that those processes which have been operating on the site during the Quaternary Period will continue to operate. Therefore, it will be necessary to evaluate the hydrogeologic, geochemical, geomorphic, structural and tectonic stability, and climatologic conditions since the start of the Quaternary Period.

It is not likely that an ideal rock type or an ideal site will be found. Each must be analyzed and evaluated with an open mind and any conditions, natural or man-made, which detract from isolation must be judiciously weighed against those which favor the site.

## GEOTECHNICAL FACTORS

The unsaturated zone in alluvial-filled valleys of the Basin and Range province offers several factors that would help inhibit the migration of radionuclides from discrete sites into and through the hydrologic system: (1) the flux of moisture into and through the alluvium is extremely low; (2) the components of the alluvium at specific locations favor the sorption of radionuclides if they should escape from waste canisters; (3) at some locations the alluvium is underlain by tuffaceous rock above the water table, which also provides a lower-permeability, sorptive barrier to downward migration of radionuclides (This is illustrated in an idealized geologic cross section in Figure 2.); (4) the thickness of the alluvium exceeds 1000 m at many localities and the thickness of the vadose zone exceeds 500 m. These factors, together with the presence of a tuff aquitard below the alluvium, provide a substantial hydrologic flow path between a repository at a depth of a few hundred meters and the underlying saturated zone.

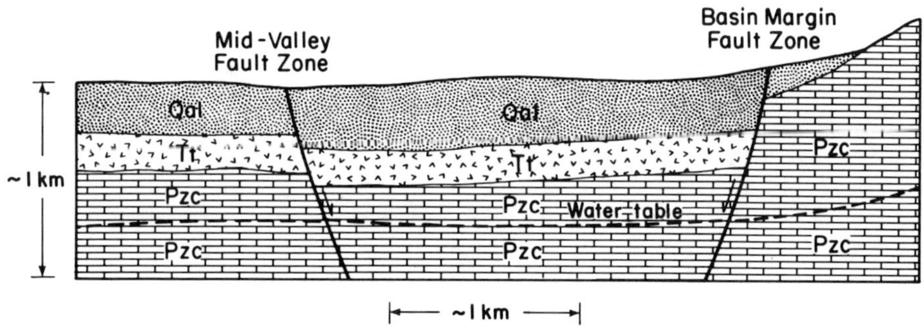


Figure 2. Schematic geologic cross section through an idealized Basin and Range valley, showing alluvium (Qal) overlying tuff (Tt) and Paleozoic carbonate rock (Pzc), with the water table at depth in the Paleozoic rock (based on [1]). [XBL 821-1621]

Several of the advantages of waste isolation in the vadose zone of alluvium also pertain to isolation in tuffaceous rock--low moisture flux, high sorptive capacity, and substantial thickness of rock above the water table. Specific to tuff is the attribute of having competent rock units, amenable to machine mining and requiring minimal ground support in underground workings, sandwiched between less competent but highly sorptive zeolitized units.

It is emphasized that even though tuffs underlie highland areas, the depths at which a repository would be located should be determined primarily by the occurrence of a favorable stratigraphic setting. Whether the repository is in the tuff sequence comprising a topographic high (a mesa or range) or at a depth well below the local topographic relief, is less important than its location above the water table in a competent tuff unit encompassed by zeolitized units. It is conceivable that a favorable setting might be a tuff unit overlain by alluvium beneath a Basin and Range valley [1].

### Tectonics, Volcanism and Seismicity

Properly chosen sites in alluvium of tectonically active areas may take advantage of local tectonism--situation of a repository on the basinward side of an active normal fault would insure that an increasing thickness of material would accumulate above the repository with time, alleviating the concern of future tectonic-erosive exhumation of the site [1]. Unless a site in tuff is unfortunately located in or closely adjacent to a caldera which will have volcanic activity associated with it in the near geologic future (a possibility which should be laid to rest by appropriate geological and geophysical investigations in the course of site selection) the concerns of volcanism and seismicity are similar to those for the case of an alluvial repository. Obviously, the location of the repository astride an active Basin and Range fault must be avoided. The relatively high seismicity of the Basin and Range region requires that the response of surface and underground facilities to ground motion associated with earthquakes of given magnitude and epicentral distance be assessed for given site locations, and these should be taken into account in the design of the facilities.

### Climatological Considerations

Evidence of the climate in the Basin and Range province during Pleistocene pluvial periods, when precipitation was 50 to 100% greater than today's [7, 8], indicates that future pluvial periods would result in increases in moisture infiltration rates. However, the saturation conditions and water table elevations at alluvial sites away from obvious locations of playa lakes would not differ significantly from those of today [9] insofar as affecting a repository in alluvium. Climatological considerations for a site in tuff are essentially the same as those for alluvium; significantly increased moisture infiltration rates in a future pluvial period may affect the position of the water table. However, as with alluvial areas in arid regions, the effect of even doubling the infiltration rate should be insignificant because the rates are so low and the additional vegetative cover would enhance moisture transpiration.

### Thermal Response

A principal concern of waste isolation in the unsaturated zone is the response of the host rock to heating. For a given waste form and thermal loading, the maximum temperature rise in the plane of a repository due to the introduction of

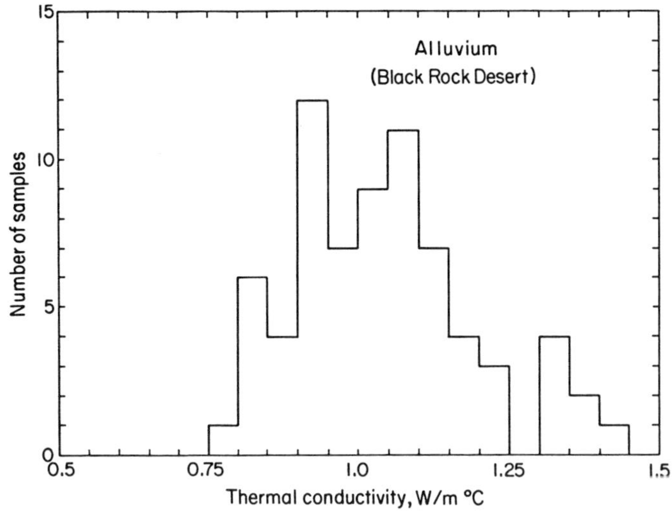


Figure 3. Thermal conductivities measured in situ in the southern Black Rock Desert (average value = 1.05 ± 0.15 W/m°C, 71 samples over 12 holes, data from [14]). [XBl 8111-4849]

the waste varies with $1/\sqrt{K\rho c}$, where K is the rock's thermal conductivity, ρ its density, and c its specific heat [10]. The thermal conductivity of alluvium covers a large range, as shown in Figure 3, and depends strongly on the degree of saturation. The influence of thermal conductivity on repository temperature increase has been calculated [9] based on procedures developed by Wang et al. [10] and is illustrated in Figure 4. Calculations were based on a repository depth of 150 m, an area of 2000 acres (radius 1605 m), and a thermal loading of 40 kW/acre with 10-year-old spent fuel. The range of thermal conductivities shown is that which could be expected for dry, poorly indurated alluvium, 0.5 W/m°C, to saturated indurated alluvium, 1.5 W/m°C. Peak repository temperature increases are 130°C and 75°C, respectively, and occur 58 years after emplacement of the waste package. These maximum temperatures are an average for the repository as a whole; the actual maximum local temperature will strongly depend on the emmplacement geometry and the thermal loading of each canister.

Repository temperatures in excess of 100°C can result in accelerated drying of saturated or partially saturated alluvium, and a corresponding decrease in thermal conductivity. Given a material with 30% porosity, a saturated thermal conductivity of 1 W/m°C, and a geometric mean model for the dependence of conductivity on water saturation, the increase in repository temperature with reduced water content was

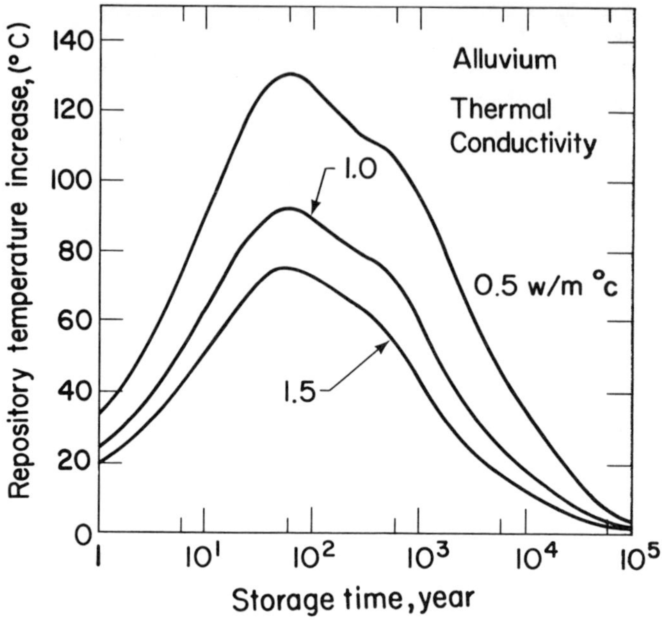


Figure 4. Effects of thermal conductivity on alluvial repository temperature, spent fuel, 10 years out of the reactor. [XBL 825-1391]

determined. Figure 5 presents the temperature rise, $\Delta T$, normalized with respect to the temperature increase for a fully saturated (wet) or dry alluvium. As indicated, the repository in dry alluvium can have a temperature increase of 1.6 times that of a repository in saturated material.

The ramifications of drying saturated or partially saturated alluvium have led to a consideration of the areal thermal loading needed to keep the maximum repository temperature at or below 100°C. For a thermal conductivity of 1 W/m°C, this is 33 kW/acre, and for a dry alluvium of conductivity 0.38 W/m°C, loading would be limited to 20 kW/acre, assuming an ambient temperature of 25°C.

These results and those in Figure 5 are conservative as they are based on constant material properties for all the material surrounding the repository. They do not consider the beneficial effect of heat transport by vapor diffusion. In reality, the properties will vary with distance from the repository heat source, and only the ground nearest the repository openings will experience a large decrease in thermal conductivity due to drying. To establish a realistic areal thermal loading for alluvium, it is necessary to be site- and material-specific and to employ improved predictive models with variable thermal properties.

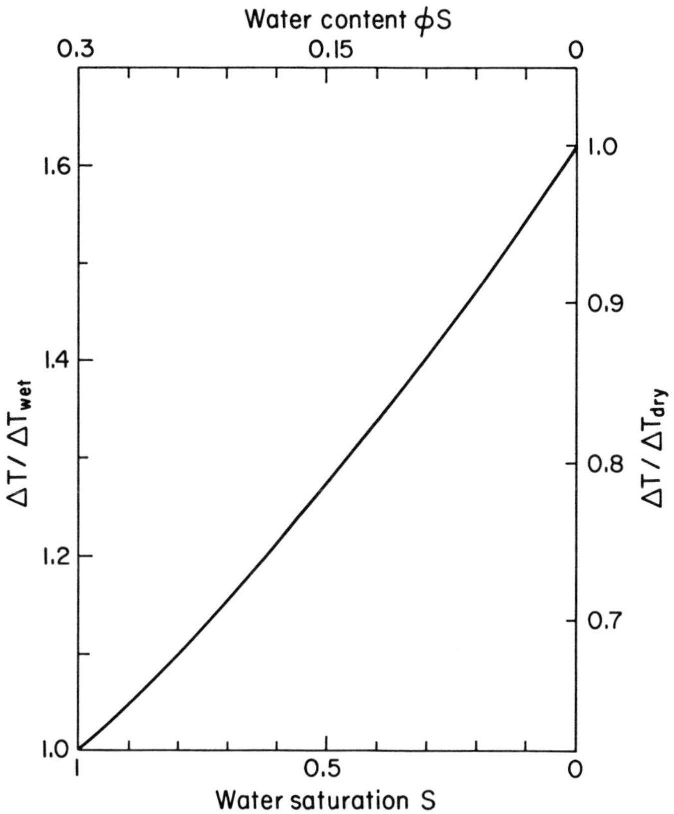


Figure 5. Variation of temperature rise with water saturation. [XBL 8111-4883]

Other factors affecting repository temperature in alluvium are waste form and repository depth, while repository area has little effect on temperature. Disposal of reprocessed waste results in a maximum temperature increase of 76°C, 35 years after emplacement, compared with 92°C for spent fuel, assuming a thermal conductivity of 1 W/m°C and 40 kW/acre thermal loading. Varying the repository depth from 150 to 600 m has negligible effect on the maximum temperature, but results in a reduction in temperature for repositories at shallower depths after 300 years of storage.

Similar calculations to those for alluvium have been made for the thermal response of a repository in tuff. Based on a repository depth of 800 m, an area of 2000 acres (radius of 1605 m), and a thermal loading of 40 kW/acre with 10-year-old spent fuel, the influence of thermal conductivity on repository temperature increase has been calculated, curves are shown in Figure 7. (The range of conductivity in tuff is shown in Figure 6.) For a conductivity of 0.9 W/m°C,

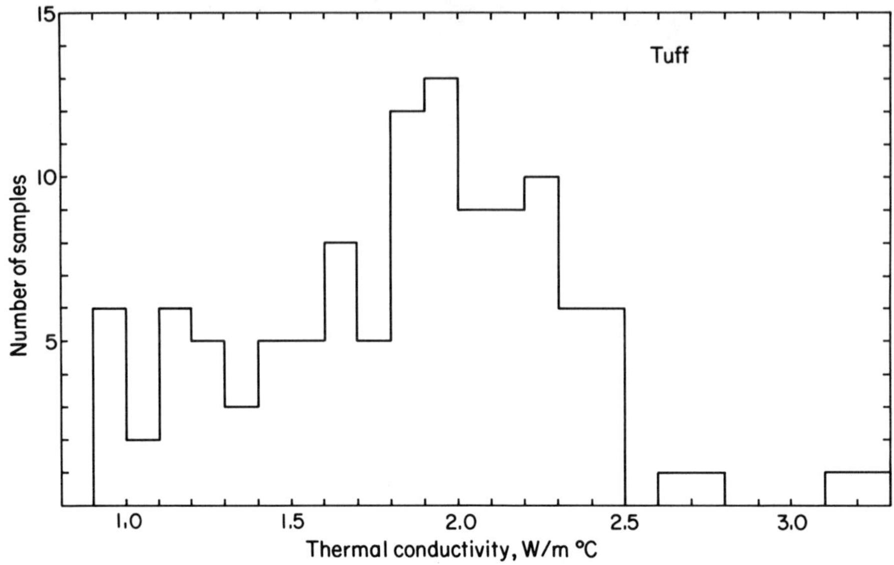


Figure 6. Histogram on the thermal conductivity of tuff; data from [15]. [XBL 8111-4850]

representative of a dry, nonwelded tuff and 2.7 W/m°C for a saturated welded tuff, the increases in temperature are 87 and 51°C, respectively. These maxima occur 58 years after emplacement of the waste and, as for alluvium, are average temperatures for the repository.

Comparison of Figures 4 and 7 reveals the significant difference in predicted repository thermal responses for underground waste disposal in tuff and alluvium. This difference is not only related to the relatively higher thermal conductivity of tuff as compared to alluvium, but also the higher density of tuff. In both cases the maximum temperatures occur at approximately the same time.

The accelerated increase in temperature due to dehydration of tuff (especially nonwelded units) is, as for alluvium, a factor limiting the maximum repository temperature and hence areal thermal loading. On establishing a 100°C maximum temperature for the repository, the permissible thermal loading in dry tuff (thermal conductivity of 1.0 W/m°C) is 27 kW/acre; for saturated or partly saturated tuff (1.8 W/m°C) the loading is 36 kW/acre. This assumes a 44°C ambient temperature based on a 20°C surface temperature, and a 30°C/km geothermal gradient with the repository horizon at 800 m depth. To more accurately determine the allowable areal thermal loading for a prescribed temperature requires detailed modeling of the dehydration process, including the fluid and vapor pressure, heating rate, and pressure release path length.

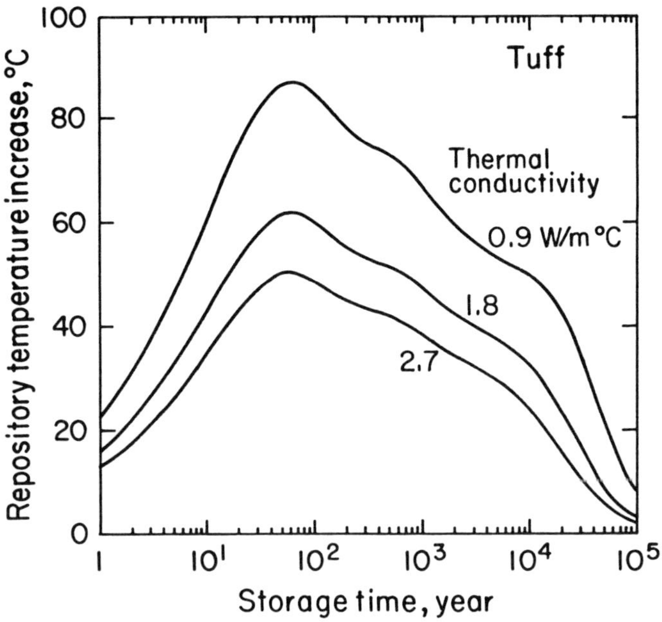


Figure 7. Effects of thermal conductivities on tuff repository temperature, spent fuel, 10 years out of the reactor. [XBL 8112-4888]

Calculations, similar to those for alluvium, were made for tuff, taking into account other factors influencing repository temperature--waste form, repository depth, and repository area. Disposal of reprocessed waste results in a maximum temperature increase of 51°C, 35 years after emplacement as compared to 62°C for spent fuel, assuming a thermal conductivity of 1.8 W/m°C and 40 kW/acre thermal loading. Altering repository depth from 400 m to 1600 m and area from 500 to 8000 acres has negligible effect on the thermal response up to 2000 years after waste emplacement. For a longer storage time, the influence is relatively minor with a greater depth and larger area, resulting in slightly increased temperature rises.

Laboratory experiments on cores of saturated tuff were performed to measure the rate of water loss on heating from 25 to 150°C, and two models were formulated to predict the measured results [12]. In one model, the transport of vapor was treated as Darcy flow driven by the higher partial pressure of water vapor at the liquid-vapor front. The other model assumes that molecular diffusion is the dominant mode of water vapor transport. A comparison of the two predicted results with the experimental measurements clearly indicated the dominance of the diffusion mechanism. Furthermore, an in situ heater experiment at G tunnel in the Grouse Canyon tuff [13] basically confirms the results of the laboratory experiment.

The thermal-hydrological properties of densely welded, strongly fractured tuff units which may contain lithophysal zones must also be considered. Considerations of rock mass permeability in tuff, in the context of waste isolation, would be similar to those of other hard rock types--principally the effect of heating on the near-field and mid-field hydrologic systems. The coupled thermomechanical-hydrologic effects of heating are also of concern, as are the presence of minerals subject to volume change in response to heating [14] and the effects of varying degrees of saturation on these responses.

## Construction Considerations

Construction of a relatively shallow repository in alluvium, either by underground mining methods if the depth exceeds 100 m or by open pit methods at shallower depths, is comparable to construction costs in harder rock under saturated conditions [4]. Based on experience at the Nevada Test Site [5], the strength characteristics of alluvium are such that extensive ground support for underground workings is not required to depths of 200 m. Open pit construction would permit installation of barrier materials completely encompassing the repository if necessary. In either case, a repository at a relatively shallow depth in alluvium, compared with a deep hard rock site, permits relatively comparable ease of emplacement and if necessary retrieval of the waste.

Only underground construction is considered applicable to a site in unsaturated tuff. This would take advantage of a competent, most likely densely welded unit, amenable to machine mining, and requiring minimal ground support of the workings [6]. The stratigraphic position of such a unit between zeolitized units to furnish sorptive barriers would be highly desirable.

## Resource Potential

The Basin and Range province has a very low population density--the vast majority of the land is under federal jurisdiction; large alluviated areas exist where there is no present mining activity. The future mineral resource potential of these areas is not considered to be significant. At sites amenable to waste isolation, the depth of the water table is excessive for development of groundwater resources.

Table I. Major Advantages and Concerns of Waste Isolation in Unsaturated and Saturated Regimes

| Major Advantages | Major Concerns |
|---|---|
| **Thermal Effects** | |
| Unsaturated Zone in Arid Regions: Relatively shallow emplacement of waste will result in lower repository temperature earlier in the life of the repository, compared to deeper burial in a given medium. Diffusion of water vapor in response to heating may result in bulk thermal conductivity comparable to that of saturated material [17]. | At relatively low temperatures, bulk thermal conductivity will depend on degree of saturation. Variations in thermal conductivity will control maximum temperature rise experienced in a repository for a given waste form and thermal loading. Data on thermal expansion of unsaturated media are scarce. Formation of vapor phase in response to heating may lead to a two-phase convective system; its ramifications for radionuclide transport are not well understood. |
| Saturated Zone: Mechanical and hydrological responses of saturated (or nearly saturated) media to heating are becoming better understood as experiments progress in various rock types. | Potential exists for formation of convective cells in the hydrologic system of saturated media in response to heating. |
| **Transport of Radionuclides** | |
| Unsaturated Zone in Arid Regions Repository not in direct contact with groundwater regime; wastes would remain hydrologically isolated as long as partially saturated conditions pertained. | Difficulty in predictive modeling of mechanisms of transport and pathways of radionuclides in unsaturated regimes. Degree of saturation may vary. |
| Saturated Zone: Deep burial affords long pathways for radionuclides to the accessible environment. Mechanisms of transport of radionuclides in saturated rocks are presently better understood than are transport mechanisms in the unsaturated zone. | Direct contact of the repository with the groundwater regime that could transport radionuclides to the accessible environment. |
| **Potential for Human Intrusion** | |
| Unsaturated Zone in Arid Regions: Generally low population density; little potential conflict with future utilization of groundwater resources | Relatively shallow emplacement of waste compared with deep burial below water table. |
| Saturated Zone: Deep burial decreases likelihood of human intrusion. | Possible future development of groundwater resources in the vicinity of the repository |

Compared with alluvial sites, there is a greater potential for conflict of a waste repository location with mineral resource potential in the tuffaceous rocks. The occurrence of precious metals is relatively widespread in ash flow tuffs associated with magmatic-hydrothermal systems of Tertiary calderas. A resource assessment should therefore be an important part of the site investigations.

## Advantages and Concerns of Waste Isolation in the Unsaturated and Saturated Zones

In comparing the attributes of waste isolation in the unsaturated zone of arid regions and saturated hydrologic regimes, major advantages and concerns are clearly identifiable in the considerations of transport of radionuclides, thermal effects, and the potential for human intrusion. These are presented in brief in Table I. Given appropriate study, beyond the scope of this paper, similar comparisons of advantages and concerns of unsaturated and saturated regimes may be made for the considerations of the effects on the waste form and on its surrounding canister and overpack material. Considerations would include the effect of saturated and unsaturated conditions at repository temperature and pressure on corrosion of the canisters, on the leaching of waste forms, and on the mechanical and hydrological integrity of overpack and backfill material.

It is concluded that the unsaturated zones in alluvium or tuffaceous rocks of the Basin and Range province are strong candidate environments for consideration as sites for nuclear waste repositories, and as such should be investigated as comprehensively as the other geologic settings presently being considered.

## ACKNOWLEDGEMENT

This work was supported by the Director, U. S. Nuclear Regulatory Commission, Office of Safeguards, Fuel Cycle, and Environmental Research, Office of Nuclear Regulatory Research under Interagency Agreement DOE 50-80-97, NRC FIN No. B 3040-0, and by the U. S. Department of Energy under Contract No. DE-AC03-76SF00098.

REFERENCES


1. Winograd, I. J. "Radioactive Waste Disposal in Thick Unsaturated Zones," *Science*, 212(4502):1457-1464 (1981).
2. Wollenberg, H. A., J. S. Y. Wang, and G. Korbin. "An Appraisal of Nuclear Waste Isolation in the Vadose Zone in Arid and Semi-arid Regions," Lawrence Berkeley Laboratory Report, in preparation (1982).
3. Nuclear Regulatory Commission. "Proposed Rules for Disposal of High-level Radioactive Wastes in Geologic Repositories," 10 C.F.R. Part 60, July (1981).
4. Smyth, J. R., B. M. Crowe, P. M. Halleck, and A. W. Reed. "A Preliminary Evaluation of the Radioactive Waste Isolation Potential of the Alluvium-filled Valleys of the Great Basin," Los Alamos National Laboratory Report LA-7962-MS, 22 p. (1979).
5. Pack, P. D., and E. H. Skinner. "Design Review of a Rectangular Shaft in Alluvium," in *Proceedings, 17th U. S. Symposium on Rock Mechanics* (1976).
6. Smyth, J. R., and M. L. Sykes. "Tuff Properties," in Johnstone, J. K., and K. Wolfsberg, Eds., in *Evaluation of Tuff as a Medium for a Nuclear Waste Repository: Interim Status Report on the Properties of Tuff*, Sandia National Laboratory Report SAND80-1464 (1980).
7. Snyder, C. T., and W. B. Langbein. "The Pleistocene Lobe in Spring Valley, Nevada and its Climatic Implications," *Jour. Geophys. Res.*, 67:2385-2394 (1962).
8. Mifflin, M. D., and M. Wheat. "Pluvial Lakes and Estimated Climates of Nevada," Nev. Bur. Mines Geology, Bull. 94 (1979).
9. Winograd, I. J., and G. Doty. "Paleohydrology of the Southern Great Basin, with Special Reference to Water Table Fluctuations Beneath the Nevada Test Site During the Late(?) Pleistocene," U. S. Geol. Survey Open-File Rept. 80-569 (1980).
10. Wang, J. S. Y., and C. F. Tsang. "Thermal Calculations for Radioactive Waste Repositories in Alluvium and Tuff," Appendix to Wollenberg et al. (1982) Reference #2 above (1982).
11. Wang, J. S. Y., C. F. Tsang, N. G. W. Cook, and P. A. Witherspoon. "A Study of Regional Temperature and Thermohydrologic Effects of an Underground Repository for Nuclear Wastes in Hard Rock. *Jour. of Geophys. Res.*, 86(B5):3759-3770 (1981).
12. Hadley, G. R. "Water Loss by Drying from Welded Tuff," in Johnstone, J. K., and K. Wolfsberg, Eds., *Evaluation of Tuff as a Medium for a Nuclear Waste Repository: Interim Status Report on the Properties of Tuff*, Sandia Nat. Lab. Rpt. SAND80-1464 (1980).

13. Johnstone, J. K., and K. Wolfsberg, Eds. *Evaluation of Tuff as a Medium for a Nuclear Waste Repository: Interim Status Report on the Properties of Tuff*, Sandia Nat. Lab. Rept. SAND80-1464 (1980).
14. Lappin, A. R. "Thermal Properties," in Johnstone, J. K., and K. Wolfsberg, Eds., *Evaluation of Tuff as a Medium for a Nuclear Waste Repository: Interim Status Report on the Properties of Tuff*, Sandia Nat. Lab. Rept. SAND80-1464 (1980).
15. Sass, J. H., J. P. Kennelly, Jr., W. E. Wendt, T. H. Moses, Jr., and J. P. Ziagos. "In Situ Determination of Heat Flow in Unconsolidated Sediments," U. S. Geol. Survey Open-File Rept. 79-595 (1979).
16. Sass, J. H., and R. J. Munroe, Eds. U. S. Geol. Survey Open-File Rept. 74-9 (1974); data reproduced in: Touloukian, Y. S., and C. Y. Ho. "CINDAS Data Series on Material Properties (Vol. II-2)," *Physical Properties of Rocks and Minerals*, McGraw-Hill, New York (1981).
17. DeVries, D. A. "Thermal Properties of Soils," in Vanwijk, W. R., Ed., *Physics of Plant Environment*, North Holland Pub. Co., Amsterdam (1963).

HYDROGEOLOGICAL INVESTIGATION
HAZARDOUS WASTE SITE
ATLANTIC CITY, NEW JERSEY

Dennis Pennington
International Exploration, Inc.
577 Sackettsford Road
Warminster, PA 18974

ABSTRACT

Hydrogeological studies of an abandoned hazardous waste site, alleged to threaten the water supply of Atlantic City, New Jersey, were utilized to identify a leachate plume. Also, several additional sources of pollution in the vicinity of the hazardous waste site were found. Resistivity surveys, a stream sediment survey and soil investigations which included soil profiles were part of the site investigation.

Resistivity measurements, utilizing the Schlumberger array, identified a leachate plume originating from the disposal site. Observation wells and boreholes were drilled for geological control and correlation of the resistivity data. An unknown clay layer also was identified which separated two aquifers and acted as a barrier to downward movement of leachate.

INTRODUCTION

A hydrogeological investigation was initiated on January 5, 1981 at one of the most publicized abandoned hazardous waste sites in the United States. This site, known as the Price Landfill, is located west of Atlantic City, New Jersey, in Pleasantville (Figure 1). The Price Landfill was alleged to threaten the water supply of Atlantic City. Some governmental agencies alleged that twelve wells of Atlantic City were threatened. Ten of these wells are located in the Cohansey Formation. The other two wells are located in the Kirkwood Formation. Our firm, International Exploration, Inc., was retained to (1) determine the potential for contamination of the water supply of Atlantic City, (2) identify the areal extent of groundwater pollution, including the existence of a leachate plume, (3) determine the direction of movement of a leachate plume (if identified), (4) plan an effective monitoring program and (5) develop an abatement plan. As part of the investigation, the following studies were to be completed: resistivity surveys, soil surveys, geochemical surveys of stream sediments, supervision of pump tests, monitoring of groundwater and surface water (including drilling wells), and evaluation of contaminant transport through computer models. Due to the magnitude of the potential damage awards, legal constraints from litigation were a major factor in limiting some studies which were proposed for the site. For example, computer modelling of contaminant transport was not initiated due to time constraints of the court preceedings.

Although the site had been studied for more than 2 years by others, little was known prior to 1981, regarding the areal extent of pollution including the identification of a leachate plume. As a result of severe time constraints by the court, resistivity surveys in addition to geochemical surveys in the vicinity of the landfill site were initiated immediately. The primary purpose of these investigations was identification of a leachate plume and direction of movement of the leachate plume.

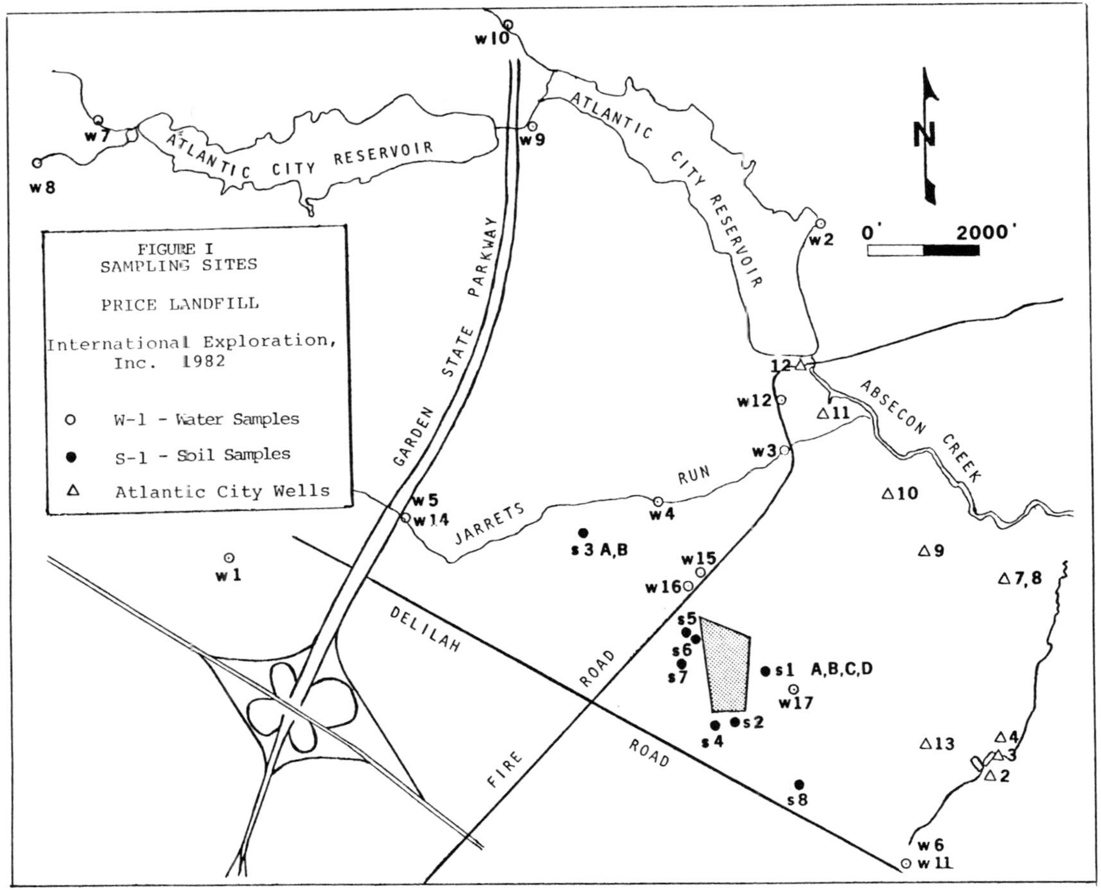


Figure 1. Site location map.

## HYDROGEOLOGY

The landfill site is underlain by unconsolidated sediments of the coastal plain of New Jersey. In the vicinity of the landfill, a thin veneer of Cape May Formation is often found. Underlying this formation is the Cohansey Formation, one of two major aquifers in the Atlantic City area. Below the Cohansey Formation is the Kirkwood Formation, the other major aquifer. The Cohansey Formation consists of unconsolidated quartz sand, gravel, and large amounts of clay. Sediments generally are medium to coarse grained and occur in long, thin, gently dipping beds. Sands range from angular to well rounded. Beds of clay occur locally and can be greater than 45 feet thick and may extend laterally over several hundred feet to several hundred acres. Gravel beds are generally less than one foot in thickness. Porosity of the sediments ranges from 27 percent to 42 percent with an average value in the vicinity of the landfill of 35 percent.

The Cohansey Formation is a shallow water table aquifer and one of the most productive aquifers in the coastal plain of New Jersey. Wells capable of yielding 1,000 gallons per minute (gpm) can be located throughout the Cohansey Formation. Depth to the permanent water table is usually less than 15 feet. For the landfill area, the depth to the water table was often less than 10 feet. Hydraulic conductivities generally ranged from 90 ft./day to 250 ft./day. Undisturbed samples of the clay layer at the landfill were obtained and are characterized by hydraulic conductivities of 0.6 to 1.0 ft./day.

Strata of the Kirkwood Formation consist of two major layers. An upper layer of silty blue clay, which is 300 to 400 feet in thickness, separates the water bearing zones of the Cohansey Formation from the water bearing zones of the Kirkwood Formation. Water bearing zones are present in the lower layer of the Kirkwood Formation. This lower layer consists predominantly of light, very fine quartz sands. Also, a high yielding aquifer, the Kirkwood Formation is artesian. Because the landfill is located in the Cohansey Formation and due to the presence of the thick clay layer of the Kirkwood Formation, the Cohansey Formation was primarily emphasized in the investigation.

A major goal of the hydrogeological investigation was identification of a potential clay layer in the Cohansey Formation. A clay layer in the Cohansey Formation, depending on depth, would have major implications in (1) an abatement plan, (2) a monitoring program, and (3) which public water supply wells might be threatened by contamination. An analysis of geological logs of Atlantic City wells indicated a high potential for the presence of a clay layer. These well logs indicated a thick clay layer in the Cohansey Formation near depths of approximately 140 to 160 feet. It was felt that any well screened below this clay layer may not be threatened by contamination if the clay layer was continuous and existed below the landfill. Surprisingly, the possibility of the clay layer at the landfill was not investigated previous to this investigation initiated in 1981.

Two general directions of groundwater flow in the vicinity of the landfill were identified based on data from about 14 wells. These directions were east and northeast from the landfill. Although no sufficient information was available to the north of the landfill, some limited movement to the north was also expected.

## GEOCHEMICAL INVESTIGATION

To determine background concentrations of contaminants, soil and water samples were collected in the vicinity of the landfill. Water samples were collected in the streams draining the areas, and from surface reservoirs of Atlantic City located north of the landfill. The nearest of these reservoirs is approximately 4,700 feet north of the landfill. In addition to surface waters, geochemical data were collected from existing wells near the landfill including 3 existing monitoring wells of the landfill and several observation wells east of the landfill installed by governmental agencies. Because no monitoring wells were located west of the landfill (upgradient), no data were available to characterize background groundwater quality. Several monitoring wells were proposed west of the landfill. However, the court involved would not allow time to drill these wells. As a result, groundwater samples were obtained from existing homes west of the landfill.

Representative contaminants present in the landfill wells and other monitoring wells were selected to characterize water quality and identify zones of contamination. These contaminants are:

Organic Contaminants

| | |
|---|---|
| Benzene | Toluene |
| Chloroform | Trans-1,1 Dichloroethylene |
| 1,2 Dichloroethane | 1,1,1 Trichloroethane |
| Methyl Chloride | 1,1,2 Trichloroethane |
| Methylene Chloride | Trichloroethylene |
| Phenols | Vinyl Chloride |

Trace Elements

| | |
|---|---|
| Arsenic | Mercury |
| Cadmium | Nickel |
| Chromium | Zinc |
| Lead | |

Chemical analyses for other contaminants were completed for some wells, surface water samples, and soil samples. Also, many of the observation wells east of the landfill were analyzed for priority pollutants. These analyses were provided by governmental agencies. All chemical analysis were conducted according to prevailing Federal and State standards.

For surface waters, concentrations for many of the contaminants indicated high background levels. Pollution was encountered in areas expected to be at background levels. Streams such as Jarrets Run, both branches of Absecon Creek, and an unnamed tributary (W-6) of Absecon Creek were identified as being contaminated with several chemicals. For example, Jarrets Run (2,100 feet northwest of landfill) contained concentrations of the following contaminants:

| | |
|---|---|
| Benzene | - traces |
| Chloroform | - traces |
| 1,2 Dichloroethane | - 0 to 25 ppb |
| Methylene Chloride | - 40 ppb to 90 ppb |
| Phenols | - 6 ppb to 20 ppb |
| Toluene | - traces |
| Trans-1,2 Dichloroethylene | - traces |
| Trichloroethylene | - traces |
| Vinyl Chloride | - traces to 25 ppb |

Streams draining into the reservoirs also had concentrations of some contaminants (Table 1).

TABLE I

Contaminants in Surface Waters (ppb)

| | W-7 | W-8 | W-9 | W-10 |
|---|---|---|---|---|
| Chloroform | - | 30 | Trace | - |
| 1,2 Dichloroethane | Trace | - | Trace | - |
| Methyl Chloride | - | 55 | - | 50 |
| Methylene Chloride | Trace | 40 | 50 | 48 |
| Phenols | 26 | 6 | 12 | 12 |
| Toluene | - | Trace | - | Trace |
| Vinyl Chloride | - | Trace | - | 35 |

Water samples in the only reservoir sampled, the eastern reservoir, identified at least trace concentrations of benzene, chloroform, 1,2 dichlorothane, methyl chloride, methylene chloride, vinyl chloride, trichloroethylene and 22 ppb of phenols.

A major spring of more than 150 gpm was sampled which entered an unnamed tributary of Absecon Creek. Water samples are identified as W-6 and W-11 on Figure 1. This spring discharged into the creek approximately 1,900 feet upstream of Atlantic City wells located at the Atlantic City water treatment plant. The stream flows through this well field. Major contaminants identified in the spring included chloroform, 5 ppb; vinyl chloride, 25 ppb; phenols, 12 ppb; and methylene chloride, 35 ppb to 75 ppb. Also, traces of toluene, benzene, and trichloroethylene were found. During the field investigation, three wells west of the landfill were located which were accessible and available for sample collecting. These were utilized to obtain some information on background water quality of the aquifer. One well was located approximately 1,200 feet west of the landfill (upgradient of the landfill) while the other two wells were located between 560 to 720 feet northwest of the landfill. Not all of the representative organic contaminants were analyzed in the western well. Of the contaminants analyzed by the laboratory, chloroform, bis-2 ethyl hexyl phthalate, and methylene chloride (traces) were identified. Concentrations of chloroform and bis-2 ethyl hexyl phthalate were 4 ppb and 9 ppb, respectively. In the two wells northwest of the landfill, concentrations of chloroform, 3 ppb to 7 ppb, and methylene chloride, 5 ppb to 15 ppb were found.

Surface water quality data collected in areas not expected to be affected by the landfill (west and northwest of the landfill) indicated relatively high background concentrations for several contaminants. These contaminants are listed in Table II.

TABLE II
Range of Concentrations in Water Samples

| Contaminant | Concentration |
|---|---|
| Benzene | Traces |
| Chloroform | Traces to 35 ppb<br>average = 9 ppb |
| 1,2 Dichloroethane | Not Detectable to 25 ppb<br>average = 3 ppb |
| Methyl Chloride | Trace to 55 ppb<br>average = 12 ppb |
| Methylene Chloride | Trace to 90 ppb<br>average = 45 ppb |
| Phenols | 6 ppb to 26 ppb<br>average = 14 ppb |
| Toluene | Traces |
| Trichloroethylene | Traces |
| Vinyl Chloride | Not Detectable to 25 ppb<br>average = 10 ppb |

Some of these concentrations, however, indicate sources of pollution other than the landfill. As a result of the relatively high concentrations of contaminants found in the vicinity of the landfill, interpretation of chemical data for leachate plume identification was much more complex. For example, concentrations of chloroform in most water samples not affected by any pollution sources in the vicinity of the landfill were greater than 3 ppb. Therefore, concentrations less than 3 ppb were not considered anomalous.

During the reconnaissance survey of the site area, several potential contamination sources (e.g. illegal chemical dumps) were identified. Soil samples and in some cases samples of stream sediments were collected. Collection of samples was conducted according to federal procedures and standard soil sampling procedures. A minimum of four major pollution sources in addition to the landfill were discovered. Because of time constraints imposed by the courts, additional potential pollution sources could not be investigated. These contamination sources (west to east) were located at S-3A, S-3B, S-5, S-6, and S-1, and upstream of S-8. Soil samples at S-3 and S-4 confirmed presence of landfill materials. Potential contaminants selected at each site for analysis differed depending on site conditions. All soils at the sites except S-8 were discolored when compared to natural soils in each of the areas. The alleged dumping sites were distinct and had sharp boundaries separating contaminated and uncontaminated soils. Some of the sites were known as illegal dumping sites by nearby residents.

Soil samples at sites 3A and 3B were taken in a trench dug across a red and black discolored zone which was about 200 feet in diameter. In addition to high concentrations of several metals including arsenic, chromium, copper, lead, and mercury, many organic contaminants were identified. Some of these contaminants are vinyl chloride = 75 ppb, methylene chloride = 1,300 ppb, trans-1,2 dichloroethylene = 165 ppb, chloroform = 75 ppb, 1,2 dichloroethane = 100 ppb, benzene = 100 ppb, and toluene = 100 ppb. Because surface runoff from this site enters Jarrets Run, the concentrations of some of these contaminants found in the stream may be due to this contamination source.

Some contaminants were found in stream channels and drainage areas down gradient from an equipment storage site west of the landfill (S-5 and S-6). Drainage from this area was conveyed to the northeast and east of the Price Landfill. In addition to petroleum hydrocarbons (up to 1.7 percent) found in the soils down gradient of the site, some of the major contaminants which were identified are listed in Table III.

TABLE III
Selected Contaminants at Local Pollution Sources (ppb)

| Contaminant | Site S-6, S-5 | Site S-8 |
|---|---|---|
| Vinyl chloride | up to 650 | - |
| Methylene chloride | Trace to 425 | Trace |
| Trans-1,2 dichloroethylene | Trace to 65 | 125 |
| Chloroform | - | 100 |
| 1,2 dichloroethane | 175-290 | 190 |
| Trichloroethylene | 65-220 | Trace |
| Benzene | - | - |
| Toluene | - | - |
| PCB (collectively) | 4,000 | NS |
| Arsenic | 1,200-15,000 | 280 |
| Lead | 9,400-29,000 | 12,000 |
| Mercury | - | 17 |

Also, listed in Table III are the major contaminants identified at S-8. Stream sediment was collected at this sample site. The stream drains into the tributary which flows through the well field at the Atlantic City water treatment plant. Because of time constraints, the actual location of the pollution source was not identified.

An illegal dump site (S-1) was also found approximately 300 feet east of the landfill. A thick residue coated the soil surface at the site. This site was located between the landfill and several of the monitoring wells. Two soil profiles were analyzed. One soil profile was excavated upslope away from the dump area and was observed to be clean, representing natural soils. At the other soil profile, located where the thick residue was found, several soil samples at different soil horizons were collected and analyzed. Results of the analysis are listed in Table IV.

TABLE IV

Concentrations Found at an Illegal Dump Site (S1)

| Soil Sample # | S1A | S1B | S1C | S1D |
|---|---|---|---|---|
| Depth | 0-9" | 1'3"-2'3" | 3'7"-3'11" | 4'8"-5'2" |
| Trace Organics (ppb) | | | | |
| Vinyl Chloride | ** | 380 | <75 | <75 |
| Methylene Chloride | ** | 765 | 250 | <75 |
| Trans-1,2-Dichloroethylene | ** | <100 | <75 | <75 |
| Chloroform | ** | <100 | <75 | <75 |
| 1,2 Dichloroethane | ** | 570 | <75 | <75 |
| Trichloroethylene | ** | 150 | <75 | 130 |
| Benzene | ** | 190 | 100 | <75 |
| Toluene | ** | ** | 230 | 100 |
| Percent Petroleum Hydrocarbons | 32 | 0.5 | 0.9 | 1.1 |

** Indicates the presence of large quantities of volatile organic compounds that interfere or mask the presence of the target compounds.

Trace metals identified in the soil profile included arsenic, cadmium, chromium, lead, mercury, nickel, and zinc. Concentrations for these metals were highest at shallow depths. For example, the concentrations of arsenic, chromium, lead, and mercury are 4,500 ppb, 25,000 ppb, 100,000 ppb, and 17 ppb, respectively.

In addition to the four contamination sources identified from the soil surveys, three sources of contamination were identified by water samples. One of these sources was discussed previously at site W-6. A major source of contamination upstream of both branches of Absecon Creek before the streams enter the two surface reservoirs was also discovered. Another pollution source located 8,400 feet west of the landfill, which directly entered the groundwater system, was located near W-1. Contaminants found at this site were chromium and traces of vinyl chloride, chloroform, benzene, toluene, and 1,2 dichloroethane (20 ppb).

Evaluation of these seven known contamination sources as well as general water quality data collected from nearby streams is critical to assessing data from monitoring wells and the public water supply wells in the vicinity of the landfill. Some of these contamination sources may be affecting water quality found at some monitoring wells and public water supply wells. Without down playing the potential magnitude of pollution problems from the landfill, the source of contamination located at W-6, for example, may be the more immediate threat to water quality of wells located at the water treatment plant. The Price Landfill, itself, represents a more long term threat to these wells. Data from some monitoring wells could be influenced by these 7 additional sources of contamination. Additionally, some wells which were previously considered contaminated by the landfill, for example, by chloroform and methylene chloride, were found to be within the normal range of background values for these two contaminants. However, several homes east and northeast of the landfill were determined to be affected by the Price Landfill.

## RESISTIVITY SURVEY

One relatively rapid method of identifying a leachate plume is the use of resistivity surveys. Supplementing monitoring well data, the method has been utilized by the author for investigating several abandoned hazardous waste sites and landfills. The resistivity survey of the Price Landfill site was conducted as part of a more comprehensive hydrogeological investigation.

The plan was to conduct a survey using vertical

electrical soundings (VES) to identify a leachate plume and determine if a clay layer was present at the landfill site. Although the geology of the site area was known, no information existed on the presence or absence of a clay layer. Presence of the clay layer was critical because it would significantly influence the abatement plan and would be a barrier to downward movement of leachate. After soundings were completed, horizontal profiling was planned to delineate the leachate front in more detail and depth of the contaminated zone. Correlation test holes, monitoring wells, and pump tests were also part of the investigation.

Planning and execution of the resistivity survey was complicated due to location of the landfill in a developed area. A large number of man-made obstacles were present including numerous homes, a power substation, major power transmission lines, smaller power transmission lines, large areas which contained scattered metal debris, areas of salt marsh and freshwater swamps. The landfill is located in an abandoned sand and gravel pit with several other landfills in the vicinity of the study area.

As a result of the site conditions, the Schlumberger array was utilized for the resistivity survey. The equipment for the survey was a Soil Test D.C. Earth Resistivity Meter-Model R-65. This meter system has a separate transmitter and receiver with a power input to the earth up to 810 watts, delivery of voltage up to 810 volts, and providing current up to 1 ampere. A preliminary sampling grid was developed after groundwater quality data was evaluated from several monitoring wells in the vicinity of the landfill. Several traverses were located perpendicular to the two major directions of groundwater flow (north and east). More than 100 geophysical stations were completed during the resistivity survey. As part of quality control background, geophysical stations were located, resistivity measurements were repeated at the same geophysical station on different days, sounding lines were repeated perpendicular to each other at other geophysical stations, and correlation holes drilled. Five correlation holes were drilled in addition to monitoring wells.

Interpretation method utilized was curve matching using the master curve set of Orellana and Mooney (1). Results of curve matching were compared to a computer program of Zohdy (2).

Resistivity measurements identified the presence of an impermeable layer (clay) of relatively high resistivity. The clay layer was confirmed by test holes. One test hole was specifically designed to penetrate the layer and obtain water samples of the underlying aquifer and a sample of the clay layer for laboratory analysis. All five test holes confirmed the presence of the clay layer around the landfill. Thickness of the clay layer was approximately 45 feet (hole located 800 feet east of the landfill) and was determined by drilling. The clay layer became thinner southwest of the landfill. Because the clay layer was less than 3 feet thick in this area, more work was planned to determine if the clay layer was discontinuous. Correlation holes were located at the corners of the landfill and immediately north of the landfill. Presence of the clay layer (35 feet thick) was also confirmed in a water supply well located within 200 feet of the landfill in an easterly direction.

Water samples from the monitoring well at W-17, screened below the clay layer, indicated no contamination. However, additional samples were recommended to confirm this result. The approximate areal extent of the leachate plume was defined by the resistivity survey. Results of the resistivity survey are depicted in Figure 2. A significant contrast between the contaminated zone and background groundwater quality was observed. The leachate plume was confirmed by groundwater samples. Generally representing the extent of the leachate plume, the identified contamination zone requires refinements to more accurately define the leachate front. Additional resistivity surveying (horizontal profiling) and the drilling of additional observation wells on the approximate edge of the leachate front was initiated but was ceased due to time constraints imposed by the court. Drilling monitoring wells in a designated pattern could result in an interpretation which would move the leachate front to the east. Homes which were identified as contaminated and all monitoring wells were within the general boundar-

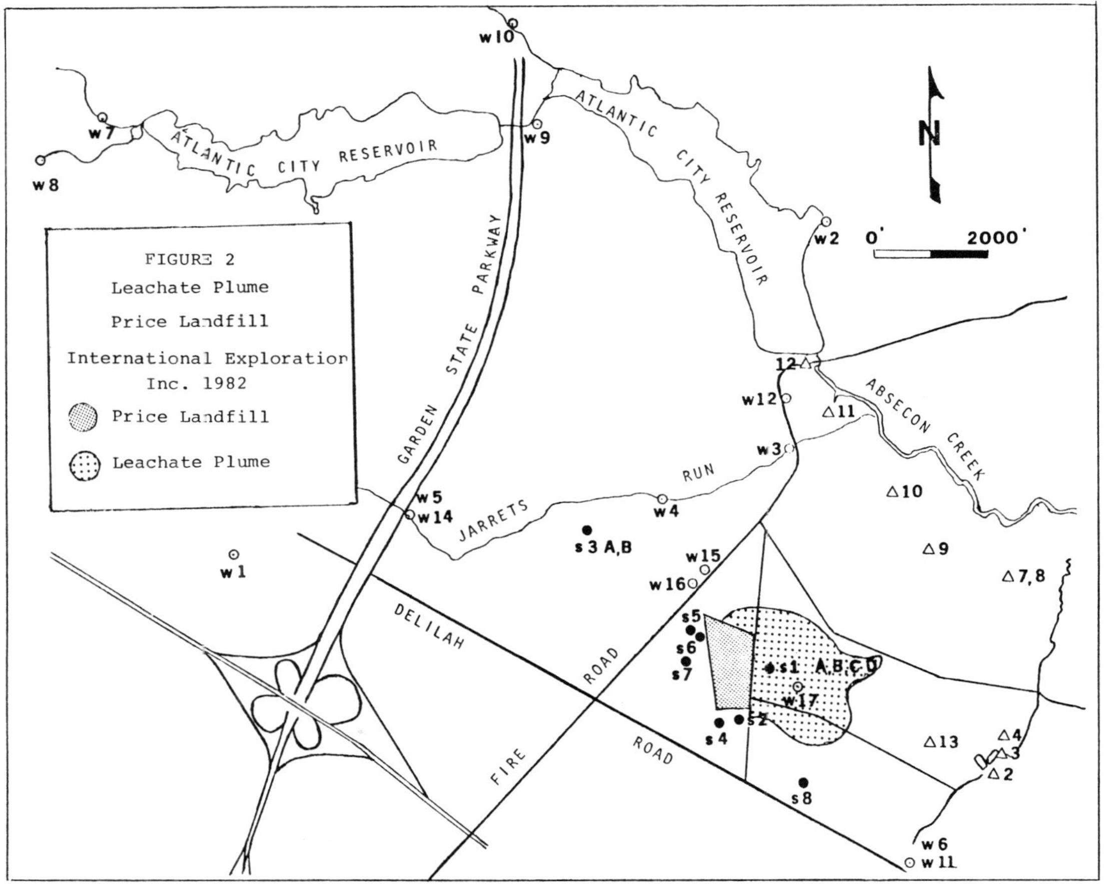


Figure 2. Leachate plume at Price Landfill.

ies of the identified leachate plume. A traverse of resistivity stations parallel to Rt. 9 and located west of the nearest Atlantic City well did not indicate the presence of leachate. One monitoring well at the equivalent distance east of the landfill also did not indicate contamination. For example, chloroform was found but at a concentration of 1 ppb, well below the normal background concentration. Only one compound bis-2 ethyl hexyl phthalate was high but could have been due to a more local source of contamination or laboratory error. Additional sampling in the area was recommended. Orientation of the plume indicates that the leachate from the landfill is moving in a slightly northeastward direction.

## CONCLUSIONS

The hydrogeological investigation of Price Landfill resulted in several important findings which were being incorporated into a detailed abatement program. This abatement plan and some parts of the investigation were not completed as mentioned previously due to time limits imposed by the court on the investigation. For example, pump tests were scheduled but cancelled because of court requirements. As a result, investigation of site specific aquifer characteristics was not complete. However, in the four months of field work, several critical facts were ascertained. A summary of this information, which is based on data available through the end of the investigation (June 1981), follows:

1. A leachate plume was identified. However, the location of the leachate front is only approximate and requires better definition.

2. The leachate plume has not reached the Atlantic City wells at the time of our investigation. Because of the direction of movement of the plume, several Atlantic City wells will not be endangered.

3. Resistivity surveys were successful and were utilized to identify the leachate plume, a clay layer, and in development of a more representative and comprehensive monitoring network.

4. The clay layer separating the Cohansey aquifer into two separate aquifers was confirmed through a drilling program around the landfill after initial identification by the resistivity survey.

5. Additional pollution sources are recommended to be considered in the development of both the monitoring program and abatement plan. These pollution sources are influencing water quality in areas near monitoring wells and residential homes. Without consideration of these sources, abatement programs for the landfill may not be as comprehensive and effective in abating pollution.

6. More investigation is recommended northeast of the landfill where residential homes are directly affected.

7. Additional studies in the area southwest of the landfill where the clay layer thins is needed to determine if the clay layer is intact and/or if contamination is leaking to the lower aquifer of the Cohansey Formation. Present data available to this investigation indicates the lower aquifer is not contaminated.

# CASE HISTORY STUDY OF WATER FLOW THROUGH UNSATURATED SOIL

Stephen J. Trautwein and David E. Daniel
Civil Engineering Department
The University of Texas at Austin
Austin, Texas

Michael W. Cooper
Underground Resource Management, Inc.
Austin, Texas

## INTRODUCTION

Evaporation ponds are commonly used in arid regions for the disposal of large amounts of liquid waste. In order to maximize evaporation, the ponds are generally shallow (only one to three feet deep) and are constructed to cover large areas. Seepage from evaporation ponds has not usually been a major concern. Besides locating ponds at sites with clayey soils, little additional effort has usually been made to control leakage. Presently, however, due to increased awareness of possible damage to the environment and stricter government regulations concerning the pollution of ground water, special design considerations are needed to control seepage from these ponds. Many newly constructed ponds are required to be lined with compacted clay or a synthetic material. In addition, some type of leak detection system is generally required to monitor the quality of ground water in the vicinity of a pond. Monitor wells are most commonly used for this purpose.

While more stringent regulations will hopefully minimize pollution of ground water caused by leakage from newly constructed ponds, ground-water pollution due to leakage from older ponds that are still in use is a problem of great concern. Many of these older ponds are being investigated to determine if leakage has occurred and whether remedial measures are necessary. The investigation of one of these older ponds is the topic of this paper.

## BACKGROUND INFORMATION

The pond in question has been used by an industrial facility for disposal of liquid waste which contains organic compounds. The pond covers approximately 48 acres and has been in use for 22 years. It was constructed on a thick deposit of soil consisting of alternating strata of sand and clay. The regional ground-water table is located at a depth of 390 feet beneath the pond. The soils above the water table had degrees of saturation ranging from 20 to 70 percent prior to construction of the pond.

Because of the natural clay strata and the presence of unsaturated soils, there had been little concern that wastewater leaking from the pond would ever reach the underlying aquifer. However, in response to State regulatory programs, the industrial facility commissioned an investigation to address the following questions:

1. Is the pond leaking?

If so

2. How far have contaminants migrated?
3. How and when will the wastewater impact the aquifer?
4. Are remedial measures necessary?

Underground Resource Management, Inc. was retained to direct the investigation. Analytical studies for the project were performed at the University of Texas. This paper is based on work performed in the initial phase of the investigation.

## FIELD STUDIES

### Sampling and Boring

The field studies began with definition of the regional geology and ground-water resources at the site. Information from oil, gas, and water wells drilled in the vicinity of the site was used to define the general soil conditions. This general knowledge was then used as a guide for conducting a preliminary program of boring and sampling to gain more specific information about the subsoils in the vicinity of the pond. Ten shallow borings (<40 feet deep) and 7 deep borings (>100 feet deep) were drilled. Soil samples for identification and laboratory testing were obtained using thin-walled sampling tubes. Gamma logs were obtained from the deep borings to aid in identification of individual strata. Several borings were drilled in areas unaffected by the pond in order to obtain baseline data for comparative purposes. A plan view of the site with boring locations is given in Figure 1. The generalized stratigraphy is shown in Figure 2. The upper surface of the saturated aquifer was found to be at a depth of 390 feet beneath the pond.

### Installation of Monitor Wells and Lysimeters

In conjunction with the boring program, a number of monitor wells and lysimeters were installed in order to obtain ground-water samples for chemical analysis. Two deep monitor

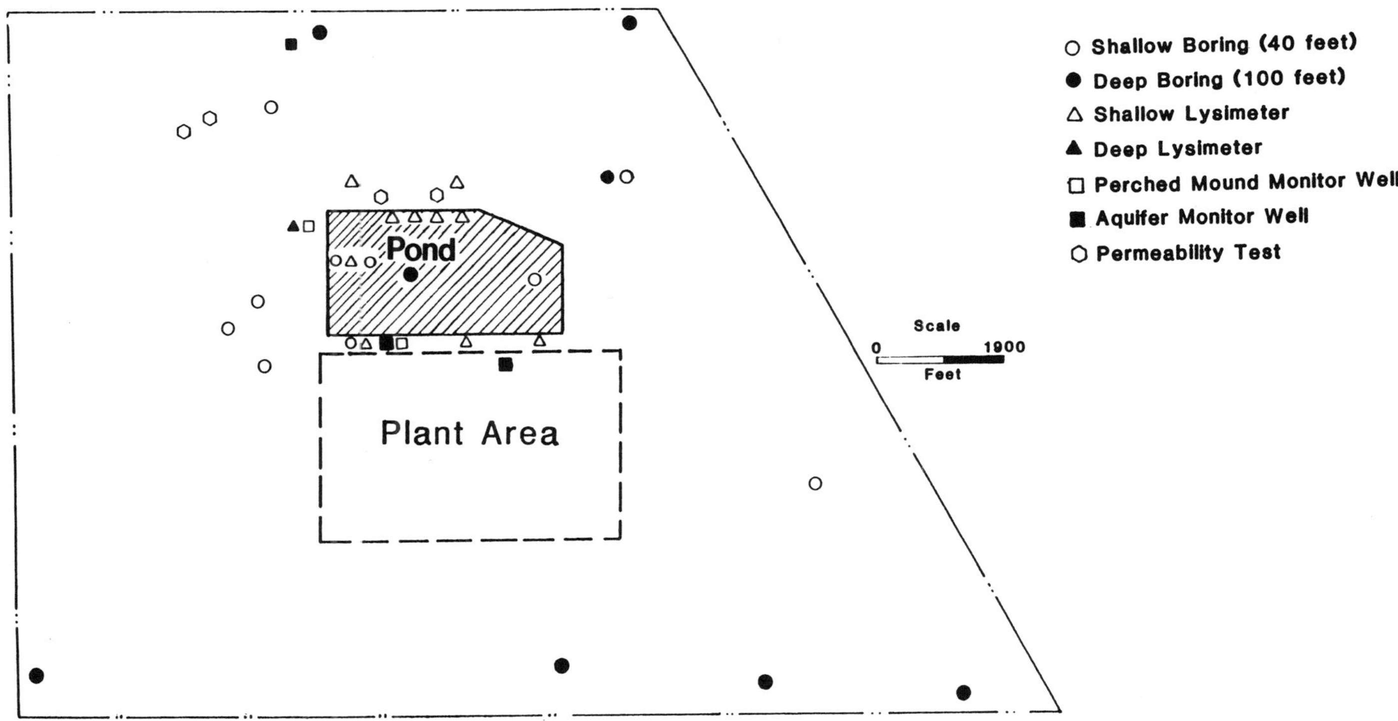


Figure 1. Plan View of Site

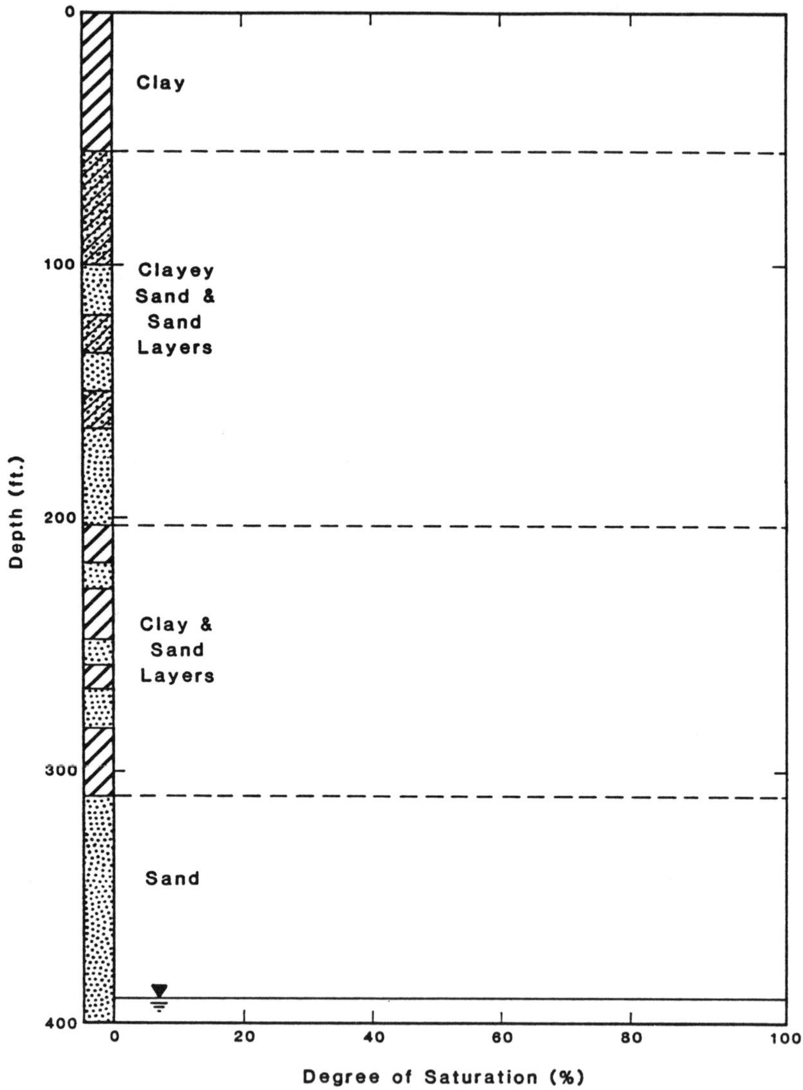


**Figure 2. Generalized Stratigraphy at the Site**

wells were installed into the aquifer. A perched mound of water was found to exist between depths of 300 and 140 feet. Two monitor wells at depths of 180 and 185 feet were installed to sample water from this perched mound. Three deep lysimeters were installed at depths of 308, 324, and 339 feet to sample ground water in the sand zone beneath the perched mound. Ten other lysimeters were installed at shallow depths (five to twenty feet) to sample the ground water in the upper clay zone beneath the pond.

The configurations of the monitor wells and lysimeters are shown in Figures 3a and 3b. Appropriate measures were taken to minimize the possibility that installation processes would cause contamination of the borehole near the lysimeter or the screened zone of the monitor well. Boreholes were cased and grouted to prevent contaminated water from upper zones from leaking down to lower zones. Drilling mud that was free from organic additives was used. Where possible, compressed air was used instead of drilling mud to remove cuttings from boreholes in the zones where lysimeters were to be installed. Lysimeters were constructed in such a way so that the various parts were not held together with organic solvents. Such measures were necessary so that the presence of organics in the ground water was not masked by contaminants added during the drilling process.

### Permeability Testing

Four field permeability tests were conducted in the upper clayey zone directly beneath the pond. The details of the permeability apparatus are shown in Figure 4. Permeant flowed out of the borehole through a slotted casing that was five feet in length. The average depth from the ground surface to the center of the slotted casing for the four tests was 5, 8, 11 and 17 feet. Falling head tests were performed intermittently over a period of several months until a constant value of permeability was obtained. The results of the tests will be discussed subsequently.

## LABORATORY TESTING

### Soil Testing

Standard index tests were performed to classify the soils. Moisture content determinations were made in numerous samples in order to develop subsurface moisture content profiles. Profiles were constructed for soils beneath the pond and for soils in areas unaffected by the ponds.

Permeability tests were performed on disk shaped speci-

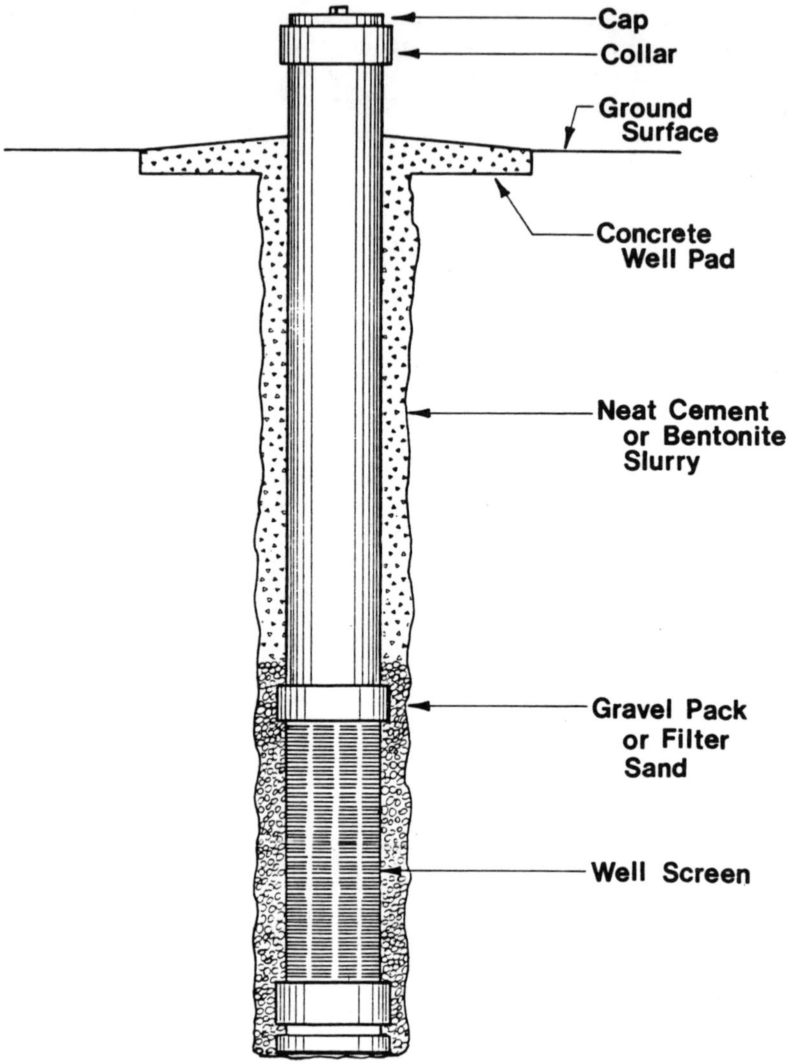


**Monitor Well Installation**

Figure 3a.

# Lysimeter Installation

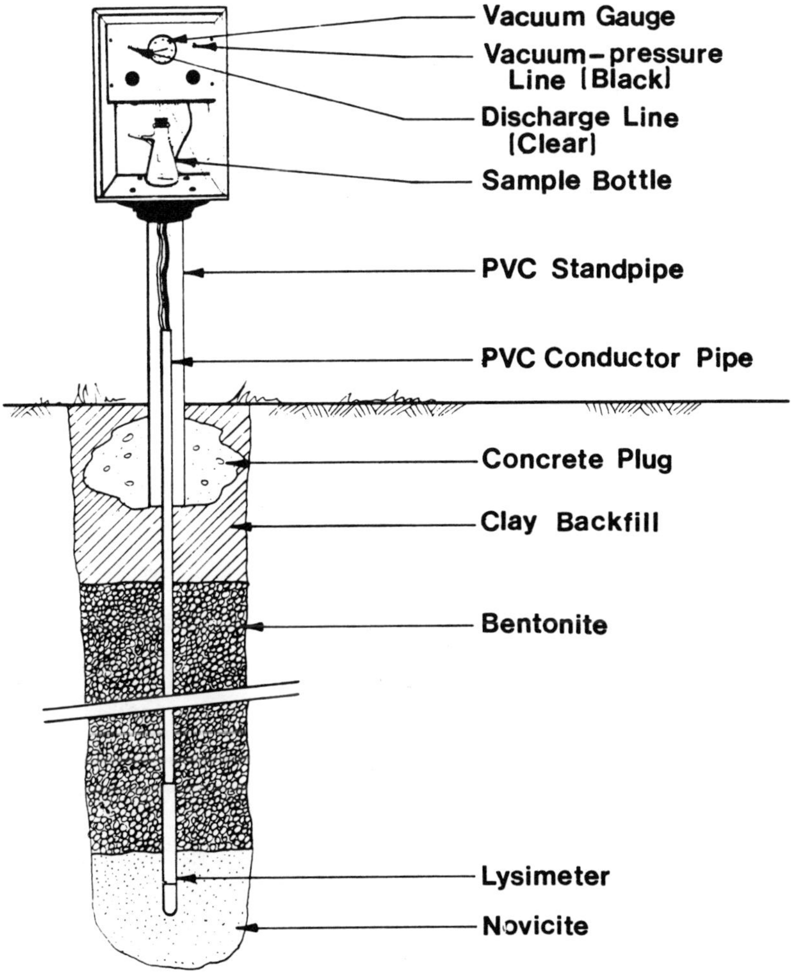


Figure 3b.

## Figure 4.

Field Permeameter Installation

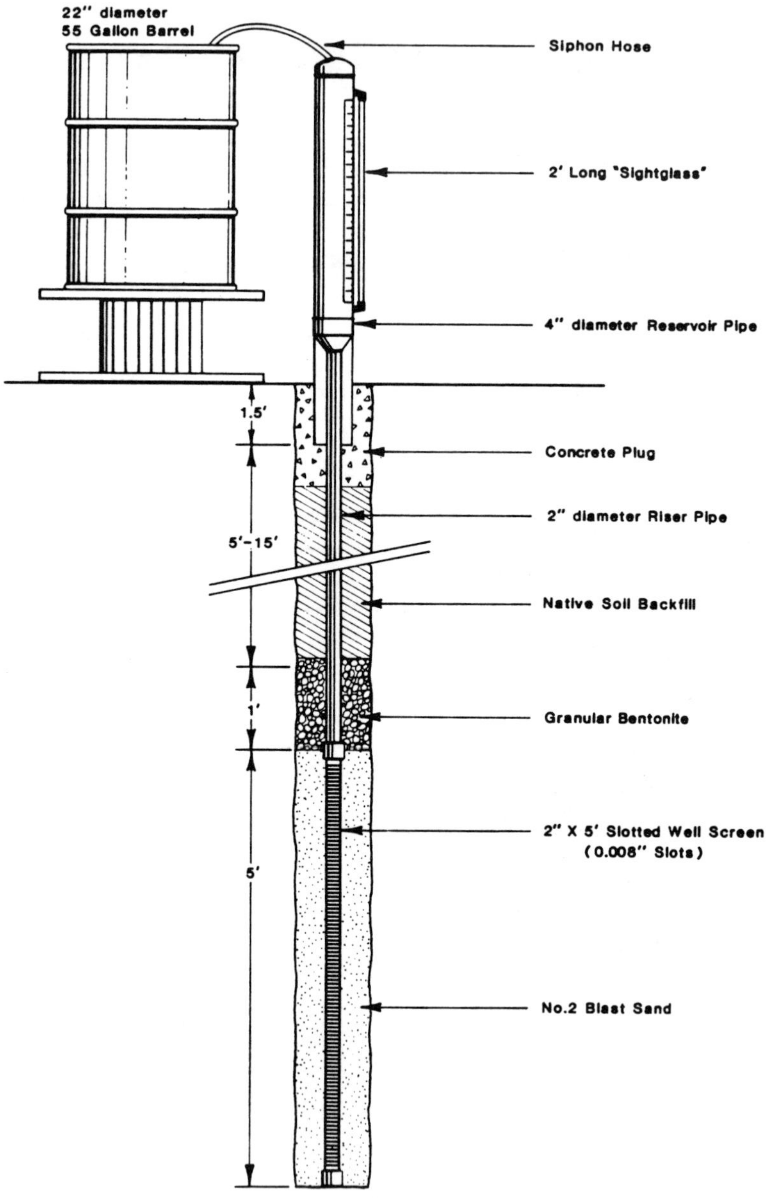

mens trimmed from samples obtained using thin-walled tubes. Four of the permeability tests were performed on samples obtained from the boreholes used for the field permeability tests.

The tests were performed in rigid ring cells and care was taken to ensure that no channels formed between the sample and the ring. Gradients used during the testing ranged from 10 for the high permeability sand samples to 300 for some of the low permeability clays. Tap water was used as the permeant.

### Chemical Analysis

Chemical analyses were performed on soil and groundwater samples in order to determine if they were contaminated with pond water. The primary tests used for this purpose included measurements of total organic carbon, total dissolved solids, and concentration of certain metals known to be in the waste stream.

## FIELD AND LABORATORY FINDINGS

The field and laboratory studies provided enough information to answer the first two questions that were posed. Borings in the vicinity of the pond revealed the presence of a perched mound of water between the depths of 140 and 300 feet. Chemical analysis of water and soil samples obtained in this zone confirmed the presence of pond water. Also, the water contents of samples obtained in the zone above the mound were higher than water contents of samples obtained at the same depths but in an area unaffected by the pond.

Water samples obtained from the deep lysimeters indicated that the contaminant front had migrated to a depth of at least 308 feet. No contamination was found in water samples obtained from the aquifer wells.

Once it was determined that leakage was occurring and the vertical extent of the contaminant front had been defined, the remaining questions concerning future migration of the contaminants were addressed. The approach taken to answer these questions is discussed below.

## NUMERICAL STUDIES

The complex geometry, transient nature of flow, and presence of initially unsaturated soils precluded the use of closed-form analyses to predict water migration around the pond. Instead, a numerical model was used to simulate flow.

For this project, simplifications were made in order to make such an analysis economically feasible.

Because large uncertainties exist in some of the input parameters, a highly sophisticated model was considered to be unwarranted. Instead, it was decided to construct a simple model that would be economical but still adequately represent the behavior of the system.

In order to simplify the problem, it was decided to concentrate on the flow beneath the center of the pond. It was reasoned that initial aquifer impact would occur beneath the center of the pond and if flow could be modeled in this zone, the results could be used to estimate the impact of the contaminants on the aquifer.

Also, in order to simplify the solute-transport problem, it was assumed that contaminants would travel at the same rate as the wetting front. This assumption was considered to be conservative, i.e., the worst case possible, and would predict the earliest possible time for impact to occur. Actual aquifer impact would be expected to occur at a later time since the migration of the contaminants should be retarded to some degree by the clayey soils.

Because flow beneath the center of the pond was being modeled, and because the diameter of the pond (1200 feet) is so much greater than the thickness of the unsaturated zone (390 feet) it was decided to analyze the case of one-dimensional (vertical) seepage. The field investigation indicated that the soil layering is essentially horizontal and that the various soil strata are continuous across the site.

It was thought that the success or failure of this simplified analysis would help in deciding whether a more sophisticated representation of the problem was needed.

## Description of Model

A computer program called DUNSAT [1] was selected for modeling flow at the site. DUNSAT is a Galerkin-type finite element code which solves the equation of flow derived from Darcy's Law and the principle of conservation of mass. DUNSAT is used to solve for the distribution of hydraulic head, given initial and boundary conditions, throughout the region being modeled. The method of solution is described in detail Daniel [1] and by other researchers who have used the same general approach [2, 3, 4, and 5].

Some of the features of DUNSAT which are applicable to this problem are the ability to model:

- saturated - unsaturated flow
- one or two dimensional flow
- steady-state or transient flow
- non-linear soil properties
- multilayered systems.

Some of the important assumptions that were made in developing DUNSAT are:

- Darcy's Law is valid and flow results only from gradients in total hydraulic head.
- relative hydraulic conductivity and soil suction are unique functions of hydraulic head.
- volume changes in the soil skeleton are small.
- hysteretic soil behavior is not taken into account.
- vapor transport is negligible.

It was considered that the errors introduced by making these simplifying assumptions were insignificant compared to uncertainties in soil properties.

## Convergence

The computational costs for modeling saturated/unsaturated flow are frequently very high, particulary when a sharp wetting front exists. One of the main reasons for the high computational costs is the difficulty in obtaining convergence. Convergence is defined to have occurred when the difference between the solution for two successive iterations is less than some specified tolerance. If soil properties vary greatly over a given time interval, as when a wetting front passes through a soil element, a large number of iterations is needed to obtain convergence. Smaller time steps can be used to reduce the change in soil properties. However, if the time step is too small, computational costs will be prohibitively large.

Incorporated into DUNSAT are several features that facilitate achieving convergence within a reasonable number of iterations while at the same time using time steps large enough to obtain solutions at large times. These features include:

1. The ability to use backward or central differences to step through time. Backward differences lead to faster convergence when the solution is changing rapidly. Central differences provide faster convergence when the solution is relatively smooth with time.
2. The number of iterations required for convergence is used to control the time step. The time step is increased if few iterations were required and it is decreased if a large number of iterations were necessary on the previous time step.
3. Different tolerances can be used during the same run. If the program becomes bogged down at a cer-

tain time step (as when a wetting front passes a boundary between two soil layers with large differences in hydraulic conductivities) the tolerance can be increased to achieve convergence and advance to the next time step. Some accuracy is lost by increasing the tolerance.

4. The program has a stop-restart feature which allows the user to monitor the convergence rate and make changes in input parameters (e.g., time step and tolerance), if necessary.

## DEVELOPMENT OF THE MODEL

The steps used in developing the model are:

1. Definition of the stratigraphy in the flow region.
2. Construction of the finite element grid.
3. Establishment of initial and boundary conditions.
4. Estimation of soil properties.
5. Performance of computer runs to obtain a solution.
6. Comparison of computed results with field observations. Steps 3-6 are repeated until a match is obtained.

The stratigraphy used in the model is that shown in Figure 2, which is a composite of all the soil borings made in the vicinity of the pond. A one-dimensional finite element grid composed of linear elements was used. The grid was composed of 209 nodes spaced two feet apart.

The boundary conditions varied depending on the problem being analyzed and are discussed subsequently.

### Calibration of the Model

An important step in developing a model is to calibrate it as described in steps 4-6 listed above. A numerical model is only a very limited representation of the physical problem and some method should be used to determine how well the model simulates field behavior. The level of confidence one has for the model's ability to predict future behavior is greatly increased if the model can successfully predict known behavior.

In order to calibrate the model for this problem, two cases, where field behavior was reasonably well defined, were analyzed. The two cases will be referred to as 1) steady-state case; and 2) transient case.

The steady-state case represents conditions at the site 22 years ago, before the pond was constructed. A profile of degree of saturation, $S_r$, was constructed based on water content measurements from samples obtained from borings in areas unaffected by the pond. It was assumed that this $S_r$ profile is what existed in the pond area before the pond was constructed.

The transient case refers to conditions that exist at the site after 22 years of ponding. A saturation profile based on water content measurements from samples obtained in the pond area was used as the known condition for comparison with the $S_r$ profile obtained from the model.

## Initial and Boundary Conditions

Steady-State Case. A total hydraulic head of -660 feet (equivalent to 20 atmospheres of suction) was used as the average boundary condition at the ground surface. The datum for defining elevations was set at the ground surface. The boundary condition at the ground surface was selected based on measurements of suction obtained with thermocouple psychrometers [6, 7]. The measurements were made in the laboratory on samples obtained near the ground surface. The total hydraulic head at the base of the profile was set at -390 feet. This was selected to represent a water table condition at a depth of 390 feet.

For convenience, the initial heads were made to vary linearly between the two boundaries. The analysis was performed by allowing the heads at the interior nodes to vary until steady-state conditions developed. The steady-state conditions did not involve a linear variation in head because hydraulic conductivity varied with depth.

Transient Case. The total hydraulic head at the upper boundary was set to one foot to represent a pond that was one foot deep. The lower boundary condition was the same used in the steady state case.

The initial heads at the interior nodes were set to those obtained in the steady-state analysis.

## Soil Properties

Each soil layer is characterized by:

1. moisture characteristic curve (relationship between suction, p and $S_r$) (Figure 5).
2. relative hydraulic conductivity $k_r$, versus $S_r$ (the relative hydraulic conductivity is the ratio of the hydraulic conductivity, k, at any $S_r$ to the saturated conductivity, $k_s$).

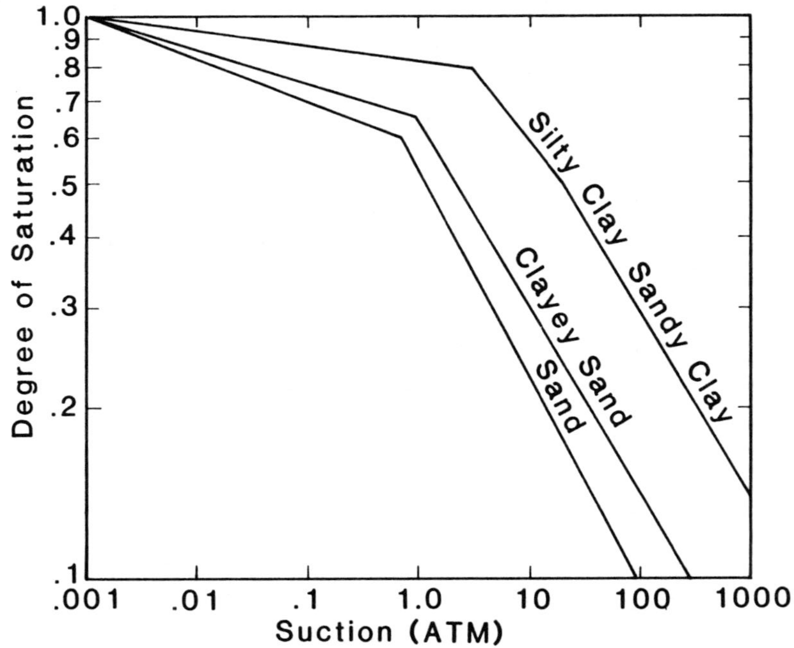


Figure 5. Moisture-Characteristic Curves Used in the Analyses

3. dry unit weight and specific gravity so that $S_r$ can be calculated from water content, and vice versa.

The dry unit weight was determined from samples obtained at the site. Values of specific gravity were assumed.

It was not in the scope of this project to measure moisture-characteristic curves or $k_r$ - $S_r$ relationships. Instead, initial estimates for these relationships were made based on curves published in the literature [6] and from the authors' experience in measuring these relationships.

The initial estimate of the soil properties was used in the model to obtain a profile of $S_r$ vs. depth for the steady-state case. The properties were adjusted until the computed $S_r$ profile matched the measured profile reasonably well.

Next, the adjusted properties were used to predict the $S_r$ profile after 22 years of ponding. Repeated runs were made, making additional adjustments in the soil properties until the model predicted the measured $S_r$ profile corresponding to 22 years of ponding (Figure 6).

The steady-state case was then reanalyzed to see if adjustments made for the transient analysis had any significant effect on the steady-state solution. The adjustments did affect the steady-state solution so further adjustments were made in the soil properties and the transient case was reanalyzed again. This process was repeated until the model was capable of predicting the correct $S_r$ profile for both cases.

The adjusted moisture characteristic and $k_r$ vs. $S_r$ relationships are shown in Figure 7. A comparison of the measured values of saturated hydraulic conductivity ($k_s$) with the adjusted values used in the model is shown in Figure 8. The controlling layer with respect to vertical seepage appears to be the clay layer located at a depth of 290 feet.

## PREDICTION OF FUTURE MIGRATION

Once the model was calibrated to simulate known behavior of the system, it was used to predict future migration of the pond water. The results of the analysis are shown in Figure 6. The entire region between the ground surface and the lower clay becomes saturated after 30 years of ponding. The degree of saturation in the lower sands increases gradually until after 60 years of ponding a constant value is reached. Steady-state flow is achieved at this point and a constant seepage rate of approximately three feet per year is obtained. The advancement of the wetting front predicted by the model is shown in Figure 9. The wetting front reaches

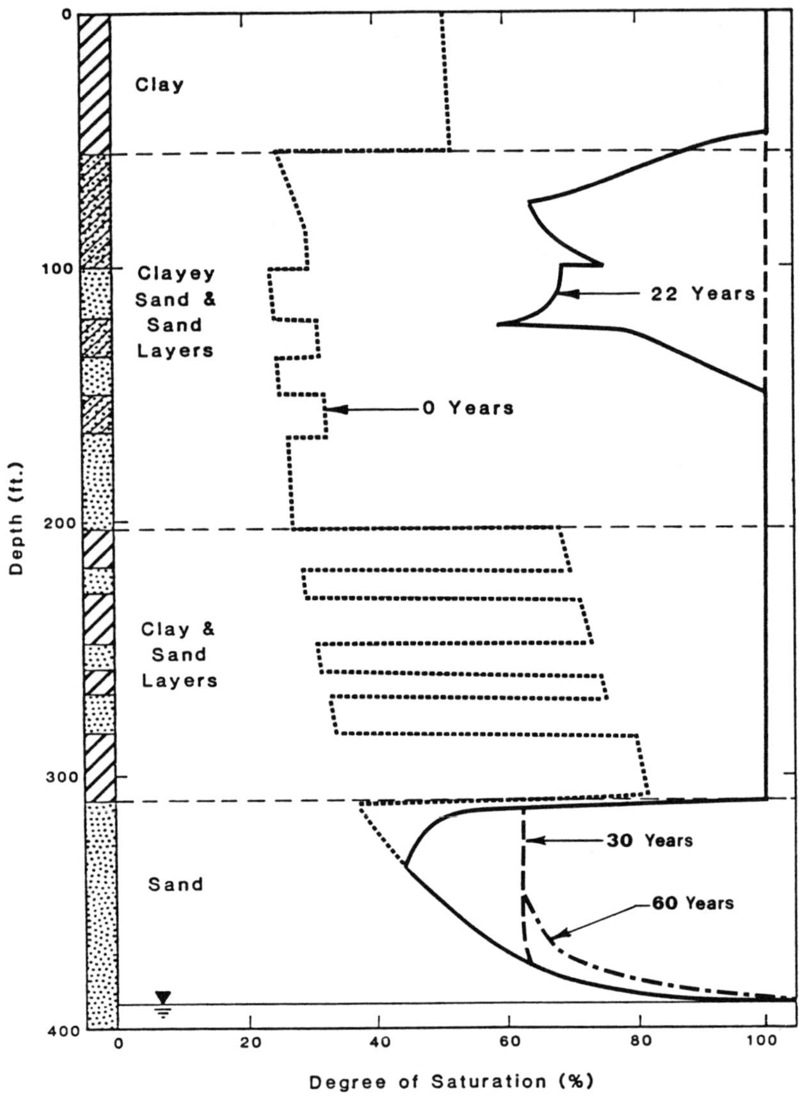


**Figure 6. Degree of Saturation Profiles for the Various Analyses**

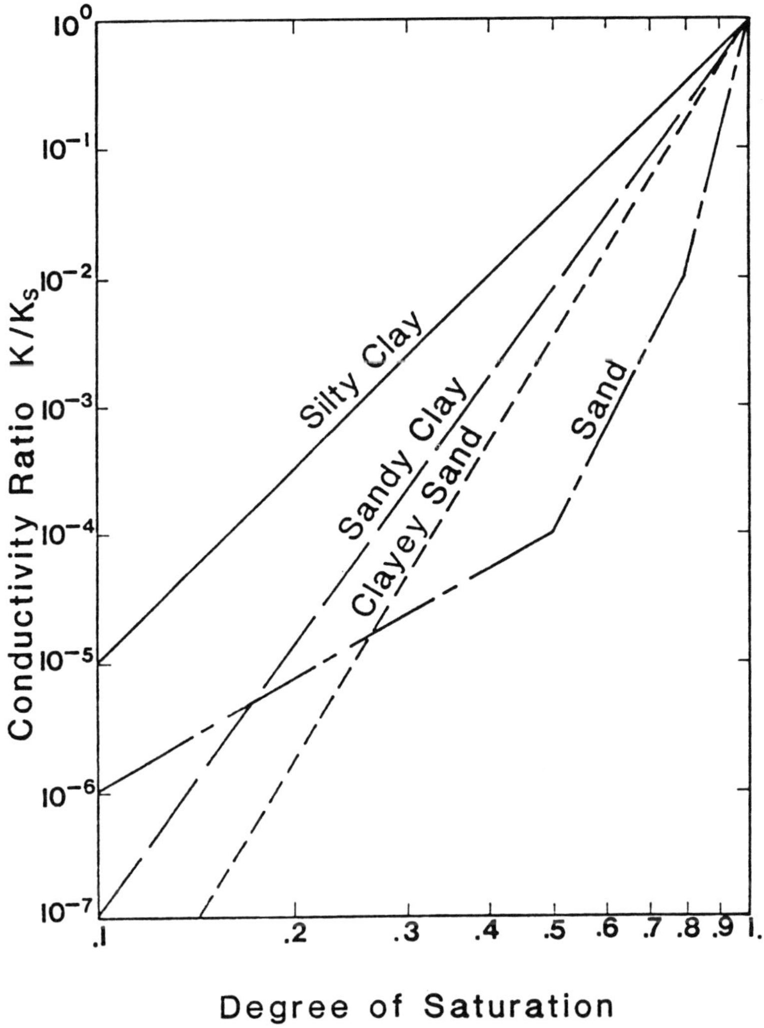


**Figure 7. Relative Conductivity Versus Degree of Saturation Relationships Used in the Analyses**

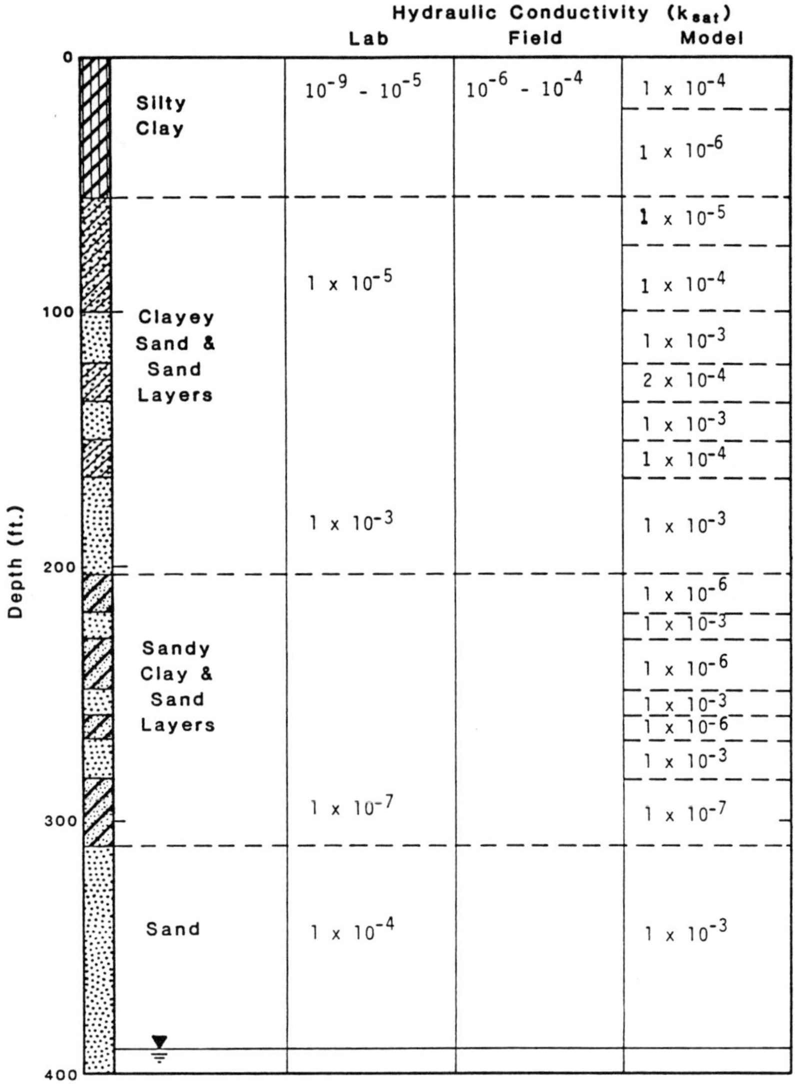


Figure 8. Comparison of Measured Values of Saturated Hydraulic Conductivity ($k_{sat}$) With Those Used in the Model

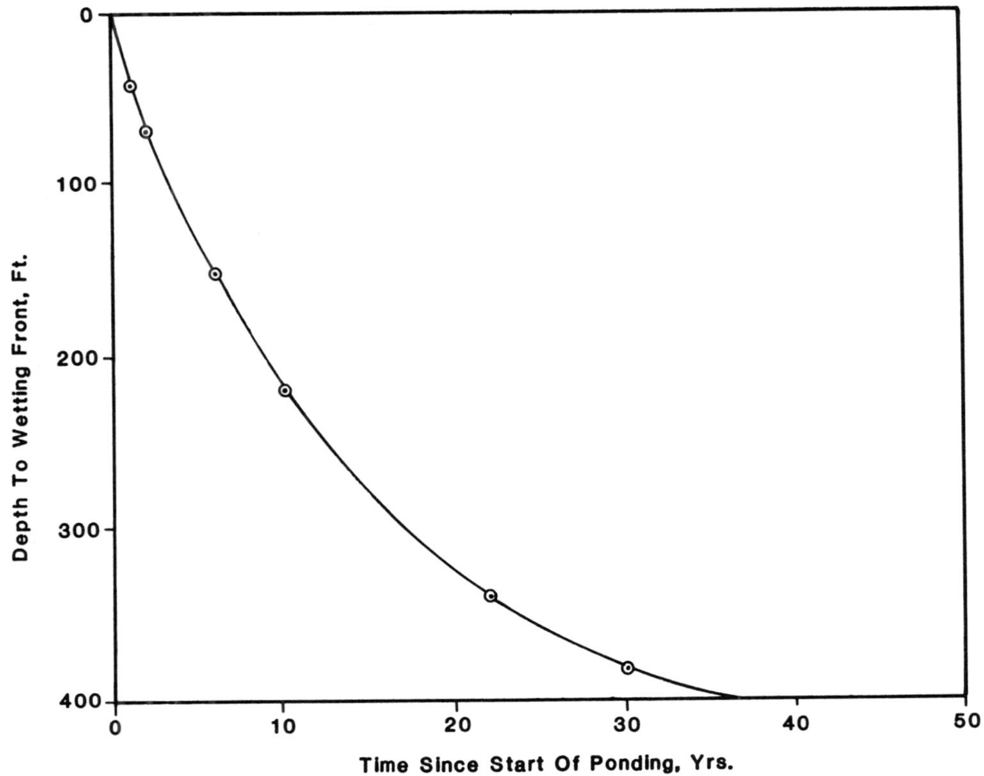


Figure 9. Predicted Advancement of the Wetting Front

the aquifer approximately forty years after ponding begins. The predicted location of the wetting front after 22 years of ponding is -340 feet. This is approximately 30 feet beyond the contaminant front found in the field investigation.

The analysis was repeated to see if elimination of the pond would stop the advancement of the wetting front before it reached the aquifer. The results of this analysis are shown in Figure 10. The wetting front continues to advance for an additional 18 years after ponding is ceased. The wetting front just reaches the aquifer as the maximum values of $S_r$ are obtained in the lower sand layer. Water flow then reverses direction and after 350 years the $S_r$ profile is close to the original steady-state case. It should be noted that the results for this part of the analysis probably contain significant error because hysterisis in soil properties was not taken into account. However, the results are useful in that a general idea is obtained about how the elimination of the pond affects the water migration. The model indicates that changes in the boundary condition at the ground surface has little immediate effect on flow at large depths. Rather, the effects are seen 10 to 20 years later.

The execution time for modeling 60 years of ponding on a CDC Cyber 170/750 computer was approximately 600 seconds. The time step ranged from 0.001 years at early times to 0.25 years at times beyond 30 years. The maximum tolerance used to achieve convergence was 10%. A more typical value for the tolerance used was 2%.

The major expense involved in performing the analysis was the calibration of the model. A record was not kept of the total amount of execution time used during this step but an estimate of 3,000 seconds would be close.

## AQUIFER IMPACT

A general assessment of aquifer impact can be obtained using the results of the one-dimensional studies and making some simplifying assumptions. After the wetting front reaches the aquifer, the model predicts the rate of infiltration to be 8 gallons per square foot per year. If flow is considered to be completely one-dimensional beneath the pond, the total seepage for a 48 acre pond would be 17 million gallons per year. The total volume of water contained in the aquifer directly beneath the pond is 470 million gallons. The seepage that occurs during the course of the year represents approximately 4% of this amount. The 4% will be diluted somewhat by the 8 million gallons of ground water that naturally flows horizontally through the aquifer in the area beneath the pond. However, because of the low rate of flow in the aquifer, the dilution of the contaminants seeping

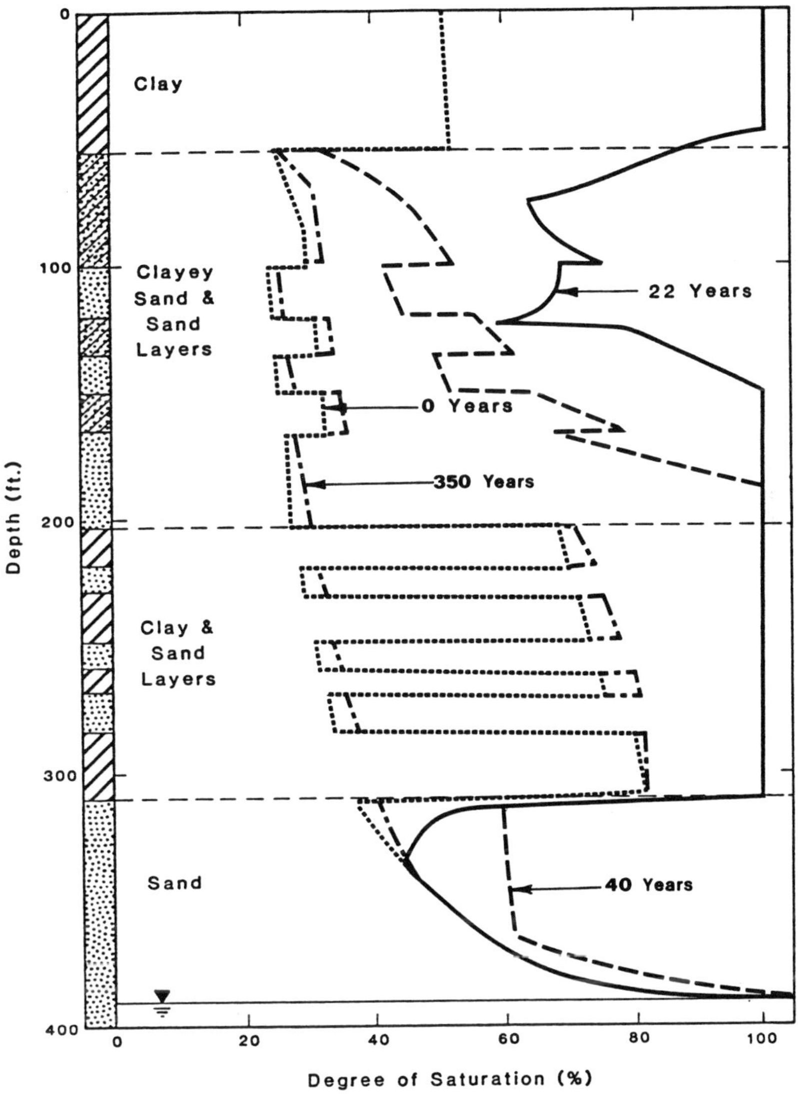


**Figure 10. Degree of Saturation Profile After Ponding is Eliminated**

into the aquifer will diminish with time. Also, a mound of contaminated water may eventually form at the level of the aquifer. While high concentrations of contaminants may develop in the aquifer beneath the pond, the regional down gradient movement in the aquifer is only 26 feet per year. The nearest user of water from the aquifer is several miles away, so the short-term impact is not considered to be significant.

During the course of the field studies, there were indications that extensive lateral migration of the perched pond water had occurred. Currently, both field and numerical studies are being conducted to determine the extent of the lateral migration.

## COMMENTS ON ADJUSTED PROPERTIES

A comparison between the measured values of saturated hydraulic conductivity ($k_s$) and the adjusted values used in the model are shown in Figure 7. Large differences exist between the measured values of $k_s$ and those required to obtain a good match with actual behavior. Discrepancies between laboratory measurements of $k_s$ and back-calculated field values has been noted by the authors on several other recent projects. The reasons for the discrepancies have been discussed [7] and occur mainly because the laboratory tests fail to account for macroscopic effects such as cracks and root holes that naturally occur in the field.

A graphical comparison of the various measurements of $k_s$ for the upper clay layer are shown in Figure 11. The plot illustrates the difficulty in selecting, for use in analyses, appropriate values of $k_s$ based on laboratory measurements. No only are laboratory measurements of $k_s$ significantly lower than field values in most cases, there is generally a wide range of values obtained. For this site, laboratory measurements of $k_s$ for the upper clay ranged over five orders of magnitude.

The importance of calibrating a model cannot be over emphasized. For instance, if the pond was just now being designed, a numerical model using laboratory measurements of $k_s$ for the clay in the upper zone would predict that the contaminant front from the pond would migrate only a few feet beyond the bottom of the pond after 20 years. In actuality, the contaminant front moved downward approximately 300 feet in 20 years.

The field permeability tests produced values of $k_s$ that were closer to the back-calculated values. This is expected since the field tests account for more of the macroscopic effects mentioned earlier.

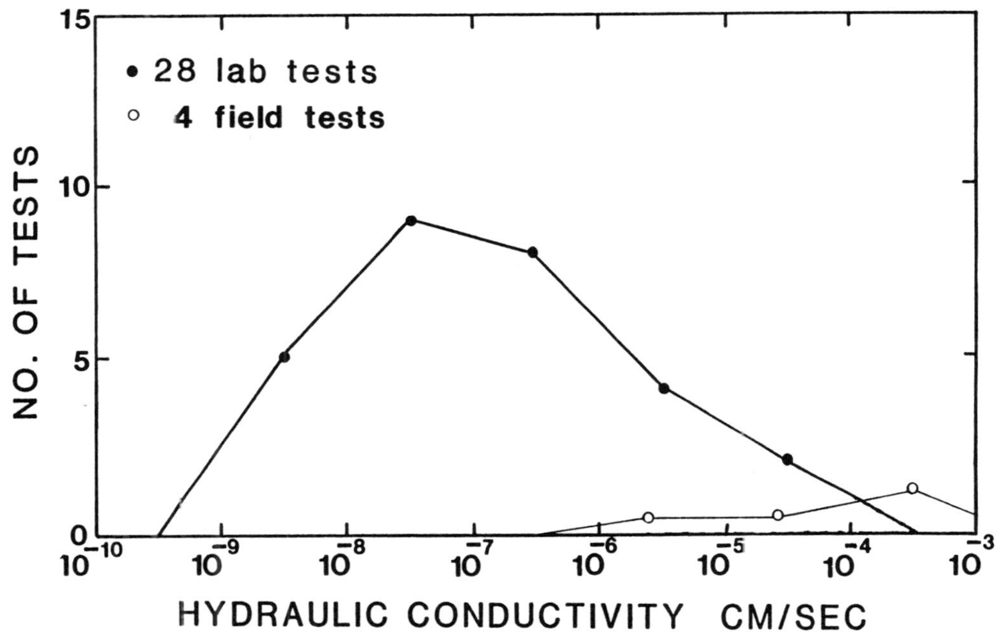


Figure 11. Comparison of Hydraulic Conductivity Measured in the Laboratory and in Field Tests for Upper Layer of Clay

## SUMMARY AND CONCLUSIONS

Seepage from the ponds occurred at a much faster rate than had been expected. This is in agreement with general experience obtained recently on other projects where the actual field value of hydraulic conductivity was an order of magnitude or more larger than values measured in the laboratory. Macroscopic features such as cracks, root holes, and fissures probably control the rates of seepage in the field, but these features are difficult to account for in laboratory tests.

The large differences between laboratory values of hydraulic conductivity and values back-calculated from the analyses emphasize the importance of calibrating numerical models with known field behavior.

This case history also provided the opportunity to determine if flow at a complex site could be simulated with a numerical model. The results obtained for the simplified problem seem reasonable even though there were significant uncertainties in some of the soil properties. The ability of the model to simulate two different known flow conditions increased the level of confidence in predictions obtained for future patterns of migration. The single most important parameter that influenced the predicted results was saturated hydraulic conductivity.

The effects of multi-dimensional flow are currently being investigated. The experience gained performing the one-dimensional studies should provide a useful background for addressing the problem of multi-dimensional flow.

## ACKNOWLEDGEMENTS

Mr. Charles Mauldin of Underground Resource Management, Inc. (URM) supervised the field studies described in this paper. Mr. David McMurtry (URM) assisted with the computer studies.

REFERENCES


1. Daniel, D. E. (1980), "Moisture Movement in Soils in the Vicinity of Waste Disposal Sites" Ph.D. Dissertation, The University of Texas at Austin, Austin, Texas.

2. Reeves, M., and J. O. Duguid (1975), "Water Movement through Saturated-Unsaturated Porous Media: A Finite Element Galerkin Model," Oak Ridge Natl. Lab., Oak Ridge, TN Rept. No. ORNL-4927, 232 p.

3. Neuman, S. P. (1975), "Galerkin Method of Analyzing Non-steady Flow in Saturated-Unsaturated Porous Media," in Finite Elements in Fluids, Vol. 1, Wiley, New York.

4. Narasimhan, T. N. (1975), "A Unified Numerical Model for Saturated-Unsaturated Groundwater Flow," Ph.D. Dissertation, Univ. of Calif., Berkeley, Calif.

5. Hamilton, J. M. (1979), "Measurement of Permeability of Partially-Saturated Soils," M. S. Thesis, Univ. of Texas, Austin, Texas.

6. Daniel, D. E. (1979), "Thermocouple Psychrometers for Measuring Suction in Unsaturated Soils," Univ. of Texas, Geot. Engr. Rept. GE79-3.

7. Hamilton, J. M., Daniel, D. E., and R. E. Olson (1979), "Measurement of Hydraulic Conductivity of Partially-Saturated Soils," Proc., ASTM Spec. Conf. on Permeability and Groundwater Contaminant Transport, in press.

8. Yong, R. N., and B. P. Warkentin (1975), Soil Properties and Behavior, Elsevier, Amsterdam, 449 p.

9. Olson, R E., and D. E. Daniel (1981), "Field and Laboratory Measurement of the Hydraulic Conductivity of Saturated and Partially Saturated Fine-Grained Soils," State-of-the-Art Paper, ASTM Symposium on Perm. and Groundwater Contam. Trans., Proc. in Press.

# A HYDROLOGIC STUDY OF THE UNSATURATED ZONE ADJACENT TO A RADIOACTIVE WASTE DISPOSAL SITE AT THE SAVANNAH RIVER PLANT, AIKEN, SOUTH CAROLINA

Paul Gruber, P.G.
Environmental Resources Management, Inc.
999 West Chester Pike
West Chester, Pennsylvania 19380

## INTRODUCTION

This paper was prepared as a description of work sponsored by the United States Energy Research and Development Administration through a Laboratory Graduate Participation Appointment awarded under its contract with Oak Ridge Associated Universities. Field work for this project was completed in August 1976 at the Savannah River Plant and Savannah River Laboratory, Aiken, South Carolina operated by E. I. duPont deNemours and Company, prime contractor with ERDA under Contract AT (07-2)-1. Laboratory studies were completed in January 1977 at the soil physics laboratories of the School of Forest Resources, University of Georgia, Athens.

Disposal facilities of low-level solid radioactive wastes in the form of landfills or burial grounds is a major concern to those generating these wastes as well as state compacts involved in the development of new sites. Regulations for the design, licensing, and operation of low-level radioactive waste disposal sites are currently being formulated by state and federal agencies to insure that these disposal sites are maintained properly and do not present health and safety hazards to communities surrounding them. A complete understanding of the interaction of radioactive materials with the surrounding geologic environment is necessary to design properly and safely new radioactive waste disposal sites and to upgrade substandard facilities currently in operation.

In order to monitor and predict the movement of radioactive nuclides in the environment, if leakage of radioactive materials from a disposal site should occur, data must be collected to meet specific objectives. These objectives include:

1. a definition of the geometry and geology of the disposal site;
2. an evaluation and description of the regional geologic and tectonic environment of the study area to include its lithology, stratigraphy, structure, and seismicity;

3. an evaluation and description of the regional ground water flow system underlying the disposal site including, ground water flow rates and direction, hydraulic conductivities, and recharge rates;
4. an evaluation of the relationship of ground water to surface water, and;
5. an evaluation of the nature of the interaction of radioactive nuclides with the soils at the disposal site.

In order to predict accurately ground water recharge rates and understand the interaction of radioactive nuclides with the soils, the hydrologic properties of the unsaturated zone above the ground water table should be determined. The hydraulic conductivity, hydraulic gradients, soil-water fluxes, and storage capacity of the soils under varying soil-moisture conditions in the unsaturated zone are important factors governing the rate of recharge into the underlying water-table aquifers. Additionally, water is an important medium for the transport of materials through the soil and will actively transport radioactive isotopes through the unsaturated zone to the saturated zone. The nature of soil-water properties in many existing solid radioactive waste disposal sites has been cited as lacking detailed and comprehensive study.

Two low-level radioactive waste disposal sites are currently in operation at the Savannah River Plant in Aiken, South Carolina (Figure 1). Solid radioactive wastes from plant operations, typically contaminated equipment, are buried in two centrally located disposal sites between the F and H production areas (Figure 2). The older of the two disposal sites is completely filled with waste. A newer and larger disposal site, adjacent to the first, is now in operation. Solid wastes contaminated with plutonium and other transuranium nuclides are encased in retrievable containers for possible removal to a long-term repository in the future. Wastes are buried in slit trenches and backfilled to grade. Each burial trench is approximately twelve feet wide with twelve feet of undisturbed soil separating each trench. Wastes are buried at a depth of twenty feet, nearly twenty feet above the elevation of the local water table.

Investigators at the Savannah River Plant have undertaken extensive studies and collected data designed to evaluate the geologic and hydrologic characteristics of the study area and determine the nature of the interaction of radioactive nuclides with the soils at these disposal sites, but few data are available on the movement of soil-water in the unsaturated zone above the water table.

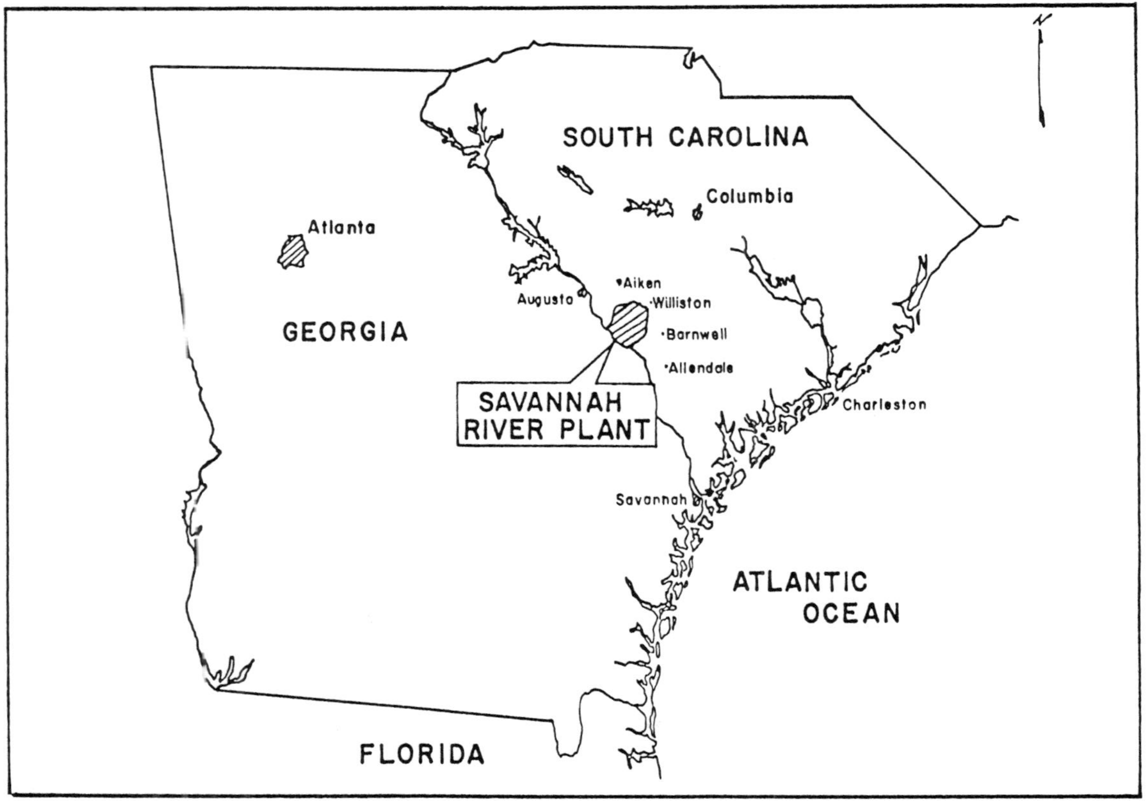


FIGURE 1. LOCATION MAP OF THE SAVANNAH RIVER PLANT.

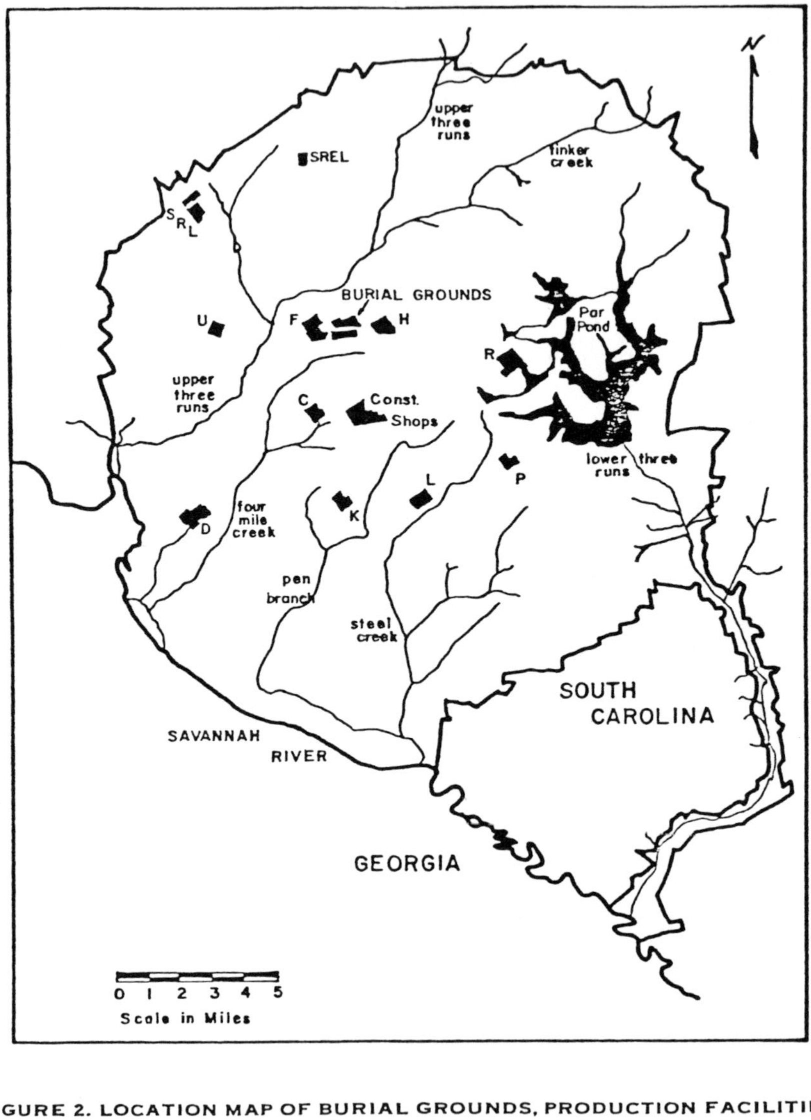


FIGURE 2. LOCATION MAP OF BURIAL GROUNDS, PRODUCTION FACILITIES, AND MAJOR STREAMS OF THE SAVANNAH RIVER PLANT.

## Purpose of Investigation

Experimental procedures, designed in conjunction with researchers at the Savannah River Plant, were aimed at characterizing the hydrologic nature of the unsaturated zone in order to prepare a more comprehensive model for the movement and transport of water at this disposal site. The ability to monitor and predict the movement of contaminants in the environment surrounding the disposal site will not be possible without such a model. The purpose of this investigation was to provide additional data and the conceptual framework necessary to understand the hydrologic character of the unsaturated zone in the vicinity of the low-level radioactive waste disposal site at the Savannah River Plant.

## Objectives of Study

In any radioactive waste disposal site containing slit trenches, the hydrologic properties of disturbed versus undisturbed areas will be different. Additionally, inhomogeneities in the characteristics and composition of undisturbed soils in the study area will also affect the hydrologic characteristics of the soil.

The objectives of this study were to furnish investigators at the Savannah River Laboratory with:

1. reasonable estimates of the hydrologic properties of the unsaturated zone, using accepted field and laboratory techniques;
2. an assessment of the magnitude of the spatial variability among these hydrologic properties in disturbed and undisturbed soils, and;
3. a comparison of field and laboratory estimates of hydraulic conductivity calculated for these soils.

The hydrologic properties investigated included saturated and unsaturated hydraulic conductivity, soil-water flux, soil-water storage, porosity, and grain-size distribution. Recommendations are to be made regarding which of these experimental and computational procedures can be most useful in similar ongoing studies at the Savannah River Plant.

## Previous Work

Few investigations regarding the nature of spatial variability of hydrologic properties of field soils, especially of disturbed versus undisturbed soils, have been made. The most extensive investigations of spatial variability of field soils were made by Nielson and others (1, 2). Klute (3) examined the various computational techniques currently available for determination of unsaturated hydraulic conductivity of field soils, as did Kunze (4). Mualem (5) proposed a simplified analytic method for predicting the unsaturated conductivity of soils by using the soil-water characteristic curve and measured values of the hydraulic conductivity at saturation. Mualem's model provides better agreement between the calculated and observed values of conductivity than several older computational techniques. The calculation of unsaturated conductivities in this study's laboratory investigations are based on his equations and assumptions.

## GEOLOGY OF THE SAVANNAH RIVER PLANT

The Savannah River Plant is located near the inland margin of the Atlantic Coastal Plain. The climate of the Plant is typical of the southeastern United States. Mean annual temperature, recorded by the U.S. Weather Bureau at Augusta, Georgia (on the western edge of the Plant), for the 86-year period ending in 1960, is 62.4°F.

Precipitation is normally abundant and evenly distributed throughout the year. Average annual precipitation, recorded at the Savannah River Plant during the 23-year period, 1952 through 1975, is approximately 48 inches. Normal monthly precipitation ranges from 2.21 inches in November to 5.10 inches in March, during the period of record.

The Savannah River Plant is located within the Savannah River drainage basin. The Savannah River flows southeastward and forms the western boundary of the Savannah River Plant. Principle tributaries to the Savannah River that drain the plant site are: Four Mile Creek, Pen Branch, Steel Creek, and Upper and Lower Three Runs (see Figure 2). Upper Three Runs, the largest tributary to the Savannah River, drains the major portion of the Savannah River Plant. All tributaries flow in a southwesterly direction, almost at right angles to the direction of flow of the main river channel, and have low stream gradients.

The Aiken Plateau, a physiographic subdivision of the Atlantic Coastal Plain, as characterized by Cooke (6), is the major landform of the plant site. The surface of the plateau is highly dissected with broad interfluvial areas sloping to the southeast at about eight feet per mile. Streams draining the Savannah River Plant have narrow, flat-bottomed channels which have cut steep-sided valleys into the plateau.

## Stratigraphy and Water-Bearing Characteristics of the Geologic Units in the Study Area

Late Cretaceous, Teritary, and Quaternary sediments form the sequence of consolidated and unconsolidated deposits underlying the Savannah River Plant. These sediments were deposited unconformably on a basement of igneous and metamorphic rocks of Precambrian and Paleozoic Age and sedimentary rocks of Triassic Age. Strike of the sediments is approximately N 60°E and dip is six to twenty feet per mile to the southeast. These sediments form a wedge-shaped mass increasing in thickness to more than 1,200 feet on the southeast or downdip side of the plant site (7).

Aquifers underlying the Savannah River Plant that are extensively used as a source of water are sands of the Cretaceous Ellenton and Tuscaloosa Formations. Recharge to these aquifers is directly from precipitation falling on the outcrop area north of the Plant and as downward leakage from overlying Tertiary and Quaternary aquifers. Recharge to these Tertiary and Quaternary aquifers is directly from precipitation. Discharge from all of these aquifers occurs principally as base flow of streams which flow through the outcrop, or are incised into these formations, and to wells which penetrate them.

Within the Savannah River Plant, ground water in the Ellenton and Tuscaloosa Formations occurs under artesian conditions. Ground water in Tertiary and Quaternary sediments occurs under water table conditions in their outcrop area and under artesian conditions downdip (7).

Radioactive wastes at the Savannah River Plant are buried above the water table in deposits of the Hawthorn and Barnwell Formations. These formations do not transmit water readily to wells and do not recharge appreciable amounts of water to underlying aquifers. However, if leakage should occur from these disposal sites, a complete understanding of the recharge-discharge relationship of these formations to the underlying aquifers and the rates and directions of soil-water movement within these formations are important considerations in predicting the movement of radioactive wastes into the ground water system. Figure 3 shows the areal distribution of the geologic units at the Savannah River Plant.

## Description of the Study Area

The study area selected for the field investigations is on the west side of the older burial ground, between F and H areas of the Savannah River Plant, and measures 40 feet by 20 feet (Figure 4). Tertiary sediments cropping out in the vicinity of the burial grounds include the Congaree (?),

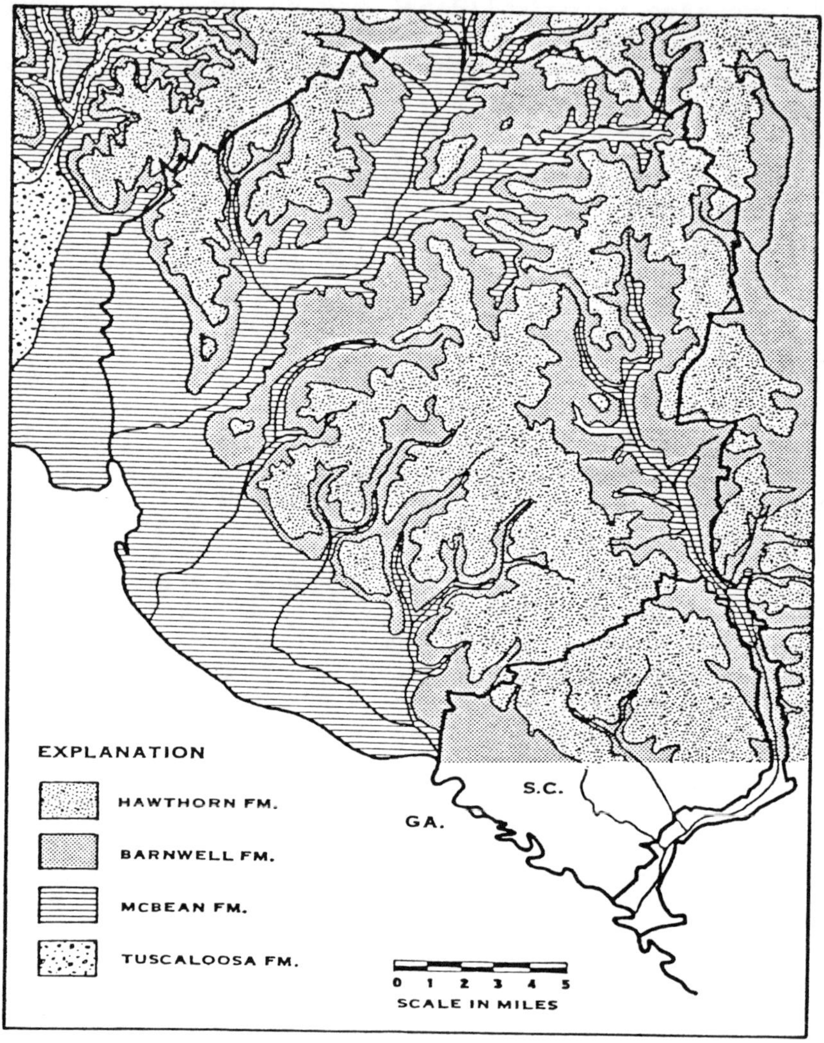


REFERENCE: MODIFIED FROM SIPLE, 1967

FIGURE 3. OUTCROP MAP OF SELECTED GEOLOGIC UNITS AT THE SAVANNAH RIVER PLANT.

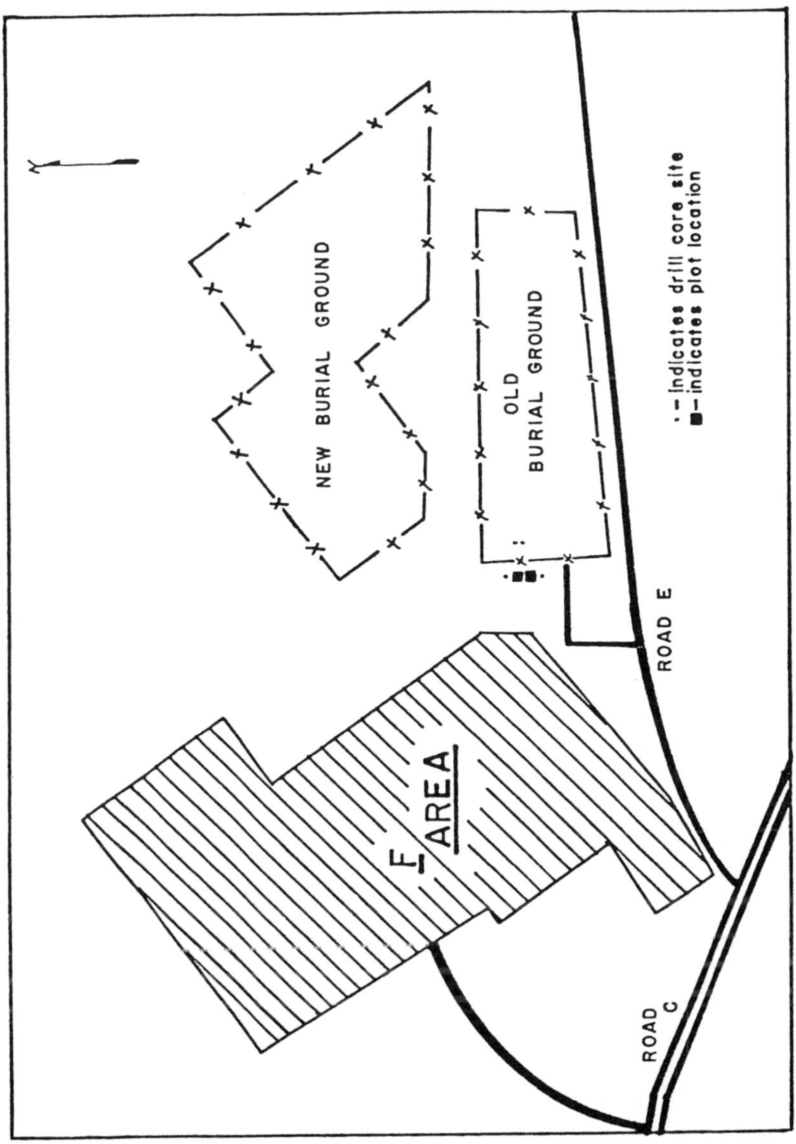


FIGURE 4. LOCATION MAP OF PLOTS IN RELATION TO BURIAL GROUNDS AND F AREA. DRILL CORE SITES ARE ALSO INDICATED ON MAP.

McBean, Barnwell, and Hawthorn Formations. Siple (7) characterized these formations as having poor water-transmitting properties, owing to their relative abundance of clay, silt, and fine-grained sand. These formations can be locally permeable and, in fact, between the F and H production areas, the McBean Formation yields moderate amounts of water to wells.

Directly overlying the Hawthorn Formation in the study area are late Teritary alluvial deposits which occur most frequently on interstream divides. Upper Three Runs and Four Mile Creek bound F and H areas and the interstream divide, or plateau, is composed mainly of coarse gravel and poorly sorted sand 20 to 40 feet thick. These deposits probably represent material reworked from the Tuscaloosa Formation and transported from its outcrop area northwest of the Savannah River Plant by the Savannah River and its tributaries.

Generally, the poorly sorted sand and gravel that constitutes the alluvial deposits are considerably above the water table and, consequently, they have little importance as a source of ground water. Nevertheless, these deposits are fairly permeable, and hence are capable of transmitting and storing water. The quantity and rates of water movement through these deposits are governed by the infiltration capacity of the soils at their interface with the atmosphere. Water from precipitation is absorbed by these deposits at the soil surface and is transported through the soil profile, either to plant roots or out of the profile and eventually into the ground water. It has already been stated that the underlying McBean Formation in the study area is capable of producing water. The possibility of radioactive materials being transported through the alluvial deposits into the ground water by percolating soil-water exists, and a quantification of the hydrologic properties of and types of soils overlying and in the vicinity of the burial grounds are important parameters necessary to estimate accurately the rate of ground water recharge in the study area.

## Description of the Soils in the Study Area

In 1963, the U.S. Forest Service completed a soil mapping survey of the Savannah River Plant. The Forest Service identified the soils in the vicinity of the burial grounds and classified them as belonging to the Dothan and Norfolk Soil Series. The dominant surface texture of these soils is sandy.

In order to assess more accurately the soil characteristics of the study area (an area less than the minimum mapping unit used by the U.S. Forest Service), a more detailed soil survey was conducted by a qualified soil scientist prior to the installation of testing equipment at the site. In addition to the soil survey, four, 15-foot continuous cores from the study area and the burial grounds were recovered and analyzed visually to determine the grain-size distribution of the soils and to assess the homogeneity and variability of soil characteristics in disturbed and undisturbed sites. Two cores were recovered from the burial ground, one from a burial trench site (disturbed site) and a second from a non-trench area (undisturbed site). Two additional cores were recovered from undisturbed sites on the west side of the old burial ground. Figure 4 shows the location of these continuous soil cores.

The core obtained from the burial trench area was markedly different from the remaining cores, exhibiting little soil profile development and more homogeneous textural and color characteristics than the cores from the undisturbed sites. The undisturbed soils were classified as belonging to the Ochlockonee Soil Series; a coarse loamy, siliceous, acid thermic family of Typic Udifluvients. These soils have brown sandy loam A horizons and stratified acid sandy loam C horizons.

## FLOW IN THE UNSATURATED ZONE

The cyclic movement of water in the field begins with its entry into the soil by the process of infiltration, continues with its temporary storage in the unsaturated soil, and ends with its removal from the soil by drainage, evaporation, or plant uptake.

Unsaturated flow processes are, in general, complicated and difficult to describe quantitatively, since they often entail changes in state and content of soil-water during flow. Such changes involve complex relations among the variable soil-water content, soil-water suction, and hydraulic conductivity.

### Saturated Flow

Water flow under saturated conditions in the soil can be treated as a steady-state phenomenon. The flux of water under these conditions is proportional to the gradient of water potential and the hydraulic conductivity. Darcy's law states that the flow of a liquid through a porous medium is in the direction of, and at a rate proportional to, the force acting upon the fluid (hydraulic gradient) and also proportional to the ability of the conducting medium to transmit the liquid (hydraulic conductivity).

## Unsaturated Flow

Under unsaturated flow conditions, negative potentials exist and an air phase is present which drastically modifies the flow characteristics. The forces causing flow in the unsaturated zone are now associated with the attractive forces of soil-particle surfaces for water and the surface tension in the air-water interface within the pores. With the introduction of an air phase in the porous system, the flow equation must be modified to take into account the fact that the cross section available for liquid flow is no longer equivalent to the total porosity (8).

The moving force of water in saturated soil is the gradient of a positive pressure potential. Water in an unsaturated soil is subject to a subatmospheric pressure, or suction, and the gradient of this suction now constitutes the driving force. The soil-water suction (attraction of the solid soil matrix to water) is due to the physical affinity of water to the soil-particle surfaces and the capillary pores. Water tends to flow from where suction is low to where it is high. When suction is uniform along a horizontal column of uniform soil, the column is at equilibrium and there is no soil-water movement. If a suction gradient exists, water will flow in pores which remain at the existing suction along hydration envelopes over particle surfaces until the potential is equilibrated.

The most important difference between unsaturated flow and saturated flow is the hydraulic conductivity. When soil is saturated, all of the pores are filled and conducting fluid, so that the conductivity is at its maximum. When soil becomes unsaturated, some of the pores become air-filled and the conductive portion of the soil's cross-sectional area decreases correspondingly.

Darcy's law, though originally conceived for saturated flow only, was modified by Richards (9) for unsaturated flow in three dimensions, with the provision that the conductivity is now a function of the soil-water suction head.

## Calculation of Unsaturated Hydraulic Conductivity

Several investigators have attempted to calculate the unsaturated conductivity of the soil using a pore-size distribution estimated from the soil-water characteristic curve. Childs and Collis-George (10) approached the problem from a statistical standpoint. Their method is based upon the probability of the continuity of the pores in adjacent places within the porous medium. Many investigators, most notably Marshall (11), Millington and Quirk (12), Nielson (13), and

Kunze (4) have modified the equation of Childs and Collis-George in an effort to improve the agreement between the calculated values of unsaturated conductivity with experimental data. Maulem (5) and earlier investigators devised a simple analytic model for predicting unsaturated conductivity by using the soil-water characteristic curve and the measured value of the hydraulic conductivity at saturation as a matching factor in his equation. It is similar to the Childs and Collis-George model but uses a modified assumption concerning the hydraulic conductivity of the pore sequence in order to take into account the effect of a larger pore section. The soil-water characteristic curve is divided into a larger number of pore classes than in the previous approaches. The equation used by Mualem for the calculation of the unsaturated conductivity is as follows:

$$K_r\ (\theta) = \left\{ S_e^{\beta} \sum_{i=1}^{n} \frac{[2(n-1) + 1]}{\psi_i^2} \right\} \Bigg/ \left\{ \sum_{i=1}^{m} \frac{[2(m-i) + 1]}{\psi_i^2} \right\}$$

m represents the total number of intervals into which the water content domain is divided, n is the number of intervals up to a prescribed value of $\theta$. By changing $\beta$ (a multiplier related to pore size distribution), the equation yields solutions for Childs and Collis-George if $\beta = 0$, Millington and Quirk if $\beta = 4/3$, and Kunze if $\beta = 1$.

The multiplier $\beta$, forces the predicted curve obtained from laboratory determined soil-water characteristics to fit the experimental or field data at one point. This means that at least one value of hydraulic conductivity must be known. The hydraulic conductivity of the saturated soil can be measured easily in the laboratory. The analytic solution is computed by matching K at saturation and is the method used in this study. Mualem's equation is a generalization of Kozeny's approach for saturated and unsaturated porous media, according to which $K_r$ is a power function of the effective saturation $S_e$.

## METHODS AND MATERIALS

In order to determine variability of hydrologic properties of soils in the vicinity of the burial grounds at the Savannah River Plant, field and laboratory experiments were designed and conducted in a manner similar to those described by Nielson and others (2).

Criteria used for selection of the study area included the following:

1. the site should be nearly level and near the burial grounds;
2. the soils at the site should be similar to the soils found within the burial grounds and have uniform profiles;
3. the site should be topographically higher than the surrounding area which will permit water to drain readily through the vertical soil profile and prevent re-infiltration of water from other sources after the field experiments have begun;
4. disturbance of the surface soils by human activity should be minimal; and
5. vegetal cover should be minimal.

No field experiments were conducted within the burial grounds. Undisturbed soil cores from burial trench and non-trench areas were recovered from the burial grounds and analyzed, along with undisturbed cores from the study area, in the laboratory.

The site selected for field investigations was on the west side of the older burial ground and is shown in Figure 4. Two 12-foot square plots were established on the surface of the site. Mercury tensiometers were placed in duplicate on opposite sides of each plot at depths of 12, 24, 36, 48, 60, 90, and 120 inches to monitor soil-water pressure changes during redistribution and initial wetting of the soil.

Narrow trenches, 36 inches deep, were excavated around the perimeter of each plot. These trenches were investigated for their soil characteristics before backfilling. Plywood sheets, four feet by twelve feet, were inserted in each trench after installation of thin polyethylene liner. In backfilling each trench, soil was compacted around the plywood sheets in order to insure that good contact was made with the walls of the trenches around the perimeter of the plots. This enclosure provided 12 inches of freeboard above the surface of each plot within which water could be ponded to any desired depth to initiate infiltration. The polyethylene liner and compaction of the soil in each plot around the plywood sheets prevented the leakage of water down the sides of the plot.

Prior to the installation of the field equipment, the vegetation covering the surface was destroyed by frequent applications of a herbicide, thereby limiting the disturbance of the soil surface and insuring the maintenance of the natural infiltration characteristics of the site. After removal of surface vegetation, four, four-inch diameter holes were drilled with a hollow-stem auger, two in each plot, to a depth of 12 feet. Each hole was encased with 12 feet of two-inch O.D. aluminum tubing, sealed at the bottom, to serve as access tubes for measurement of the in situ soil-water content by a neutron moisture probe.

## Steady-State Infiltration

Infiltration was initiated by ponding water on each plot until steady-state flow was established in the profile at all depths to 120 inches (i.e., until mercury levels in the tensiometers remained constant). Soils in the study area were dry prior to flooding and nearly ten days were required to reach steady-state conditions. The rate of subsidence of the ponded water defined the steady-state infiltration rate and was measured by a hook gauge.

After infiltration ceased, the surface of each plot was covered with six-mil black plastic to prevent evaporation of water from the soil. The plastic was covered by a light-colored tarpaulin to: (1) hold it in place, (2) prevent extreme temperature variations of the surface, (3) further reduce any evaporative losses, and (4) prevent subsequent movement of water in the vapor phase. Substantial movement of water in the vapor phase leads to errors in the calculation of unsaturated hydraulic conductivity (14). Each plot was also covered by a tent canopy which prevented rainwater from falling on the plots and disturbing the infiltration profile.

Tensiometer readings were taken hourly for the first 24 hours following infiltration and then less frequently as time passed for a period of 28 days. At the end of this period, tensiometer readings showed no changes and water movement through the profile had essentially ceased.

Concurrent with the measurements of soil-water pressure head throughout the profile by the tensiometers, the soil-water content of the soil was measured by a neutron moisture probe at depths of 12, 24, 36, 48, 60, 90, and 120 inches in each of the four aluminum access tubes. The probe consists primarily of a source of radiation and a detector. The detector is connected by a cable to a recording instrument or scaler. In order to obtain the soil-water content by volume of the tested material, the count ratio is matched to a calibration curve provided by the probe manufacturer (15).

## Core Collection

Two weeks after completion of the infiltration experiments, cores were collected from the field plots and the burial ground for laboratory analysis. Two undisturbed soil cores were collected from opposite sides of each plot at depths corresponding to the tensiometer depths. Duplicate cores were also collected from trench and non-trench areas within the burial ground.

Initially, a 4.5-inch O.D. bucket auger was used to remove the soil above each specified depth at which the soil cores were to be taken. Holes were cleared of all loose soil before samples were taken. A Soil Moisture Soil Core Sampler was used for extraction of the cores. The soil core sampler consisted of a one cm long brass ring at the top and bottom and two, three cm long brass rings in the center of the core barrel. The sampler was carefully lowered into the hole and inserted into the soil until the core barrel was filled. Upon removal of the filled core sampler, the entire core, still in its brass ring enclosures, was removed from the core barrel. Each brass cylinder, measuring three cm in length and 2.5 inches O.D. was carefully trimmed to remove excess soil from the top and bottom. A porous ceramic plate 0.25-inch thick, to be used in conjunction with the laboratory analysis, was placed on the bottom of each core, inserted into a plastic bag, sealed to prevent evaporation of soil-water, and labeled for shipment.

## Laboratory Investigations

Several separate investigations were performed on each core at the soils physics laboratory of the University of Georgia's School of Forest Resources. Each core removed from the field was slowly saturated over a period of several days so that disturbance of original pore geometry of each core was minimal. After complete saturation, each core was weighed and placed in a five-bar pressure plate extractor. Pressures of 0.1, 0.2, 0.33, 0.5, 0.75, and 1.0 bars were applied to the cores in the pressure plate extractor. Each core was weighed after equilibrium was reached in each successive pressure step (16).

Upon completion of the pressure plate analysis, all cores were resaturated. Water moves through the soil in response to various forces acting upon it. Among these are the pressure gradient, gravitational, adsorptive, and osmotic forces. Taking advantage of these forces, a constant head method for the determination of saturated conductivity in one-dimensional flow, designed by Klute (17), was utilized. The saturated conductivity of each core was calculated using Darcy's law.

After oven drying each core at 105°C temperature, the bulk density was calculated from the ratio of the saturated weight to the oven dried weight (18).

To complete the laboratory analysis of the cores, the particle-size distribution was obtained utilizing standard hydrometer techniques. Relative fractions of sand-, silt-, and clay-sized particles of the cores provided information and insights useful in the interpretation of the soil-water characteristic curves derived from the field and laboratory investigations (19).

## RESULTS AND CONCLUSIONS

Experimental procedures were designed to meet the specific objectives necessary to fully characterize the hydrologic nature of the unsaturated zone in the vicinity of the low-level radioactive waste disposal sites at the Savannah River Laboratory. Comparisons of detailed analytic laboratory methods and approximate field methods should indicate the applicability of utilizing a simplified field method for the determination of the hydrologic properties of the unsaturated zone.

### Particle-Size Distribution and Soil-Bulk Density

Values of percent sand, silt, and clay were obtained for each core at each depth of each plot and for cores removed at corresponding depths from the burial grounds. A total of 84 cores were recovered from the study area and analyzed by standard hydrometer techniques for particle-size distribution. The soil from the field plots and from the burial grounds are typically coarse sands with the percent of silt and clay decreasing with soil depth. Continuous cores from the field plots and the burial grounds were visually described. The soils appear to be very coarse-grained, poorly sorted, and generally loosely packed. Below a depth of approximately 120 to 130 inches the soil becomes finer grained and more densely packed, resulting in a decrease in permeability and porosity. The soil profile between 12 and 24 inches shows a noticeable increase in clay content which could influence the rate of soil-water movement through the entire profile. Similarity in soil texture and hydrologic character of the soils in the study area is evident after examination of the soil-water characteristic curves derived from the cores and is discussed in the following section.

Bulk density expresses the ratio of the mass to the bulk volume of soil particles plus pore spaces in a sample and is related to the total porosity of the soil. In coarse sandy soils which are loosely packed, the bulk density can be as high as 1.6 to 1.7 grams per cubic centimeter. As the pore volume decreases, grain size, total porosity, and bulk density also decrease (18). The average bulk density of the soils in the study area is 1.59 grams per cubic centimeter with a standard deviation of 0.76. These soils are fairly coarse-grained and homogeneous, with the most notable divergence from this trend being between depths of 12 to 24 inches, where the bulk density is markedly lower, owing to the increase in clay content of the soil.

## Soil-Water Characteristic

The relationship between the water content of soil and the soil-water pressure or suction, with which that water is held, can be graphed as the soil-water characteristic curve and may be interpreted as being related to pore-size distribution within the soil (20). Undisturbed soil cores from the field plots and burial grounds were analyzed in the laboratory to derive a soil-water characteristic curve for each depth over the profile tested. The mean and standard deviation of the soil-water contents obtained from the pressure plate analysis were tabulated. Figure 5 gives the average of soil-water characteristic curves obtained from the field plot cores and their relationship to the cores from the trench and non-trench sites within the burial ground. Each curve resembles the others inasmuch as they have the same general shape and differ only by the magnitude of the soil-water content. The soil-water characteristics of the cores from the burial ground follow the same trend as the field cores, with little variation in shape or magnitude. This indicates that, although the soils within the burial ground were disturbed to varying degrees, the pore-size distribution of those soils, and thus their hydrologic properties, do not change markedly from the undisturbed field soils as a result of this disturbance.

Water contents at saturation range from 0.47 to 0.52 $cm^3/cm^3$, while those at a pressure head of -75.0 mm (-1.0 bars) range between 0.05 and 0.10 $cm^3/cm^3$. Values of the standard deviation of the mean generally decrease with depth and soil-water pressure, which corresponds to increasing homogeneity of the soil and decreasing clay content with depth.

From the shape of the soil-water characteristic curve, it can be inferred, according to Hillel (21), that the soils in the study area are sandy with only minor amounts of clay. Soil-water is held in large pores of the sandy soil until the suction increases and then the soil drains rapidly over a short range of suction values. Childs (20) proposed the use of soil-water characteristic curves as a way of interpreting the pore-size distribution within the soil and it has been shown by other investigators, most notably Millington and Quirk (12), Childs and Collis-George (10), and Kunze (4), that a relationship between the shape of the curve and pore-size distribution does exist. A reliable prediction of unsaturated conductivity, from basic soil properties, has been proposed by Gardner and others (22), but no fundamentally based equations of general validity are available for relating unsaturated hydraulic conductivity to soil-water content or to soil-water pressure, since the soil-water characteristic curve is

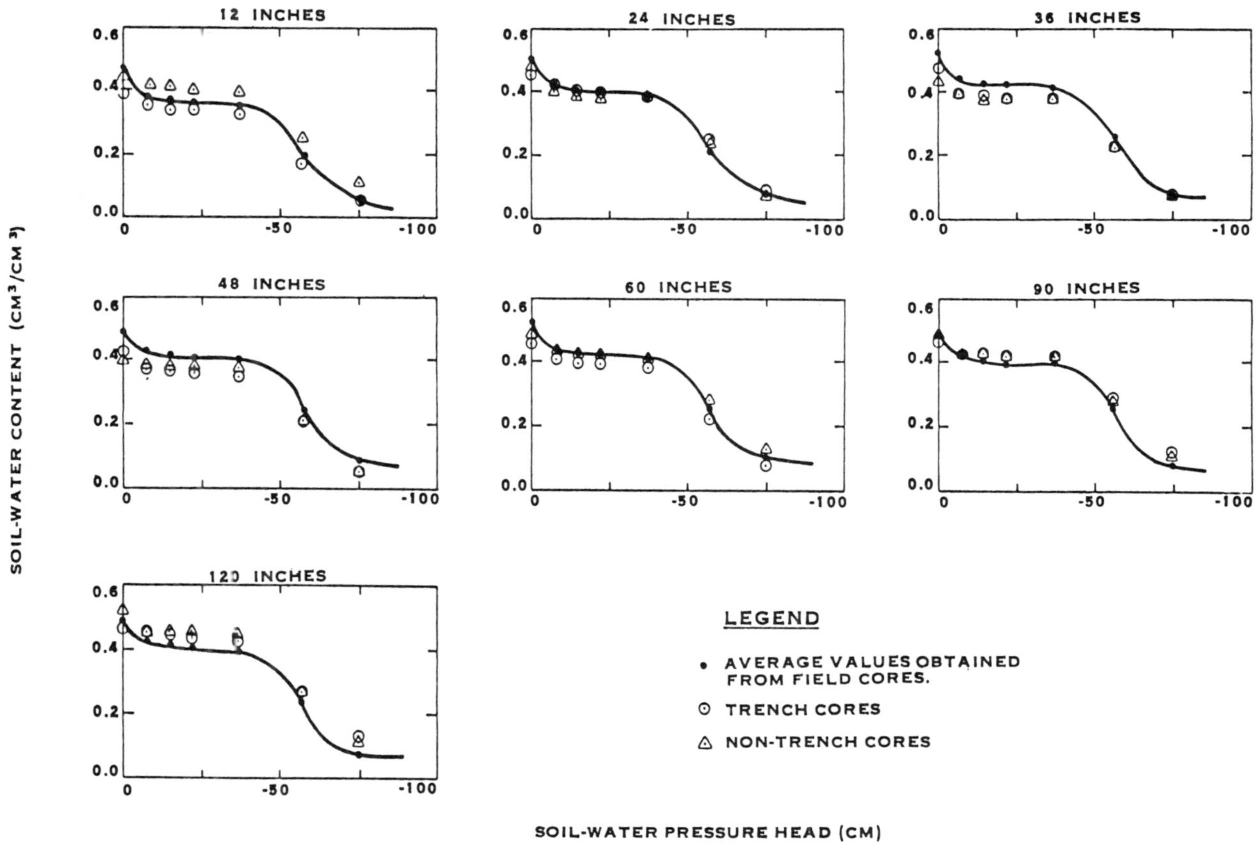


FIGURE 5. LABORATORY DERIVED SOIL-WATER CHARACTERISTIC CURVES FOR THE SOIL CORES.

strongly affected by soil texture and structure. A statistical method can produce reliable estimates of hydraulic conductivity derived in the laboratory from soil-water characteristic curves.

## Soil-Water Storage

Soil-water storage is the term used to describe the amount of water temporarily retained in the soil profile above a particular depth at any time (2). Its implications are most important in field soils in relation to crop production and retention of water in the producing root zone, but soil-water storage can also be an important consideration in siting and designing landfills and low-level radioactive waste burial grounds. If substantial quantities of water remains in storage in the soil profile for an extended period of time, during periods of heavy rainfall, recharge to water table aquifers will be high. As a result of increasing recharge, the elevation of the water table will rise and can possibly impair the integrity of the storage facility by having waste in contact with or very close to the water table. Therefore, it is important to understand the storage characteristics and soil-water fluxes (rate of water movement) of soils in the vicinity of the Savannah River Plant burial grounds in order to evaluate the recharge characteristics of the soils overlying the local water table aquifer.

The values of soil-water content with depth and the mean values of soil-water storage, soil-water content, percent saturation, and hydraulic conductivity for each depth in the field over time for the period during which the field studies were ongoing were calculated.

Figure 6 shows that the amount of water in the profile decreases with time for depths 12 through 60 inches. Soil-water storage, for depths 90 and 120 inches, after an initial decline for the first 15 days, increases following steady-state infiltration. Recalling the soil core descriptions of the continuous cores, the soil below 130 inches tends to become finer grained and more compact, resulting in a reduction in porosity and hydraulic conductivity. As the water drains through the coarse sands above 130 inches, a discontinuity boundary is encountered in the finer grained sediments (less permeable to water) and the water transmitted through the more permeable soil begins to accumulate above the fine-grained layer. A coarse-grained soil has a high saturated conductivity, but it ceases to transmit water at relatively low suctions. Thus, water content and hydraulic conductivity also increase at depths of 90 and 120 inches after 15 days of drainage. If enough water accumulates above this discontinuity, positive hydraulic pressures may occur and be expressed

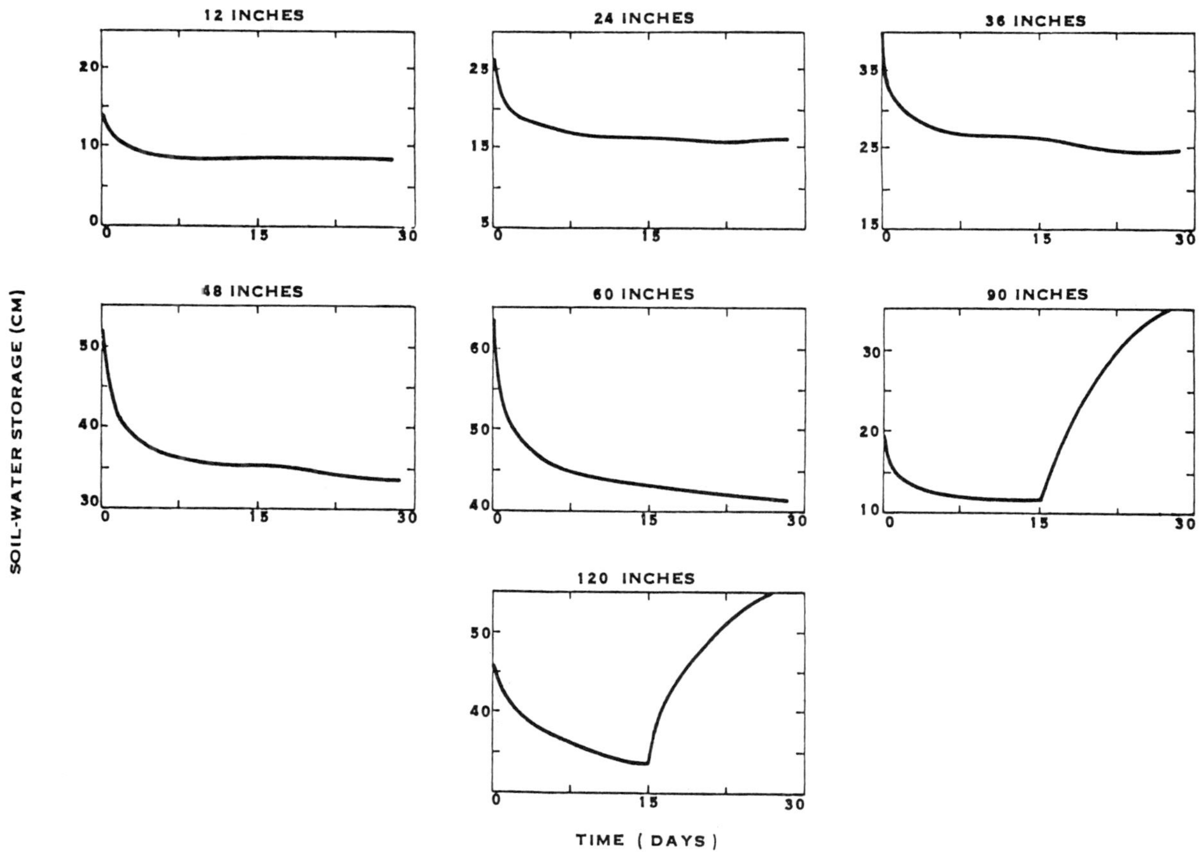


FIGURE 6. MEASURED VALUES AND SOIL-WATER STORAGE AS A FUNCTION OF TIME AFTER STEADY-STATE INFILTRATION HAS CEASED. VALUES GIVEN ARE THE MEAN OF ALL PLOTS.

as a perched water table, which results in a decrease in recharge to the underlying water table aquifer and reduces the possibility of the local water table being in direct contact with the waste materials found in the Savannah River Plant burial grounds.

Corresponding to the above changes in storage, the soil-water flux decreases with depth and time following steady-state infiltration. As can be seen from Figure 7, the water content decreases with depth and time following infiltration. Soil-water flux is the amount of water represented by the area between the lines indicating days. The soil-water content at the 36-inch depth increases following infiltration as a result of the increase in clay content at depths between 12 and 24 inches. This increase in water content at the 36-inch depth relates to the fact that when a coarse layer of soil exists below a finer grained layer, the coarser layer will not transmit significant amounts of water until many of the pores are filled with water. The movement of water from the coarse layer into the fine layer will occur only at suctions much lower than those at which the pores in the soil above are filled. As a result, soil-water content increases and soil-water flux decreases in the underlying coarse layer.

## Hydraulic Conductivity

According to Nielson (2), who did extensive statistical studies on the relationship of hydraulic conductivity to water content, hydraulic conductivity of field soils is more meaningfully correlated to percentage of saturation. Therefore, evaluation of spatial variability of hydraulic conductivity due to variation of soil-water characteristics can be more readily analyzed. Figure 8 shows the relationship of hydraulic conductivity to percentage of saturation for each depth and represents the mean values obtained from a calculation of the hydraulic conductivity by a computer program called SOIL. The hydraulic conductivity multiplied by the hydraulic gradient yields the soil-water flux. This is the basis for the calculation of hydraulic conductivity, upon which depends the measurement of the steady-state infiltration rate at the soil surface and the rate of change of soil-water content in the profile. These measurements are the inputs to the program.

As can be expected, Figure 8 clearly shows that unsaturated hydraulic conductivity decreases with percentage of saturation for each depth and range from a high of 2.44 centimeters per day (cm/d) at a depth of 60 inches and 77 percent saturation and a low of 0.0004 cm/d at a depth of 12 inches and 59 percent saturation. It also appears that the standard deviation in hydraulic conductivity decreases with

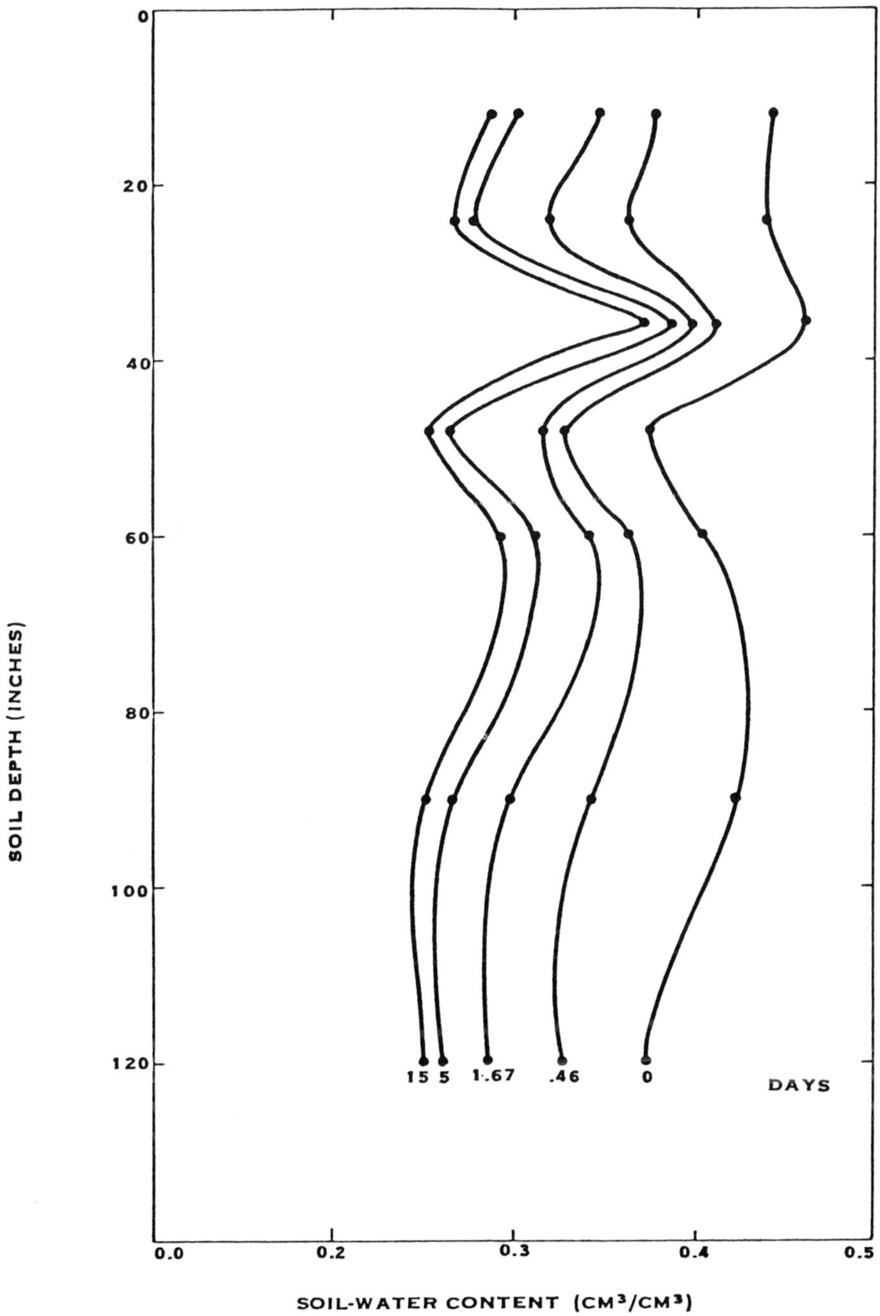


FIGURE 7. SOIL-WATER CONTENT PROFILES FOR THE STUDY AREA FOLLOWING THE CESSATION OF STEADY-STATE INFILTRATION.

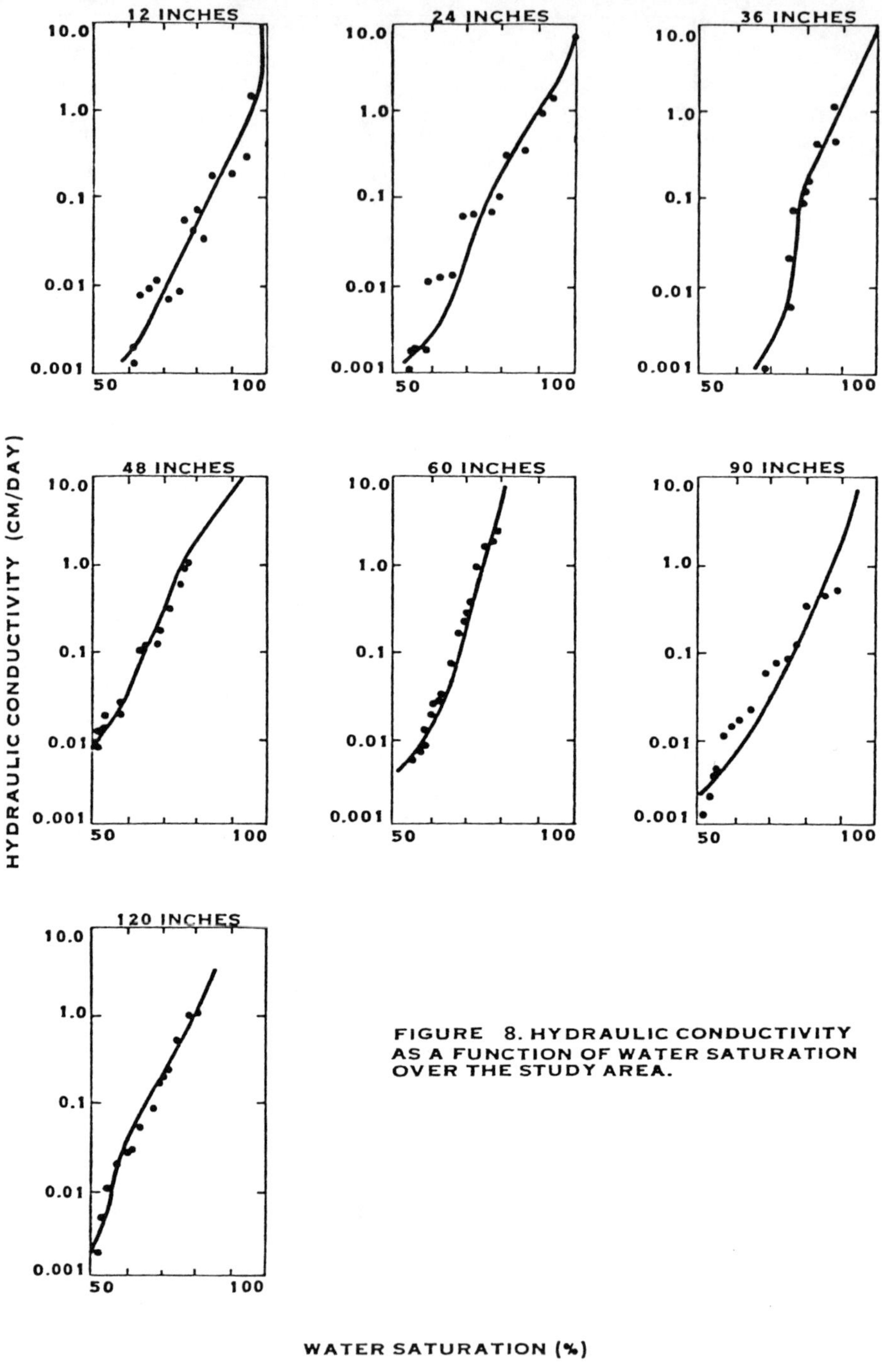


FIGURE 8. HYDRAULIC CONDUCTIVITY AS A FUNCTION OF WATER SATURATION OVER THE STUDY AREA.

depth owing to the greater homogeneity in soil characteristics. Figure 9 is the average hydraulic conductivity expressed as a function of soil-water content, from saturation, over the entire study area, and also shows that the study area has a fairly uniform profile. Figures 8 and 9 show good correlation of hydraulic conductivity with percentage saturation and soil-water content and it appears from these graphs that the hydraulic conductivity of the study area is fairly uniform with depth and percentage water saturation owing to the homogeneity of the soil profile in the study area.

Hydraulic conductivity calculated by the method devised by Mualem (5) attempts to predict the conductivity of a soil from its pore-size distribution estimated from the soil-water characteristic curve derived in the laboratory. Calculation of hydraulic conductivity using Mualem's modifications of the equations of Millington and Quirk, Childs and Collis-George, and Kunze were made for each soil core. Eighty-four undisturbed soil cores from the field plots and the burial grounds were analyzed by the pressure plate technique and soil-water characteristic curves derived are shown in Figure 5. Hydraulic conductivity values obtained by the methods of Childs and Collis-George, Millington and Quirk, and Kunze were nearly identical. A computer program called HYDRO was used to calculate the hydraulic conductivity from the soil-water characteristic. Copies of the output from programs SOIL and HYDRO are available for inspection and are on file with the Department of Geology, University of Georgia, the Savannah River Laboratory, and the author. The hydraulic conductivity of the field cores, the trench site, and non-trench site were calculated by program HYDRO using the method of Kunze, with a matching factor of 1.0 and the pore classes of the soil-water characteristic divided into 70 divisions. Figures 10A and 10B show the relationship of field versus laboratory-derived hydraulic conductivity, matched at saturation and percentage saturation, for the field, trench, and non-trench cores, over the entire profile. At shallower depths, the agreement between field and laboratory hydraulic conductivity for all cores is reasonably good. At 90 and 120 inches, however, the laboratory-derived hydraulic conductivity is consistently higher than the field hydraulic conductivity, especially at lower conductivity and water content values. This phenomenon is not related to any failure in the ability of the model to predict unsaturated hydraulic conductivity from the soil-water characteristic, but the disagreement in values is probably related to the disturbance of the soil cores at these depths. There undoubtedly was compaction of the cores upon retrieval which reduced the permeability and porosity of the cores. At lower suction values, the disturbed cores retained more moisture and hence yielded higher conductivity values than the field conductivity values.

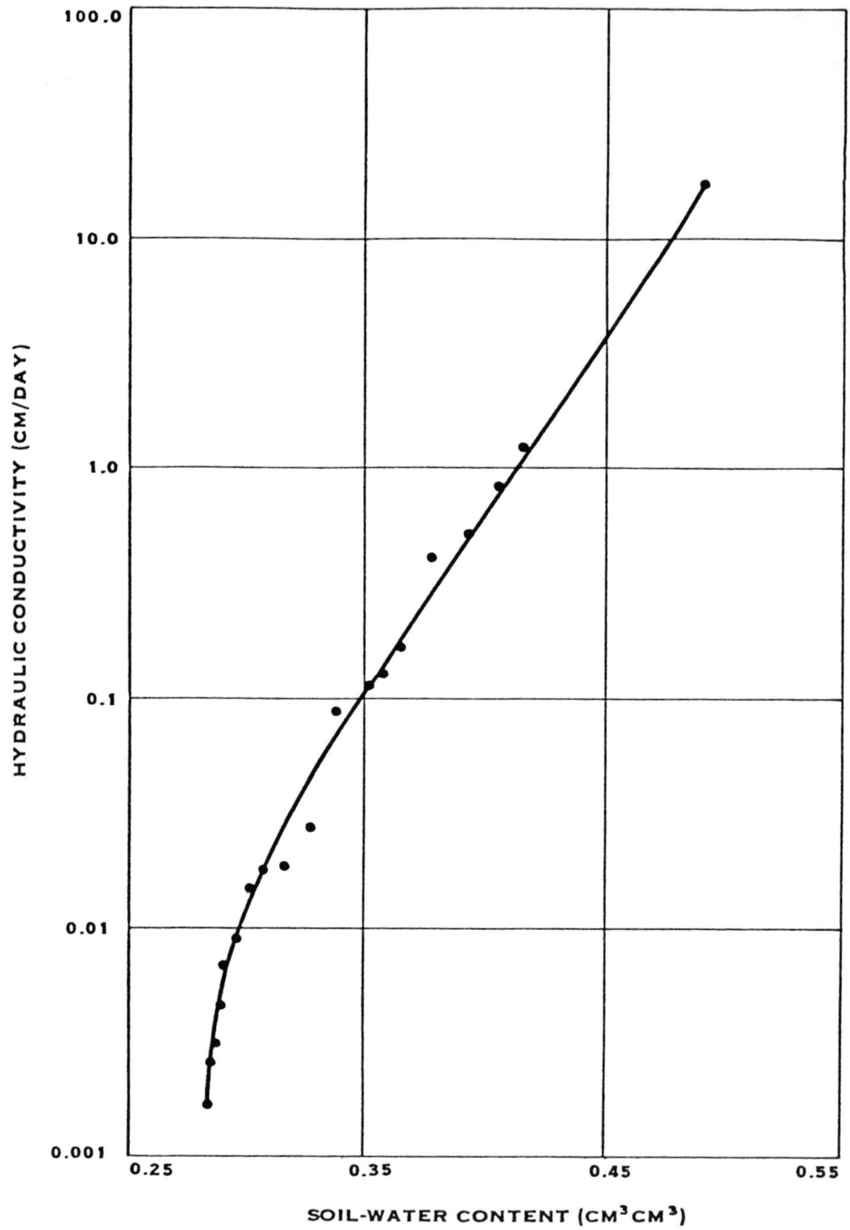


FIGURE 9. AVERAGE HYDRAULIC CONDUCTIVITY AS A FUNCTION OF SOIL-WATER CONTENT FOR THE ENTIRE STUDY AREA ASSUMING A HOMOGENEOUS SOIL TO A DEPTH OF 120 INCHES

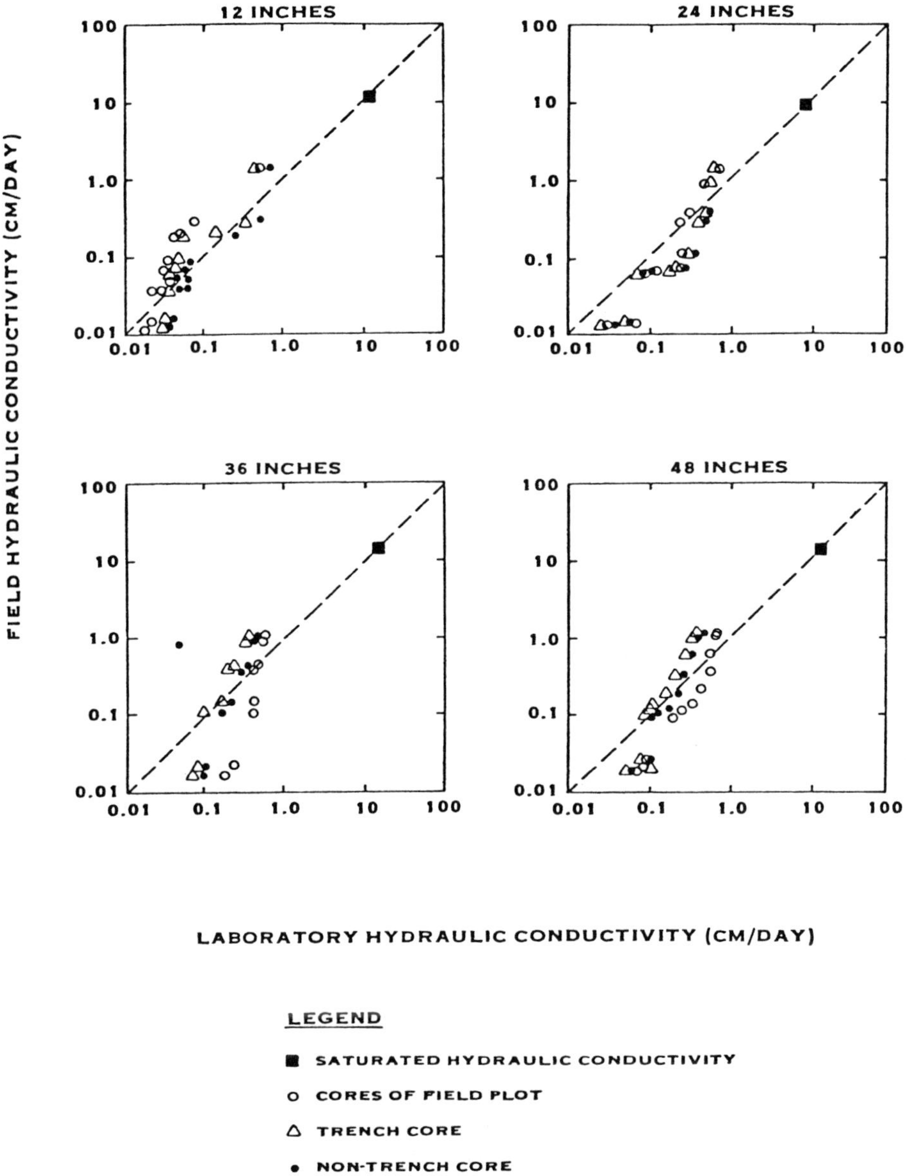


FIGURE 10A. FIELD VERSUS LABORATORY DERIVED HYDRAULIC CONDUCTIVITY. VALUES ARE MATCHED AT THE SATURATED HYDRAULIC CONDUCTIVITY VALUE.

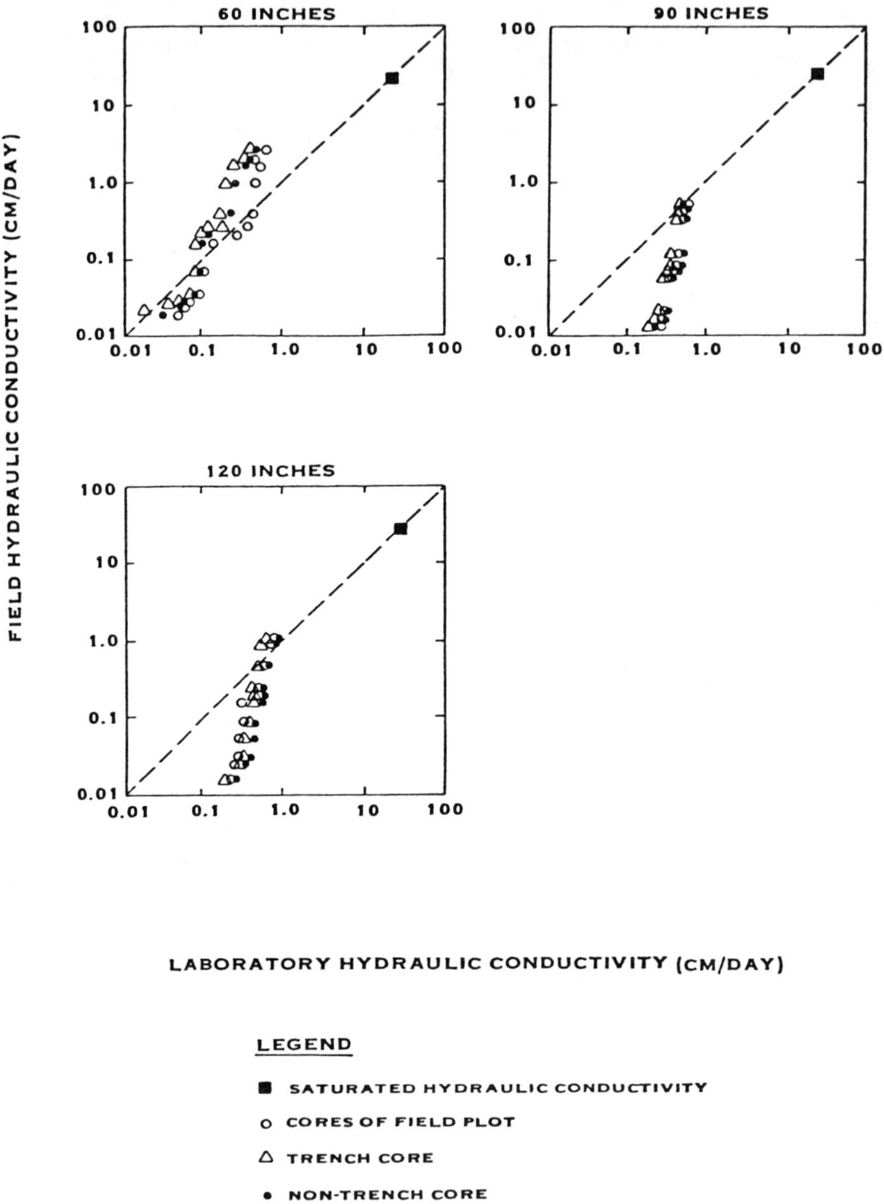


FIGURE 10B. FIELD VERSUS LABORATORY DERIVED HYDRAULIC CONDUCTIVITY. VALUES ARE MATCHED AT THE SATURATED HYDRAULIC CONDUCTIVITY VALUE.

## Summary and Recommendations

Several observations can be made about the spatial variability of the physical properties of the soil in the study area and the burial grounds. Two distinctive, well defined horizons exist in the soils in the study area at depths of 12 to 24 inches and below 130 inches. These clay-rich and compacted horizons inhibit the movement of soil-water. Throughout the rest of the profile the particle-size distribution of the soil and its bulk density is relatively homogeneous. Cores from within the burial grounds recovered for analysis also showed textural and hydrologic characteristics similar to the cores from the undisturbed field plots, as evidenced from similar soil-water characteristic curves derived for all cores and similar textural character as determined from particle-size analysis. In the non-trench areas, some disturbance of the soil is evident in the upper 24 inches of the profile. Below that depth horizonation and development was not disturbed. Owing to the coarse texture and loosely structured character of the soils, even in the trench areas where disturbance was at its maximum, the hydrologic characteristics of the soils were similar to those of the soils in the study area. The small variation in the steady infiltration rate of the field plots is further evidence of the homogeneity of the soils in the study area.

Hydraulic conductivity graphed as a function of percent saturation decreases with depth throughout the profile. If the soil profile is treated as a homogeneous soil mass, the spatial variation in hydraulic conductivity also decreases with soil depth and water content over the study area. Even though the variations in water content and hydraulic conductivity vary throughout the study area with depth, soils in the study area are fairly homogeneous. Simplified field methods for measuring field values of hydraulic conductivity, using a single or preferably two tensiometers placed above and below soil textural discontinuities and a reliable soil-water characteristic curve, can be sufficiently accurate for characterizing field conditions. It would be desirable to establish a number of field plots over a larger study area to further examine the statistical distribution of hydraulic conductivity.

Soil-water flux decreases with depth and time following steady-state infiltration, and soil-water storage at depths exceeding 90 inches increases with depth and time after an initial period of drainage. Reduction of soil-water flux coupled with increasing soil-water storage over depth and time are favorable soil hydrologic properties for disposal sites with material buried at depths below the wetting front. This relationship implies that during periods of heavy rainfall, soil-water storage in the unsaturated zone will increase, thereby reducing the potential for rapid recharge of the underlying water table aquifer. Since the water table elevation is not subject to rapid changes and its elevation remains fairly constant through time, the area chosen for the low-level radioactive waste facility at the Savannah River Plant is a suitable site from the standpoint of soil-water movement and ground water recharge.

Statistical methods available in the literature for predicting hydraulic conductivity from the soil-water characteristic provide adequate and reliable estimates of soil hydrologic parameters. The most important measurements for predicting soil-water behavior in the field are the soil-water characteristic curve and a steady-state hydraulic conductivity. These measurements, along with the simplified field approach, should provide sufficient information from which to predict the behavior of soil-water under field conditions throughout the burial grounds and the area surrounding them.

From the undisturbed sites, two well developed clay horizons were found to be present at 36 and 120 inches and are especially important in governing the rate and amount of water moving through the soil profile. In the disturbed areas of the burial grounds, however, this horizonation was destroyed by trenching during the burial operation. Although the general soil hydrologic properties did not appear to be appreciably altered, the destruction of horizonation where wastes are buried could change the rate of soil-water flux and amount of soil-water storage in the trench areas. Investigations should be conducted to determine the nature and magnitude of the differences in soil-water flux and soil-water storage that resulted from burial operations. Unless precautions were taken to recompact the material in the trenches, removal of horizonation can result in creation of efficient pathways for water movement into the trenches. To date, no reports of elevated water levels below the trenches have been made and, apparently, under natural rainfall conditions, the removal of horizonation in the trenches has not created any problems. Similarity of hydrologic properties of the soil, most notably hydraulic conductivity and steady-state infiltration rate, in disturbed and undisturbed areas also results in similar soil-water redistribution profiles.

Further investigations should be conducted to evaluate the hydrologic properties of the undisturbed soils in other areas around the burial ground to obtain a more statistically valid sample. A simplified method for measuring field values of hydraulic conductivity of the well-defined soil horizons which govern soil-water movement should be made using two tensiometers and a reliable soil-water characteristic and are sufficiently accurate for characterizing field conditions. Although variations in soil texture, bulk density, and hydraulic conductivity are minimal, for disturbed and undisturbed soils in the study area, the ability to make predictions of the hydrologic properties of the soil over a large area from a single site can range from good to unsatisfactory, depending on the particular prediction parameter of interest. Therefore, a larger sample set of field measurements should be made.

The most important laboratory measurements for predicting soil-water behavior in the field are the soil-water characteristic and a steady-state hydraulic conductivity value. These measurements provide ample information with which to approximately predict the behavior of soil-water under field conditions and eliminate the need for extensive field investigations in the burial ground.

A simplified field method of investigating undisturbed soils with well-defined soil horizons, coupled with laboratory measurements of cores from the disturbed burial ground sites should yield information statistically adequate to fully evaluate the behavior of soil-water in the burial ground in disturbed and undisturbed soil.

## REFERENCES

1. Nielson, D. R., et al., 1964, Water movement through Panoche clay loam soil: Hilgardia, Vol. 35, No. 17, pp. 491-506.

2. Nielson, D. R., Biggar, J. W., and Erh, K. T., 1973, Spatial variability of field-measured soil-water properties: Hilgardia, Vol. 41, No. 7, pp. 215-259.

3. Klute, A., 1972, The determination of the hydraulic conductivity and diffusivity of unsaturated soils: Soil Sci., Vol. 113, No. 4, pp. 264-276.

4. Kunze, R. J., Vehara, G., and Graham, K., 1968, Factors important in the calculation of hydraulic conductivity: Soil Sci. Amer. Proc., Vol. 32, pp. 760-765.

5. Mualem, Y., 1976, A new model for predicting the hydraulic conductivity of unsaturated porous media: Water Resources Res., Vol. 12, No. 3, pp. 513-522.

6. Cooke, C. W., 1936, Geology of the coastal plain of South Carolina: U.S. Geol.. Surv. Bull. 867, p. 196.

7. Siple, G. E., 1967, Geology and ground water resources of the Savannah River Plant and vicinity, South Carolina: U.S. Geol. Surv. Water Supply Paper 1841, p. 110.

8. Baver, L. D., Gardner, W. H., and Gardner, W. R., 1972, Soil Physics: 4th ed., John Wiley and Sons, New York, p. 498.

9. Richards, L. A., 1931, Capillary conduction of liquids in porous mediums: Physics, Vol. 1, pp. 215-259.

10. Childs, E. C. and Collis-George, N., 1950, The permeability of porous materials: Proc. Roy. Soc., Vol. 201A, pp. 392-405.

11. Marshall, T. J., 1958, A relation between permeability and size distribution of pores: J. Soil Sci., Vol. 9, pp. 1-8.

12. Millington, R. J. and Quirk, J. P., 1959, Permeability of porous media: Nature, Vol. 183, pp. 387-388.

13. Nielson, D. R., Kirkman, D., and van Wijk, W. R., 1961, Diffusion equation calculations of field soil-water infiltration profiles: Soil Sci. Soc. Amer. Proc., Vol. 25, pp. 165-168.

14. Richards, L. A., Gardner, W. R., and Ogata, G., 1956, Physical processes determining water loss from soil: Soil Sci. Soc. Amer. Proc., Vol. 20, pp. 310-314.

15. Holmes, J. W., 1950, Calibration and field use of the neutron scattering method of measuring soil-water content: Aust. J. Appl. Sci., Vol. 7. pp. 45-58.

16. U.S. Salinity Laboratory Staff, 1954, Diagnosis and improvement of saline and alkali soils: U.S. Dept. Agr. Handbook 60.

17. Klute, A., 1969, Laboratory measurement of hydraulic conductivity of saturated soil, in Methods of Soil Analysis, pt. 1: Amer. Soc. Agron. Monograph 9, pp. 210-222.

18. Blake, G. R., 1969, Bulk density, in Methods of Soil Analysis, pt. 1: Amer. Soc. Agron. Monograph 9, pp. 374-391.

19. Day, P. R., 1969, Particle fractionation and particle size analysis, in Methods of Soil Analysis, pt. 1: Amer. Soc. Agron. Monograph 9, pp. 545-568.

20. Childs, E. C., 1940, The use of soil-moisture characteristics in soil studies: Soil Sci., Vol. 50, pp. 239-252.

21. Hillel, Daniel, 1971, Soil and Water: Academic Press, New York, p. 288.

22. Gardner, W. R., 1960, Dynamic aspects of water availability to plants: Soil Sci., Vol. 89, pp. 63-73.

# GEOLOGIC CHARACTER OF TUFFS IN THE UNSATURATED ZONE AT YUCCA MOUNTAIN, SOUTHERN NEVADA

Robert B. Scott
Richard W. Spengler
Sharon Diehl
U.S. Geological Survey, MS 954
Denver Federal Center
Denver, CO 80225
A. R. Lappin
Sandia National Laboratories
Box 5800
Albuquerque, NM 87185
Michael P. Chornack
Fenix & Scisson
Box 498
Mercury, NV 89023

ABSTRACT

At Yucca Mountain, a potential site for a high-level nuclear waste repository on the Nevada Test Site in southern Nevada, evaluation of the geologic setting and rock physical properties, along with previous regional hydrologic studies, has provided background that can be used for construction of a preliminary conceptual hydrologic model of the unsaturated zone.

The 500-m-thick unsaturated portion of Yucca Mountain consists of alternating layers of two contrasting types of tuff. One type consists of highly fractured, densely welded, relatively nonporous but highly transmissive ash-flow tuffs. The other type consists of relatively unfractured, nonwelded, highly porous but relatively nontransmissive, argillic and zeolitic bedded tuffs and ash-flow tuffs. The contrast between these two sets of distinctive physical properties results in a stratified sequence best described as "physical-property stratigraphy" as opposed to traditional petrologic stratigraphy of volcanic rocks.

Superimposed on this layering are two sets of faults and fractures: one strikes north-northwest (N. 15° W. to N.

40° W.) and dips steeply (60°-90°) westward; the other strikes north-northeast (N. 5° E. to N. 35° E.) and also dips steeply westward. The north-northeast set constitutes the major Basin and Range style normal faults that separate the gently eastward-tilted major structural blocks of Yucca Mountain. The hanging-wall sides of these faults commonly exhibit increases in dip similar to rollover structures commonly related to listric faults. The north-northwest fracture set has the higher density, and the north-northwest fault set appears to have two directions of displacement. One is right-lateral slip perhaps related to the Las Vegas Valley and Walker Lane shear zones; the other is dip slip perhaps related to brittle drag behavior adjacent to the major north-northeast-striking faults.

The vast majority of recharge through the unsaturated zone is assumed to be vertical; the dominant migration may occur in fractures of densely welded tuffs and in the matrix of nonwelded tuff, but the mode of fluid flow in these unsaturated systems is undetermined. Limited lateral flow of recharge may occur at horizons where local perched water tables may exist above relatively nontransmissive zeolitized nonwelded tuffs. The pervasive north-northwest-striking fractures may control the direction of lateral flow of recharge, if any, in the unsaturated zone, and certainly that direction coincides closely with the observed southeasterly flow direction in the saturated zone under Yucca Mountain. Empirical evaluation of this conceptual hydrologic model has begun.

## INTRODUCTION

The Nevada Nuclear Waste Storage Investigations project, administered by the Nevada Operations Office of the U.S. Department of Energy, is examining the feasibility of emplacing nuclear wastes in southern Nevada. As part of that effort, the U.S. Geological Survey (USGS) and Sandia National Laboratories (SNL), among other research groups, have participated in investigations of the geology, hydrology, geophysics, and rock properties of the thick sequence of silicic tuffs that occur within and contiguous to the Nevada Test Site (NTS) in Nye County, Nevada (Figure 1). These efforts augment more than 20 years of geologic research by the USGS at the NTS [1], within the southern part of the Great Basin subprovince. Site-specific investigations, carried out since 1978, have concentrated on the characterization of the 1.5- to 4 km-thick sequence of ash-flow tuffs and related rocks at Yucca Mountain (Figure 2). Recently, emphasis has been placed on the unsaturated zone

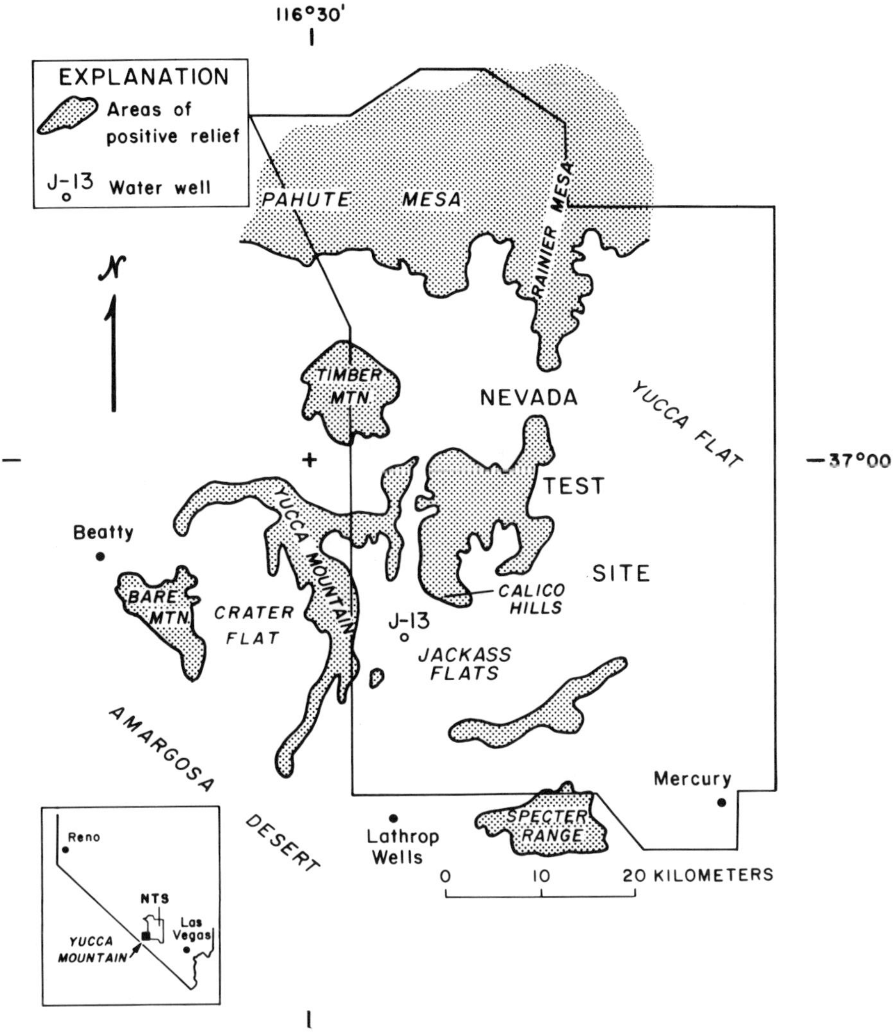


Figure 1. Location map. Index map at lower left shows the location of the Nevada Test Site (NTS) within Nevada, and the center shows an expansion of the southwest portion of the NTS.

which is some 500 to 800 m thick under Yucca Mountain. The sequence within the unsaturated zone consists of a wide range of contrasting rock types including densely welded devitrified and vitric ash-flow tuffs, nonwelded vitric ash-flow tuffs, vitric bedded tuffs, nonwelded zeolitized and argillized ash-flow tuffs, and zeolitized and argillized bedded tuffs.

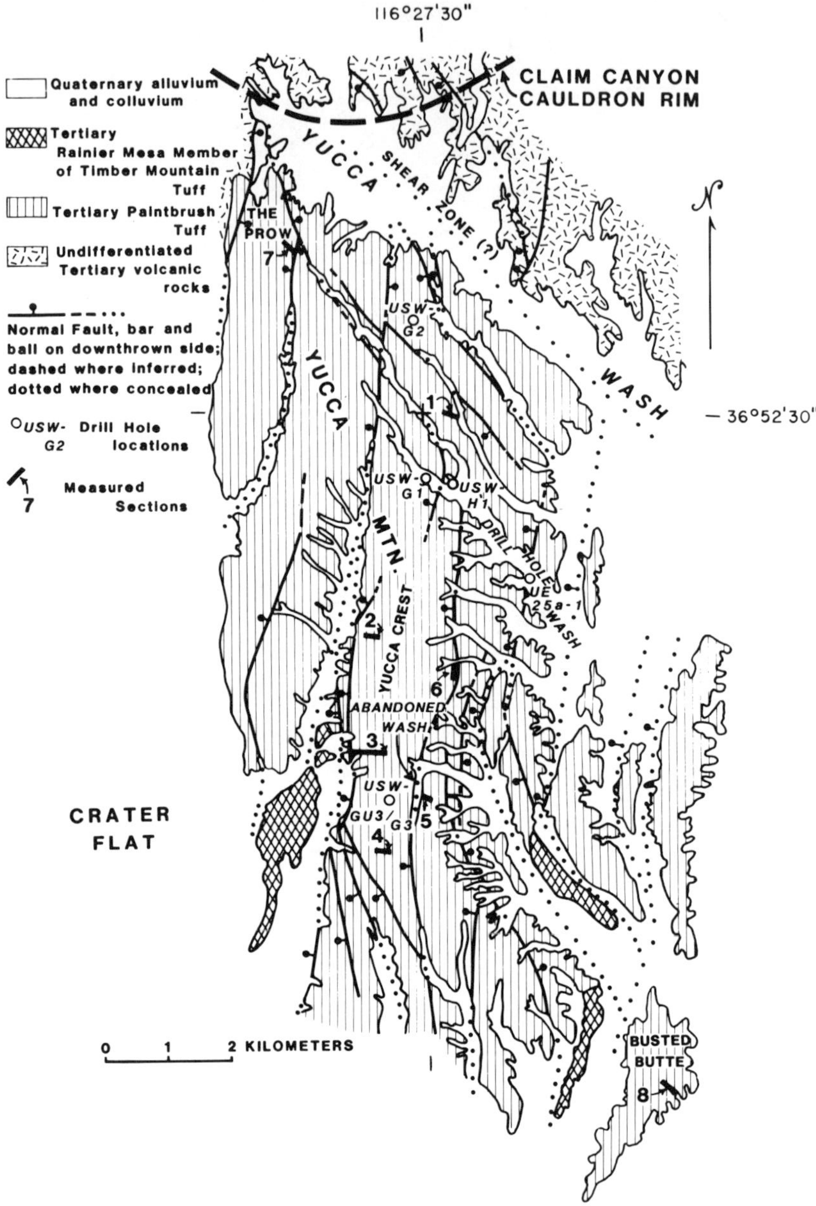


Figure 2. Geologic map of central Yucca Mountain generalized from the work of Christiansen and Lipman [3] and Lipman and McKay [4]; and modified based on results of recent detailed mapping. Exact boundaries of the area of the potential repository within this map area are not drawn because data are still being collected that may affect the most favorable boundaries.

This chapter deals principally with the geologic character and rock properties that potentially affect the hydrology of the unsaturated zone. Because the static water level at Yucca Mountain is 500 to 800 m below the surface, the unsaturated zone provides a thick sequence of welded and nonwelded tuffs for possible storage of nuclear waste. Several reasons for serious consideration of the unsaturated zone for waste storage in the Great Basin subprovince were reviewed by Winograd [2]; although the basic logic outlined by Winograd emphasized burial in thick alluvium, his logic can be extended to the unsaturated zones of fractured, densely welded ash-flow tuffs. This extension of logic includes: 1) low recharge flux in arid climates; 2) movement of recharge through highly transmissive, thermally stable, densely welded tuffs; 3) thick unsaturated zones; 4) multiple barriers to radionuclide migration including retardation by sorptive zeolitic and argillic nonwelded tuffs; 5) favorable thermal loading; and 6) economic factors of relatively shallow burial.

Among the most critical parameters of nuclear waste isolation is the transport of radionuclides by ground water. Ground-water movement in the NTS region is largely a function of effective rock conductivities [5]. Factors that govern effective rock conductivities in an ash-flow tuff are numerous; several factors produce matrix conductivity: At the time an ash flow is initially emplaced, the glass shard matrix becomes flattened by the welding process, provided the interior of the ash flow is at temperatures in excess of 500°C. This welding process creates an inverse relation between glass shard porosity and the degree of welding, but a positive relation between bulk density and the degree of welding. In the process of squeezing trapped gases from between the shards, gas-rich portions of ash-flow tuffs may develop gas pockets called lithophysal cavities. Most ash-flow tuffs consist of intercalated tongue-like sheets resulting from a series of magmatic pulses; if extruded over a short period at high temperatures, these individual sheets effectively weld into one sheet with decreases in degree of welding observable only at the margins of the sheet. Such sheets are called simple cooling units [6]. Alternatively, cooling periods of sufficient duration between individual magmatic pulses may cause decreases in the degree of welding within the interior of the cooling unit. Such ash-flow sheets are called compound cooling units. Within either compound or simple cooling units, small pore sizes (less than tens of micrometers) and assumed high tortuosities, particularly in the densely welded interiors of ash flows, create very low matrix conductivities: typical values are about $10^{-4}$ cm/s for vitric nonwelded tuffs and about $10^{-9}$

cm/s for densely welded tuffs [5].

Superimposed upon this cooling-unit framework and matrix conductivity are the effects of 1) devitrification in the slowly cooled interiors and vapor-phase crystallization in the partially welded margins, 2) vertical extension joints produced during cooling, 3) alteration to secondary phases, by reaction with pore waters to form zeolites and clays, concentrated in less welded portions of the cooling units, and 4) faults and joints related to tectonic events. The term fracture is used to refer to any crack with undetermined offset; those without offset are joints, and those with known offset are faults. A wide variety of fracture densities, attitudes, and apertures are formed as a function of the behavior of a particular rock type to its stress environment. Also, a variety of minerals may fill part or all of the fractures.

The net effect of all these processes, discussed in the paragraph above, is to alter drastically the effective rock conductivities. Highly fractured, densely welded tuffs have effective hydraulic conductivities about 5 or 6 orders of magnitude higher than their matrix hydraulic conductivities [5]. Nonwelded vitric tuffs have effective hydraulic conductivities only about 1 order of magnitude higher than matrix hydraulic conductivities, but zeolitized or argillized nonwelded tuffs have effective hydraulic conductivities 1 to 4 orders of magnitude higher than matrix hydraulic conductivities [5,7]. A note of caution should be added here because these effective hydraulic conductivities were measured in saturated rocks [5]; those for the unsaturated rocks of interest at Yucca Mountain may be significantly lower and may be related to the degree of saturation. These differences emphasize the need to evaluate rock properties in their stratigraphic and structural framework. Thus, the principal objective of this paper is to characterize qualitatively those rock-mass properties that may affect the hydrology of the unsaturated zone at Yucca Mountain.

## PREVIOUS GEOLOGIC INVESTIGATIONS

Prior to nuclear waste storage investigations, geologic research and exploration within the immediate vicinity of Yucca Mountain were limited to reconnaissance geologic quadrangle mapping [3-4]. Several summaries of the volcano-tectonic history of the NTS region that specifically address the origin of ash-flow tuffs exposed at Yucca Mountain have also been published [8-10]. Regional tectonic and structural relationships for the NTS and vicinity have been summarized by Carr [11].

## Stratigraphic Framework

Although less than 2 km of Miocene (12.5 to 14 m.y.-old) ash-flow tuffs with minor lavas and bedded tuffs have been penetrated by drilling with continuous core recovery [12-14] (Florian Maldonado, USGS, and Sarah Koether, Fenix & Scisson (F&S), written commun., 1982), geophysical evidence suggests that a sequence of tuffaceous rocks between 1.5 and 4 km thick overlies a pre-Cenozoic basement under central Yucca Mountain [15]. The volcanic sequence beneath Yucca Mountain is shown on a north-south cross section (Figure 3). Four major ash-flow tuffs have been been penetrated by

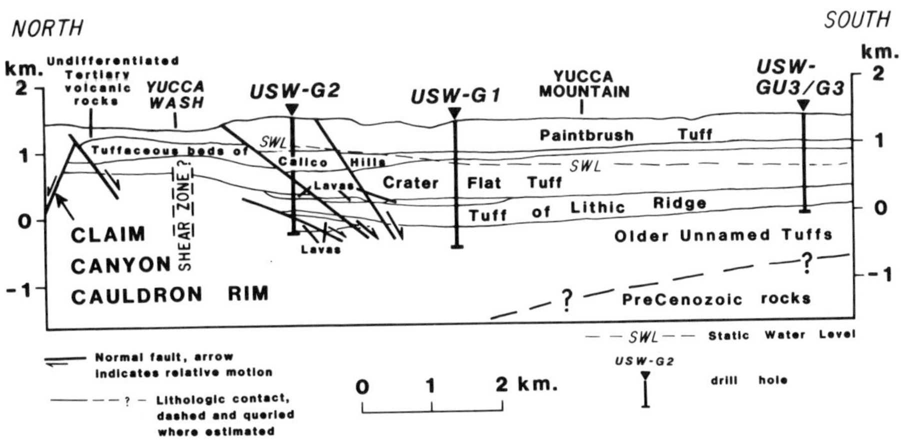


Figure 3. Generalized north-south geologic section through Tertiary volcanic strata under Yucca Mountain. SWL marks the static water level. Horizontal and vertical scales are the same. The zero depth is mean sea level.

drilling. These tuffs are listed in ascending order: 1) The informal tuff of Lithic Ridge (300±m thick) (W. J. Carr, USGS, F. W. Byers, Los Alamos National Laboratory (LANL), and Paul Orkild, USGS, written commun., 1982) consists of a partially to moderately welded, rhyolitic compound cooling unit. 2) The Crater Flat Tuff includes three rhyolitic ash-flow tuff members. The basal unit is the informal Tram unit (300±m thick), which consists of a partially to moderately welded, compound cooling unit. The middle member is the Bullfrog Member (200±m thick), which consists of a partially to densely welded, compound cooling unit. The upper member is the Prow Pass Member (100±m thick), which consists of a partially welded, compound cooling unit (W. J. Carr, USGS,

F. W. Byers, LANL, and Paul Orkild, USGS, written commun., 1982; R. B. Scott, USGS, and Mayra Castellanos, F&S, written commun., 1982). All three members are considered to have erupted from a caldera complex now buried below Crater Flat several kilometers to the west of Yucca Mountain [16] (W. J. Carr, USGS, F. W. Byers, LANL, and Paul Orkild, USGS, written commun., 1982). 3) A nonwelded sequence of rhyolitic ash-flow tuffs and bedded tuffs (30 to 275 m thick) is informally called the tuffaceous beds of Calico Hills [12] (Florian Maldonado, USGS, and Sarah Koether, F&S, written commun., 1982; R. B. Scott, USGS, and Mayra Castellanos, F&S, written commun., 1982). 4) The uppermost formation, the Paintbrush Tuff, here consists largely of two major ash-flow tuffs, the Topopah Spring Member (300±m thick) and the Tiva Canyon Member (100±m thick). Both the Topopah Spring and Tiva Canyon Members were erupted from the Claim Canyon Cauldron about 2 km north of Yucca Mountain [8]. Both the Topopah Spring and the Tiva Canyon Members largely consist of moderately to densely welded tuff, and they are both compositionally zoned from the high-silica rhyolites that form their basal and central portions to quartz latites that form densely welded caprocks near the top of each member [17,8]. In the stratigraphic interval between the Topopah Spring and Tiva Canyon Members, two minor members of the Paintbrush Tuff, the Pah Canyon and Yucca Mountain Members, are present at the north end of Yucca Mountain as distal edges of nonwelded to moderately welded sheets that pinch out toward the south and east within the central portion of Yucca Mountain [12] (Florian Maldonado, USGS, and Sarah Koether, F&S, written commun., 1982). Where these two members become too thin to map, they are lumped together with unnamed bedded tuffs. Although the Paintbrush Tuff is the youngest (12.5 to 13.1 m.y. old) ash-flow tuff to cover the entire Yucca Mountain area, the 11.3-m.y.-old Rainier Mesa Member of the Timber Mountain Tuff was locally ponded in post-Paintbrush Tuff structural depressions, covering portions of the normal faults that delineate the major structural blocks of the mountain (Figure 2).

The remainder of this chapter considers the characteristics of those units that occur above the static water level at Yucca Mountain. The static water level (Figure 3) within Yucca Mountain is located near the base of the Paintbrush Tuff at drill hole USW-G2, near the base of the tuffaceous beds of Calico Hills at USW-G1, and within the Crater Flat Tuff at USW-GU3/G3 [12] (Florian Maldonado, USGS, and Sarah Koether, F&S, written commun., 1982; R. B. Scott, USGS, and Mayra Castellanos, F&S, written commun., 1982).

Structural Framework

Paleozoic strata are not exposed within 15 km of Yucca Mountain, though seismic refraction (W. D. Mooney, USGS, written commun., 1982) and gravity data [15] suggest that Paleozoic rocks may be present under 1.5 to 4 km of volcanic strata. As no drilling has penetrated these rocks under Yucca Mountain, their character must be extrapolated from nearby exposures in the Specter Range, Calico Hills, and Bare Mountain (see Figure 1) [18]. Paleozoic rocks in these localities have been deformed by complex folding and thrust faulting during formation of the late Mesozoic Sevier orogenic belt [19-20]. Ambiguities inherent in these regionally complex structures do not allow reliable projections of either structures or lithologies under Yucca Mountain. Paleozoic carbonate or clastic rocks are likely to form the subvolcanic rocks. Regional interpretations of these Mesozoic structures vary [11,20] (M. D. Carr, USGS, oral commun., 1982), and possible structures under Yucca Mountain include upper or lower plates of regional overthrusts, low-angle gravity-controlled fault complexes, or broad folds. Because Paleozoic rocks are regionally important aquifers and aquitards in the southern Great Basin [5], penetration into subvolcanic rocks remains one of the ultimate objectives of the USGS investigations at Yucca Mountain.

Regionally, three types of Cenozoic structures exist: 1) Basin and Range style faults, 2) major strike-slip fault zones, and 3) volcano-tectonic structures. Relative to the heterogeneous and complex Cenozoic structure common in surrounding areas, the structure at Yucca Mountain appears uncomplicated. As exposures of pre-Paintbrush Tuff rocks are sparse, the apparent structural simplicity of Yucca Mountain may be misleading to some degree. However, structural data from cores to depths of nearly 2 km do not suggest the presence of structural styles or complexities not apparent on the surface. Pre-Paintbrush Tuff Tertiary structures below the 2 km depths explored by drilling probably do exist beneath Yucca Mountain, as suggested by the gravity and magnetic gradients [15] (G. D. Bath, USGS, written commun., 1982).

As shown in quadrangle maps [3,4], Yucca Mountain itself is broken by several major north-northeast-striking and generally westward dipping Basin and Range style normal faults with tens to hundreds of meters of vertical displacement, forming blocks tilted gently eastward. In general, these faults decrease in displacement and density toward the north (Figure 2). Basin and Range style normal faulting began prior to volcanism [21,11] and continued during most of the major eruptive activity; apparently, few if any faults formed after volcanism ceased, although relatively minor additional displacements have occurred locally on preexisting faults during the last few million years. Regional evidence of Quaternary faulting in response to a minimum principal horizontal stress direction of about N. 50° W. has been reviewed by Carr [11]. Direct evidence of this type of faulting at Yucca Mountain is not reported in previous studies, but has been found by recent mapping (W. J. Carr, USGS, oral commun., 1982). The original geologic mapping at Yucca Mountain [3,4] did not identify volcano-tectonic structures associated with the Crater Flat Tuff; however, such structures are now believed to exist in the vicinity of Crater Flat [15].

### Hydrologic Framework

Extensive summaries of the NTS region and local hydrology have been made [5,7,22-24], but no details were previously available concerning the Yucca Mountain area. The regional water table is commonly 500 to 800 m deep and has a south and southwest flow direction toward the Amargosa Desert in most of the region. However, in local areas considerable variation is observed such as the southeast flow direction under Yucca Mountain. The principal aquifers in the NTS region are Paleozoic carbonate rocks and highly fractured Tertiary densely welded ash-flow tuffs. The major aquitards are Paleozoic clastic rocks and Tertiary zeolitized or argillized nonwelded tuffs. For example, a few kilometers east of Yucca Mountain in Jackass Flats (Figure 1), where the Topopah Spring Member is below the water table, the densely welded tuff is an aquifer in well J-13 [24,5]. Matrix and effective hydraulic conductivities of the major rock types found within ash-flow tuffs at the NTS are given in Table I.

## GEOLOGIC CHARACTER OF THE UNSATURATED ZONE

### Physical-Property Stratigraphy

The movement of fluids through rock is fundamentally a physical process. For this reason, a stratigraphic framework designed to emphasize potential differences in the hydrologic character among strata should be based upon rock physical properties. Petrologic differences, usually identified by phenocryst ratios, are traditionally used to

Table I. Hydraulic Conductivities of Tuffaceous Aquifers and Aquitards, NTS

| Hydrologic Unit | Rock Type | Range of Matrix Hydraulic Conductivities[1] (cm/s) | Range of Effective Hydraulic Conductivities[2] (cm/s) |
|---|---|---|---|
| Aquifer | Densely welded tuff[3] | $3 \times 10^{-10}$ to $10^{-8}$ | $10^{-5}$ to $3 \times 10^{-2}$ |
| Leaky Aquitard[4] | Nonwelded vitric tuff | $10^{-4}$ | $\sim 10^{-3}$ |
| Aquitard | Nonwelded zeolitized tuff | $2 \times 10^{-9}$ to $3 \times 10^{-5}$ | $10^{-6}$ to $10^{-4}$ |
| Aquitard | Nonwelded argillized tuff | $10^{-10}$ to $2 \times 10^{-5}$ | $10^{-6}$ to $10^{-4}$ |

[1]Upper range for welded tuff from page 34, lower range for welded tuff from Figure 18, and zeolitized and argillized tuff from Table 5 from Winograd and Thordarson [5].

[2]From Table 1 of Winograd and Thordarson [5] and Thordarson (USGS, written commun., 1965), and estimated from thicknesses of nonwelded vitric tuffs at margins of cooling units.

[3]Values for both devitrified and vitric densely welded tuff are included.

[4]Term used by Winograd and Thordarson [5].

define stratigraphic breaks in sequences of ash-flow tuffs, but for the hydrologic and engineering concerns of a nuclear waste repository, physical property differences in porosity, bulk density, grain density, permeability, and degree of welding, along with the density and attitudes of fractures and faults, are the important features. To contrast these two different stratigraphic classification schemes, Figure 4 compares petrologic stratigraphy with a physical-property stratigraphy based on the degree of welding in the sequence of rocks cored at holes USW-GU3/G3, -G1, and -G2 [12] (R. B. Scott, USGS, and Mayra Castellanos, F&S, written commun., 1982; Florian Maldonado, USGS, and Sarah Koether, F&S, written commun., 1982). The physical-property subdivisions shown in Figure 4 were produced by placing bedded tuffs and nonwelded to partially welded ash-flow tuffs into one category and moderately to densely welded ash-flow tuffs into a second category. The first category consists of relatively low-density rocks that tend to be more porous and less fractured than those in the second category, which consists of relatively high-density rocks that tend to be less porous and more fractured. In combination with structural data, physical-property stratigraphy, as expressed in Figure 4, is particularly useful for construction of conceptual hydrologic models. The variability in this physical-property stratigraphy from USW-GU3/G3 in the southern part of the Yucca Mountain region to USW-G2 in the northern part of the region is considerable, over a distance of only 8 km.

Detailed investigations of the physical properties of the Tiva Canyon Member of the Paintbrush Tuff were possible because of the excellent three-dimensional exposures of this member at Yucca Mountain. The Tiva Canyon Member is very similar in overall character to the less well exposed Topopah Spring Member. This similarity is supported by the observation that the physical properties of surface samples of the welded Tiva Canyon Member are comparable to those of core samples of the welded Topopah Spring Member [26-27]. In addition to studies of the more welded units, determinations of physical properties have been made for the bedded and nonwelded tuffs [28]. These physical property studies are augmented by petrological and chemical studies made at the Los Alamos National Laboratory [29], which describe the character of clay and zeolite alteration in both welded and nonwelded tuffs.

## Mappable Zonations within the Tiva Canyon Member

Zonations within ash-flow tuffs, as defined by Smith [6], primarily reflect the effects of the thermal evolution of ash-flow tuffs; Lipman and others [17] discussed the origin of primary chemical zonations that reflect primary magmatic differences. Zonations used to study ash-flow tuffs at Yucca Mountain are recognized in the field by differences in groundmass devitrification, degree of welding, shape of eroded slopes, texture of weathered surfaces created by both surficial cracking and tectonic fracturing, lithophysal cavity abundance, lithic fragment abundance,

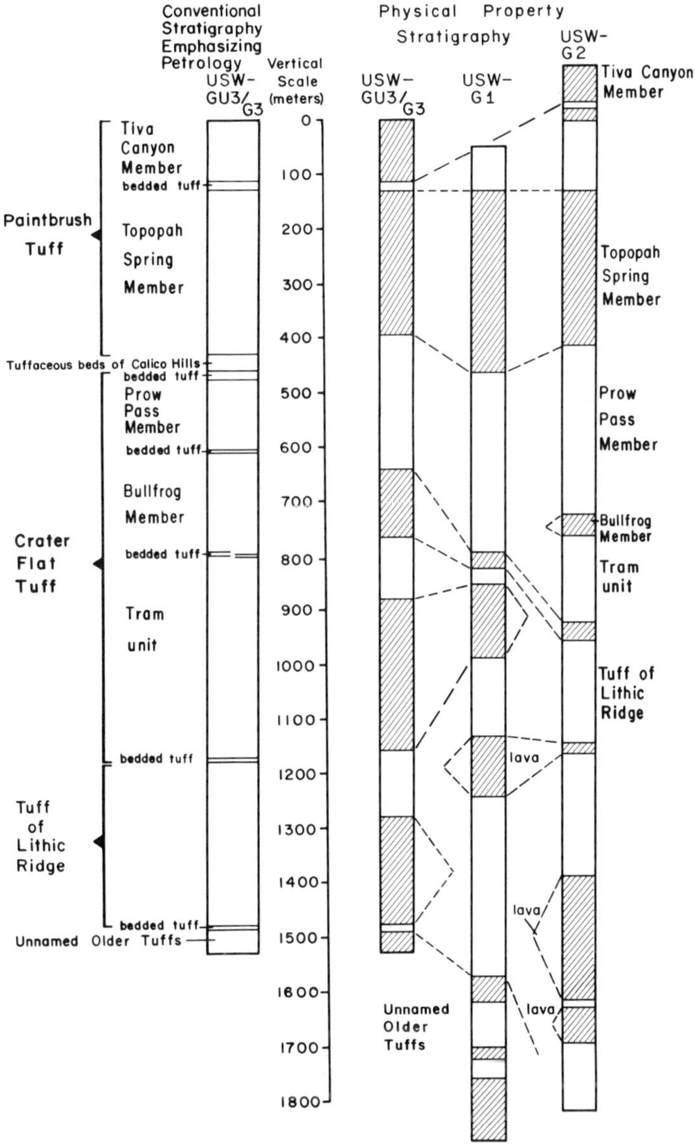


Figure 4. Comparison between stratigraphic units defined by petrologic criteria and those defined by physical-property criteria. The conventional stratigraphic nomenclature based upon petrologic criteria is given for drill hole USW-GU3/G3 on the left and the physical-property stratigraphies based on degree of welding for drill holes USW-GU3/G3, -G1, and -G2 are drawn in order toward the right. The conventional stratigraphic names

of major units are also listed on the far right for general reference but exact boundaries are not shown. The shaded zones are moderately welded to densely welded; unshaded are nonwelded to partially welded. BT = bedded tuff.

phenocryst ratios, presence of vapor-phase crystallization, and color. Besides forming stratigraphic horizons particularly valuable for recognition of small faults, these zones provide a stratigraphic framework for comparison with the physical properties of the rocks. A fence diagram of this zonation in the Tiva Canyon Member is shown in Figure 5; each station of the fence diagram is a measured section (located in Figure 2) where samples were also collected for physical- property studies. The informal field terms for these zones are used on the fence diagram.

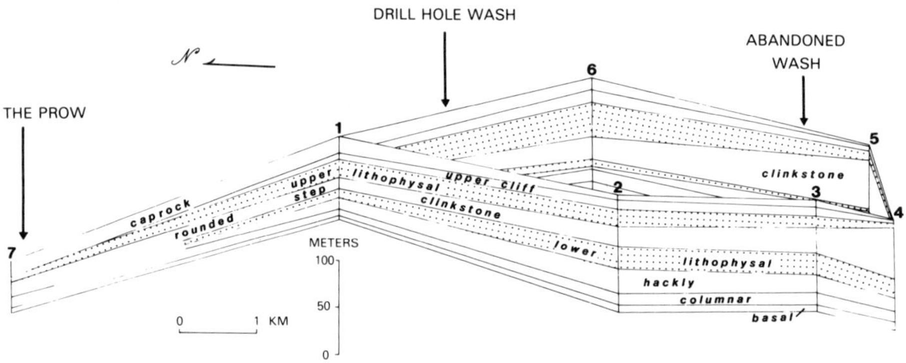


Figure 5. Generalized zonation within the Tiva Canyon Member. The rounded step zone grades laterally into the clinkstone zone. Measured section numbers are located on Figure 2.

## *Bulk Density, Porosity, and Grain Density*

Measured dry bulk densities, matrix porosities, and grain densities for different zones of the Tiva Canyon Member are shown on Table II. Similar values have been reported for these physical properties in the core of drill hole UE25a-1 [25]. The interdependence of dry bulk density and matrix porosity is shown in Figure 6; as expected, dry bulk density increases and porosity decreases as welding increases. Different degrees of welding should not be defined by specific porosity or density limits because considerable variation in both total porosity (including

Table II. Physical Properties of the Tiva Canyon Member of the Paintbrush Tuff, Yucca Mountain
(Analyses by Terra Tek, Inc., Salt Lake City)

| Measured Section Sample Number | Height above base (m) | Zone | Dry bulk density[1] ($g/cm^3$) | Grain density[2] ($g/cm^3$) | Matrix porosity[3] (percent) |
|---|---|---|---|---|---|
| YTC1-12 | 88.3 | caprock | 2.24 | 2.50 | 10.4 |
| YTC1-11 | 80.3 | caprock | 1.77 | 2.51 | 29.5 |
| YTC1-10 | 72.2 | caprock | 1.99 | 2.48 | 19.8 |
| YTC1-9 | 66.1 | upper cliff | 2.11 | 2.50 | 15.6 |
| YTC1-8 | 59.1 | upper lithophysal | 2.16 | 2.53 | 14.6 |
| YTC1-7 | 46.8 | upper lithophysal | 2.10 | 2.45 | 14.3 |
| YTC1-6 | 36.9 | rounded step | 2.20 | 2.45 | 10.2 |
| YTC1-5 | 29.2 | lower lithophysal | 2.21 | 2.47 | 10.5 |
| YTC1-4 | 18.4 | hackly | 2.29 | 2.47 | 7.3 |
| YTC1-3 | 10.8 | columnar | 2.14 | 2.46 | 13.0 |
| YTC1-2 | 6.1 | columnar | 2.17 | 2.47 | 12.1 |
| YTC1-1 | 2.3 | basal | 1.36 | 2.38 | 42.9 |
| YTC2-15 | 124.0 | caprock | 1.89 | 2.47 | 23.5 |
| YTC2-14 | 122.1 | caprock | 1.82 | 2.53 | 28.1 |
| YTC2-13 | 119.0 | upper cliff | 1.96 | 2.52 | 22.2 |
| YTC2-12 | 109.8 | upper cliff | 2.17 | 2.46 | 11.8 |
| YTC2-11 | 93.7 | upper lithophysal | 2.18 | 2.47 | 11.7 |
| YTC2-10 | 86.0 | rounded step | 2.23 | 2.44 | 8.6 |
| YTC2-9 | 78.4 | rounded step | 2.24 | 2.44 | 8.2 |
| YTC2-8 | 71.4 | rounded step | 2.18 | 2.51 | 13.2 |
| YTC2-7 | 63.0 | lower lithophysal | 2.24 | 2.44 | 8.2 |
| YTC2-6 | 49.9 | lower lithophysal | 2.18 | 2.40 | 9.2 |
| YTC2-5 | 39.2 | hackly | 2.34 | 2.52 | 7.1 |
| YTC2-4 | 23.0 | hackly | 2.29 | 2.45 | 6.5 |
| YTC2-3 | 16.9 | columnar | 2.26 | 2.44 | 7.4 |
| YTC2-2 | 10.0 | columnar | 2.01 | 2.43 | 17.3 |
| YTC2-1 | 3.8 | basal | 1.27 | 2.35 | 46.0 |
| YTC3-14B | 116.5 | caprock | 1.97 | 2.55 | 22.7 |
| YTC3-14A | 102.8 | uppercliff | 2.17 | 2.52 | 13.9 |
| YTC3-13B | 66.5 | clinkstone | 2.20 | 2.48 | 11.3 |
| YTC3-13A | 47.6 | lower lithophysal | 2.29 | 2.48 | 7.7 |
| YTC3-12B | 38.5 | hackly | 2.29 | 2.45 | 6.5 |
| YTC3-12A | 35.4 | hackly | 2.31 | 2.46 | 6.1 |
| YTC3-11A | 25.3 | hackly | 2.34 | 2.47 | 5.3 |
| YTC3-10C | 19.2 | columnar | 2.36 | 2.53 | 6.7 |
| YTC3-10B | 12.2 | columnar | 2.36 | 2.52 | 6.3 |
| YTC3-10A | 9.2 | columnar | 2.26 | 2.39 | 5.4 |
| YTC3-9D | 7.6 | basal | 2.14 | 2.43 | 11.9 |
| YTC3-9C | 3.0 | basal | 1.56 | 2.31 | 32.5 |
| YTC3-9B | 2.1 | basal | 1.32 | 2.37 | 44.3 |
| YTC3-9A | 0.0 | basal | 1.19 | 2.34 | 49.3 |

Table II. Physical Properties--Continued

| Measured Section Sample Number | Height above base (m) | Zone | Dry bulk density[1] ($g/cm^3$) | Grain density[2] ($g/cm^3$) | Matrix porosity[3] (percent) |
|---|---|---|---|---|---|
| YTC4-8B | 107.6 | caprock | 1.92 | 2.55 | 24.7 |
| YTC4-8A | 100.0 | upper lithophysal | 2.17 | 2.40 | 9.6 |
| YTC4-7A | 94.5 | lower cliff | 2.27 | 2.50 | 9.2 |
| YTC4-6C | 89.0 | clinkstone | 2.09 | 2.50 | 16.4 |
| YTC4-6B | 74.0 | clinkstone | 2.12 | 2.50 | 15.2 |
| YTC4-6A | 49.5 | clinkstone | 2.29 | 2.56 | 10.5 |
| YTC4-5A | 41.0 | lower lithophysal | 2.28 | 2.48 | 8.1 |
| YTC4-4B | 32.0 | hackly | 2.38 | 2.54 | 6.3 |
| YTC4-4A | 26.0 | hackly | 2.31 | 2.53 | 8.7 |
| YTC4-3B | 12.5 | columnar | 2.32 | 2.47 | 6.2 |
| YTC4-3A | 11.0 | columnar | 1.97 | 2.34 | 15.8 |
| YTC4-2B | 6.0 | basal | 1.51 | 2.49 | 39.4 |
| YTC6-7B | 103.3 | caprock | 1.98 | 2.51 | 21.1 |
| YTC6-7A | 78.0 | upper cliff | 2.16 | 2.49 | 13.3 |
| YTC6-6B | 52.5 | upper lithophysal | 2.14 | 2.51 | 14.7 |
| YTC6-6A | 42.4 | upper lithophysal | 2.12 | 2.51 | 15.5 |
| YTC6-5B | 30.4 | clinkstone | 2.32 | 2.50 | 7.2 |
| YTC6-5A | 16.7 | clinkstone | 2.30 | 2.52 | 8.7 |
| YTC6-3A | 10.6 | lower lithophysal | 2.26 | 2.52 | 10.3 |
| YTC6-2A | 8.5 | hackly | 2.32 | 2.53 | 8.3 |
| YTC6-1B | 3.7 | columnar | 2.34 | 2.51 | 6.3 |
| YTC6-1A | 0.0 | columnar | 2.32 | 2.48 | 6.5 |

[1]Dry bulk density calculated using volume of sample based on immersion; external lithophysal cavities are generally not evident in these data.

[2]Calculated by powdered pycnometer method.

[3]Calculated value, based on the formula: percent porosity = (grain density - dry bulk density)/grain density; therefore, lithophysal porosity generally is not included in these data.

lithophysal porosity) and matrix porosity, are common at constant degree of welding. Explanations for these variations include the presence of secondary mineralization, vapor-phase crystallization, and formation of lithophysal cavities.

Small shifts in the porosity-dry bulk density relationship shown in Figure 6 are not related to the degree of welding, but instead are related to small changes in grain density. The vitric basal portion (Figures 6 and 7) has the lowest average grain density within the Tiva Canyon Member ($2.38 \pm 0.06$ $g/cm^3$). This is the same value reported for the vitric portion of the Topopah Spring Member in drill hole USW-G1 [26]. The average grain density within the partially vitric columnar zone of the Tiva Canyon Member is $2.46 \pm 0.06$

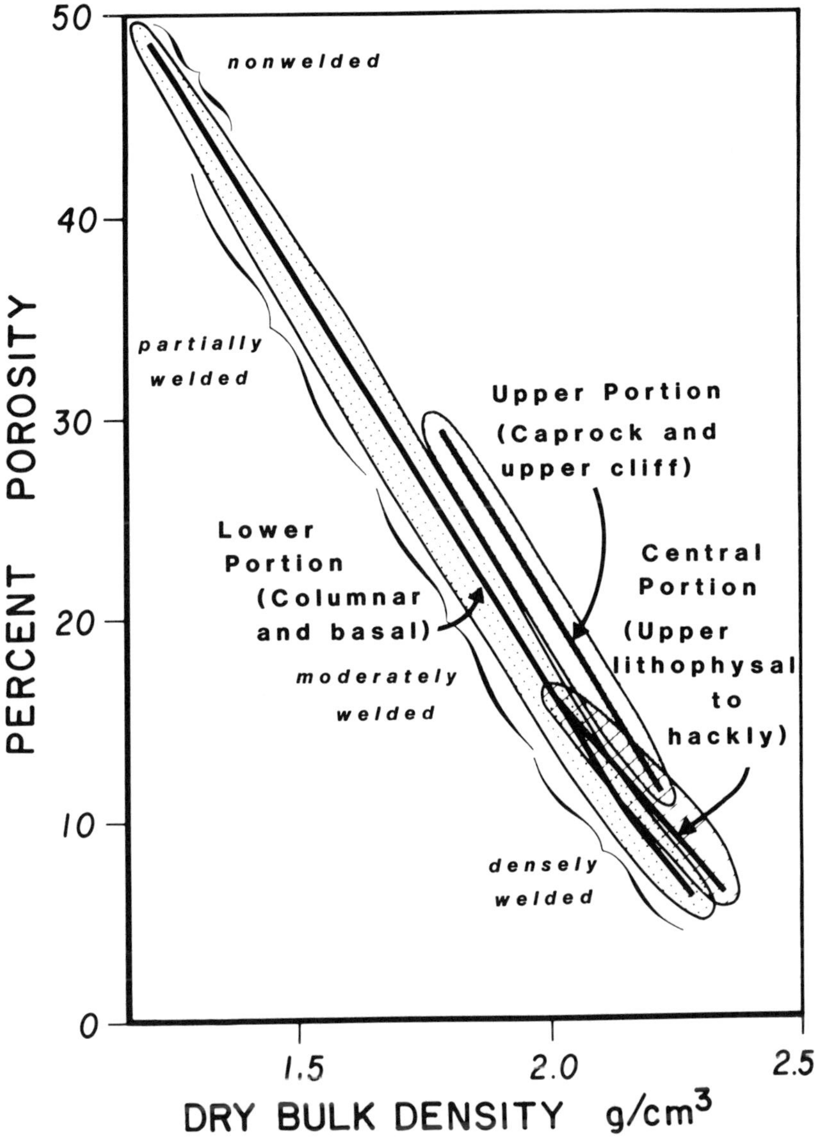


Figure 6. Relations between dry bulk density, matrix porosity, and degree of welding of the Tiva Canyon Member. The central portion includes the devitrified zones: the hackly, lower lithophysal, clinkstone, and upper lithophysal zones. Brackets show the general regions typical of rocks of different degrees of welding.

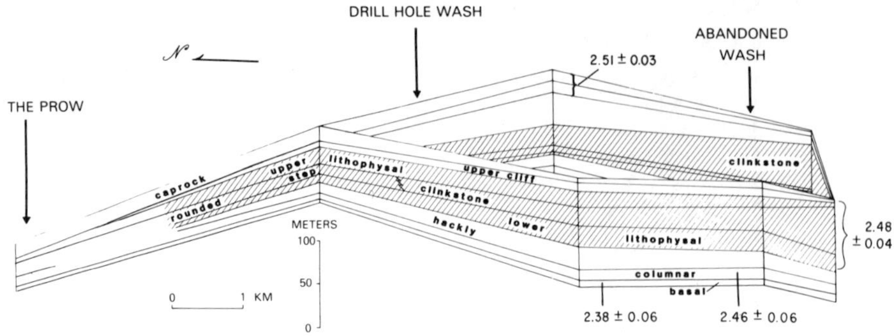


Figure 7. Grain density (g/cm$^3$) correlation with stratigraphic zones in the Tiva Canyon Member, Paintbrush Tuff.

g/cm$^3$, within the completely devitrified central portion is 2.48±0.04 g/cm$^3$, and within the upper cliff and caprock zones is 2.51±0.03 g/cm$^3$. The grain densities in the Tiva Canyon Member reported here are significantly lower than those reported for the devitrified portions of the Topopah Spring Member, which average 2.55 g/cm$^3$ [26]. Similar results were also obtained in the physical property studies of ash-flow tuffs in the Timber Mountain Caldera region (W. J. Carr, USGS, written commun., 1966).

Thus, a consistent trend of increasing grain density exists from the partially devitrified columnar zone upward through the devitrified central zones to the devitrified upper cliff and caprock zones, consistent with the offsets of curves in Figure 6. Several explanations for this trend can be given. First, the increase in average grain density of the partially devitrified columnar zone relative to the vitric basal zone is caused by the greater density of devitrification products, quartz, tridymite, cristobalite, and feldspar, compared to volcanic glass in the vitric zone. Second, higher densities in the central devitrified portion between the columnar and the upper cliff zones are also related to the greater densities of devitrified phases; the increase of grain density near the top of the central portion, specifically in the upper lithophysal zone, may be related, in part, to an increase in the phenocryst content from about 3 percent in the lower zones to about 6 percent in the upper lithophysal zone. Third, the greatest average grain densities, those found in the upper cliff and caprock zones, are probably related to two factors, an increase in phenocryst content to about 15 percent and the more mafic rock composition of those zones. Still another source of grain density variation within the devitrified portion of the unit may be the ratios of silica polymorphs and feldspar; in particular, the grain densities of cristobalite (2.33 g/cm$^3$) and tridymite (2.26 g/cm$^3$) are significantly lower than that of quartz (2.65 g/cm$^3$). However, there is almost no quartz in the devitrification products in the Tiva Canyon Member, and the feldspar/silica polymorph ratio is essentially constant (Dave Bish, LANL, written commun., 1982). The presence of abundant quartz in the devitrification products of the Topopah Spring Member is consistent with the higher densities in the devitrified portion of that unit.

*Devitrification*

The devitrification textures evident in thin sections are also related to zones mapped in the field, and probably explain many of the differences in the character of weathered surfaces. For example, the hackly zone (see Figure 7) weathers to centimeter-sized irregular fragments, in contrast to the overlying clinkstone zone, which forms meter-long, smooth, conchoidally fractured blocks; field relationships do not suggest an explanation for this difference in surface characteristics. In thin section, however, the devitrification texture within the hackly zone consists of short, finely fibrous spherulitic patches (Figure 8a), while that in the clinkstone zone consists of long, coarsely fibrous to granophyric textures which both obscure and cross glass shard boundaries, forming a more uniform three-dimensional intergrowth (Figure 8b). The devitrification textures in the upper portions of the clinkstone zone and in the upper cliff and caprock zones are increasingly coarser granophyric with gradational contacts. The columnar zone has an almost unaltered flattened glass shard axiolitic texture that gradationally becomes devitrified toward the top of the zone. The basal zone has partially flattened shards with a hydrated(?) rim. The boundaries between these devitrification textures approximately follow the mapped zone boundaries (Figure 9).

Because these differences should affect the means by which these units fracture after devitrification, a correlation between devitrification zones and the density of fractures is expected. In drill hole USW-GU3/G3 where core was recovered from the Tiva Canyon Member, the calculated density of fractures in the densely welded portion of the unit is calculated to be 14 fractures/unit m$^3$ in the upper lithophysal zone, 22 fractures/unit m$^3$ in the clinkstone zone, 22 fractures/unit m$^3$ in the lower lithophysal zone, 26 fractures/unit m$^3$ in the hackly zone, and 21 fractures/unit

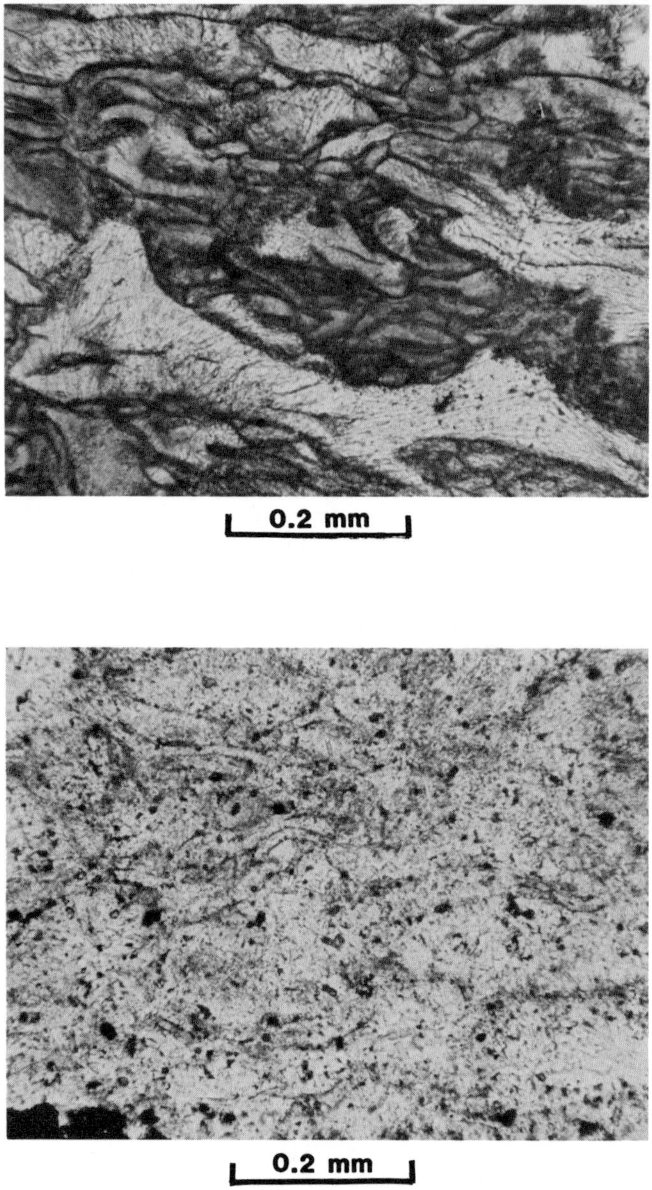


Figure 8. Photographs of thin sections of devitrified textures in the Tiva Canyon Member in the (a) hackly zone and (b) clinkstone zone. The glass shard texture is more distinct in the hackly zone but partially destroyed in the clinkstone zone.

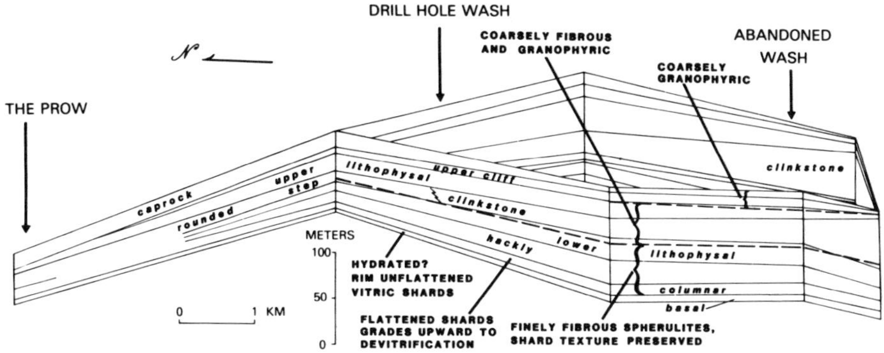


Figure 9. Stratigraphy of devitrification textures in the Tiva Canyon Member, Paintbrush Tuff. The dashed lines mark contacts between devitrification zonations that do not coincide with zone boundaries.

$m^3$ in the columnar zone. The partially welded caprock zone of the Topopah Spring Member has a fracture density of about 18 fractures/unit $m^3$ and should be representative of the unsampled portion of the Tiva Canyon caprock zone (an explanation of the fracture density units and calculation follows in a discussion of fractures). It may be significant that the hackly zone, with fine spherulitic devitrification textures, has the highest fracture density and that the upper lithophysal zone has the lowest fracture density, though it is apparent that zones of approximately the same degree of welding have approximately the same density of fracturing, in spite of variations in devitrification texture and lithophysal content. One exception to this general rule is that the density of fractures within the densely welded vitrophyres is appreciably lower than the overlying devitrified portion; for example, the Topopah Spring vitrophyre has a density of 8.8 fractures/unit $m^3$.

In contrast, the partially welded to nonwelded base of the Tiva Canyon Member has only 9 fractures/unit $m^3$, and the underlying bedded tuffs and nonwelded tuffs have less than 1 fracture/unit $m^3$. Thus, devitrification textures and lithophysal content appear to have much less impact on fracture frequency than degree of welding and vitrophyres have intermediate fracture densities.

## *Lithophysal Zones*

Zones containing abundant lithophysal cavities or gas pockets may significantly affect many of the physical properties of ash-flow tuffs. Several lithophysal zones occur in the compound cooling units of the Tiva Canyon and the Topopah Spring Members; these have been mapped and studied in measured sections of the Tiva Canyon Member (Figure 5) and studied in measured sections and core of the Topopah Spring Member (Figure 10). Field relationships of the lithophysal zones in the Tiva Canyon Member suggest that these zones are continuous sheets, and that each lithophysal zone represents a separate gas-rich eruptive pulse. Field and detailed petrographic evidence suggests that the Tiva Canyon Member actually consists of at least 8 eruptive pulses, only two of which were gas-rich. The somewhat more irregular pattern of lithophysal zones in the Topopah Spring Member shown in Figure 10 probably represents a series of

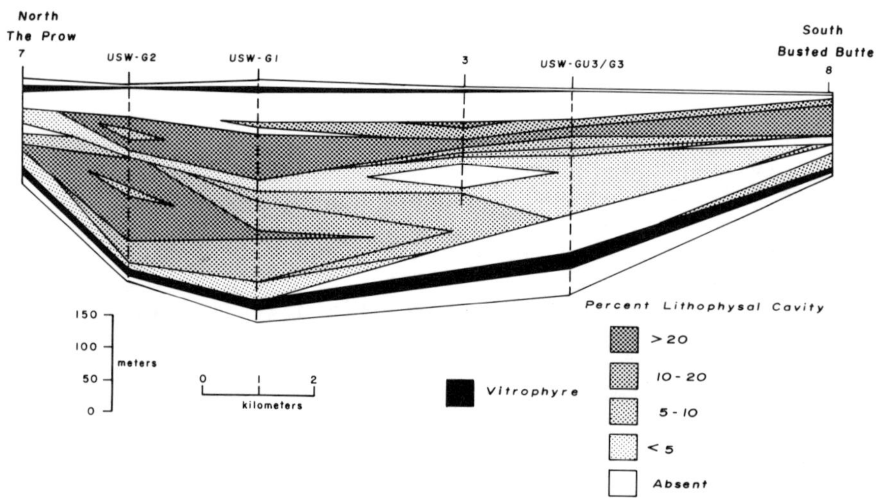


Figure 10. Lithophysal zonation in the Topopah Spring Member, Paintbrush Tuff. Locations of drill holes USW-G2, -G1, and -GU3/G3 and the measured sections 3, 7, and 8 are shown on Figure 2.

somewhat discontinuous stratigraphic tongues of gas-rich magmatic pulses that created lithophysal-rich horizons. At least 9 separate magmatic pulses can be recognized from petrographic evidence in the core from drill hole USW-GU3/G3 alone, and evidence for other stratigraphic units are found elsewhere in the field and in core [17] (R. B. Scott, USGS and Mayra Castellanos, F&S, written commun., 1982). Approximately 2 of these pulses were sufficiently gas-rich to produce lithophysal zones.

Lithophysal cavities consist of spherical to highly oblate voids ranging from less than 1 cm to as much as 30 cm in maximum diameter. Surrounding these voids is a thin (∿1 mm thick) inner rim of vapor-phase crystals; cristobalite, tridymite, and alkali feldspar are common and plates of hematite, prisms of pseudobrookite, and needles of alkali amphiboles rare. Outside the vapor-phase rim itself is an outer rim of pale-colored altered rock matrix, commonly about 1 cm thick. In general, the alteration in this rim consists of the products of high-temperature mineralogical and chemical processes, but mixed layer clays are locally reported [29]. Lithophysal porosity varies from less than 1 percent to greater than 30 percent of the rock.

Obviously these cavities will influence effective hydraulic conductivities, especially in fractured zones; hydraulic tests of these specific zones are planned. Also these cavities, which are air-filled above the static water level, decrease rock thermal conductivities and bulk densities, and will alter rock-mass engineering or mechanical properties. Therefore, determination of the actual three-dimensional distribution of lithophysae within the Topopah Spring Member is being given high priority.

## STRATIGRAPHIC CONTROL OF STRUCTURAL CHARACTER

### Expression of Fracture Density

Knowledge of the character of fractures in a rock mass is essential to understanding or predicting its hydrologic character. Therefore, it is critical to express the attitude, distribution, and density of fractures properly when interpreting data collected either along an outcrop or in drill core. A critical problem in interpretation is created by fractures that intersect cores or traverses at angles less than 90°. In such a situation, the apparent fracture frequency per unit length along core or outcrop may be significantly less than the true fracture frequency per unit length perpendicular to the fracture set [30]. First, consider a simplified example where a core is cut by a set of parallel fractures that intersect the core axis at some angle less than 90°. In this example, the actual frequency of fractures per unit length in a direction perpendicular to the fracture set can be expressed by the relation:

$$F_c = [\sin A]^{-1} \cdot F_m$$

where $F_c$ is the corrected fracture frequency, $F_m$ is the apparent fracture frequency measured along the core axis, and A is the acute angle between the core axis and the fracture set (note: if the dip of the fracture is measured, use $[\cos \text{dip}]^{-1}$). This correction, in effect, determines the number of fractures that would have been encountered if the core were cut exactly perpendicular to the fracture plane. In the real world of infinite fracture attitudes in a rock mass, fracture frequency per unit distance in all possible directions perpendicular to all fracture planes must be expressed. The locus of that unit distance in all directions geometrically generates a sphere. The sum of all the corrected fracture frequencies thus becomes the number of fractures within that spherical volume, that is, fracture density. The diameter of that sphere is the unit distance along which each fracture density is calculated. Because distance along a core or traverse increases by distance to the first power but volume increases by distance to the third power, the density of fractures must be expressed relative to a standard unit volume. In this work, the number of fractures per sphere with a unit volume of 1 $m^3$ is arbitarily chosen as the standard of comparison. Thus the unit distance, along which the fracture density is calculated, is the diameter of the sphere; for a sphere of 1 unit $m^3$, a unit distance of 1.24 meters must be used.

As an example, over an interval of 44.10 m of core in the highly fractured clinkstone zone of the Tiva Canyon Member, 179 fractures were observed. Using the angular correction described above, the sum of the calculated fracture frequencies along the entire interval would be about 800. The density of fractures within each unit volume of 1 $m^3$ is calculated by dividing the sum of the calculated fracture frequency in the entire interval by the number of standard volume diameters in the cored interval. The resulting number, 800 X 1.24/44.10 = 22, is the average number of fractures predicted within each sphere of 1 $m^3$ within the clinkstone zone.

Implicit within this statement is the acceptance of increasingly poor statistics as the angles between the fracture sets and the core axis or traverse direction become smaller. Also, the assumption is explicitly made that the total fracture density observed is representative of the entire stratigraphic zone from which data were collected. Thus, either the observed fractures or others with the same representative attitudes and densities are present throughout the sampled zone.

## Recognition of Detailed Structural Features

Among the most critical factors for understanding hydrologic flow are the distribution, density, and attitude of minor faults. At Yucca Mountain, the recognition of minor faults presents a particularly difficult dilemma because the vast majority of exposures consists of one only cooling unit, the Tiva Canyon Member. To recognize faults with less than about 3 m of displacement, detailed stratigraphic control must be available. Such control is provided by only a few horizons, such as the sharp transition at the top and base of the lower lithophysal zone and minor subhorizons within the columnar zone. In addition, the presence of tectonic breccia, gouge, slickensides, and abrupt changes in foliation attitudes were used to identify faults where insufficient stratigraphic control existed to reveal displacement. The resulting fault pattern is shown in Figure 11, and a detailed schematic east-west cross section in the vicinity of Abandoned Wash is shown in Figure 12.

Several important structural characteristics of Yucca Mountain are noted:

1) The shallow ($5^{o}$ to $7^{o}$) eastward dip of primary foliation typical of the crest of Yucca Mountain increases gradually toward Abandoned Wash, commonly to as much as $40^{o}$, and in a few instances over $70^{o}$ (Figure 12). Where dips exceed approximately $10^{o}$ to $20^{o}$, abundant small-displacement (less than 3 m), north-northwest-striking, westward-dipping, normal faults appear. These are present throughout the more highly tilted region around Abandoned Wash. The change in elevation of a stratigraphic horizon caused by the eastward dip is opposite in sense to the change in elevation caused by the offset along the west-dipping normal faults. Thus, in cross section, a single stratigraphic horizon and the connecting fault planes form a zig-zag pattern (Figure 12).

2) The abundance of north-northwest-striking (about N. $15^{o}$ W.) faults decreases from Abandoned Wash northward where the steep foliation dips are absent.

3) North of Drill Hole Wash, right-lateral strike-slip faults are essentially parallel to and lie within several of the northwest-trending (N. $25^{o}$-$40^{o}$ W.) washes. These washes are also parallel to the trend of the dip direction of ash-flow tuffs. Therefore, the trends of these washes are probably the consequence of both faulting and runoff direction.

4) Numerous fractures are present on the surface of Yucca Mountain. Fractures can originate either from extensional stresses generated within the rock mass during cooling or from tectonic stresses applied to the rock mass after cooling. If cooling joints were exclusively present, only random fracture strikes would be observed; therefore, any departure from randomness is assumed be have a tectonic origin. The fracture densities and strikes along several traverses (corrected for fracture and traverse attitude as described above) are shown in rose diagrams in Figure 13. Six traverses were made, 5 in the Tiva Canyon Member and one in the tuffaceous beds of Calico Hills near The Prow. Where

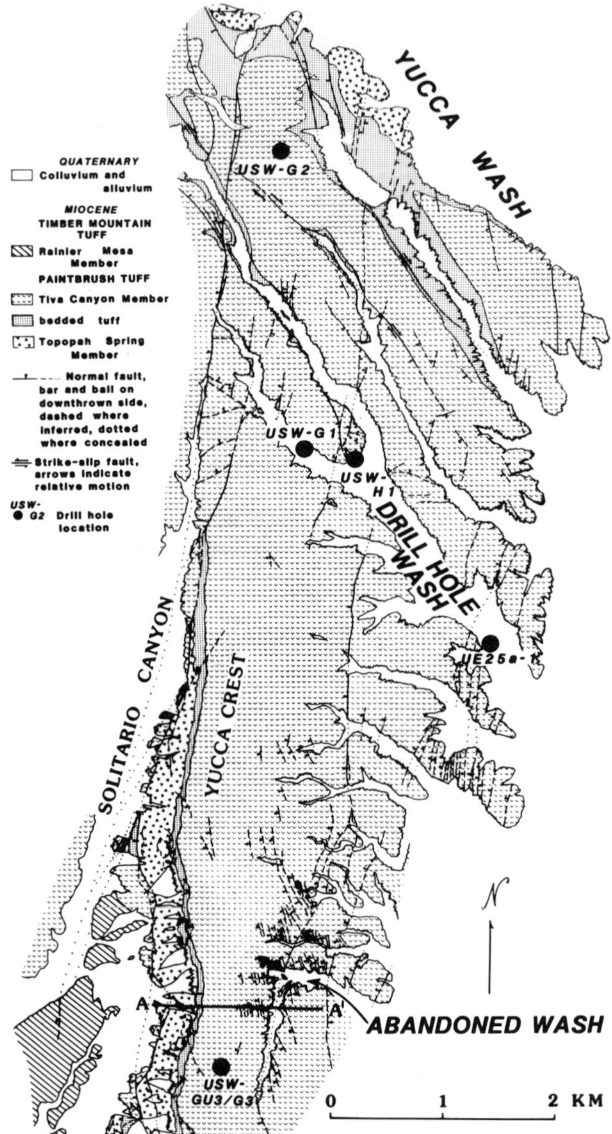


Figure 11. Detailed fault pattern at Yucca Mountain. Faults in the Abandoned Wash area are mapped only on south-facing slopes where exposures are favorable. Attitudes of primary foliation and detailed zones within the Tiva Canyon and Topopah Spring Members are not shown on this figure because of space limitations. Examples of foliation attitudes in the Abandoned Wash area are given in Figure 12.

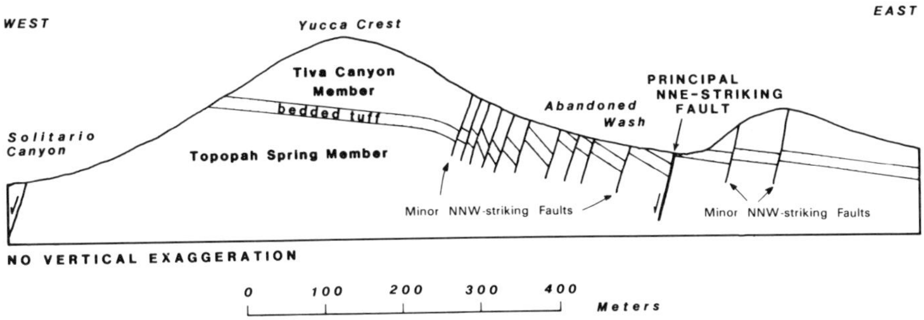


Figure 12. Structure section across Yucca Crest and Abandoned Wash showing the nature of the increase in foliation attitudes toward a north-northeast-striking normal fault and apparent dip-slip movement on representative north-northwest-striking faults.

outcrop is continuous along short traverses in the Tiva Canyon Member, fracture densities are about 6 to 8/unit $m^3$ measured along the lower lithophysal-clinkstone contact in densely welded tuff. In contrast, fracture densities in the range of only 2 to 4/unit $m^3$ were observed in the Tiva Canyon Member along the long traverses. Two factors contribute to the apparently lower fracture densities on long traverses. First, the long traverses (one just north of Abandoned Wash and one just south of Drill Hole Wash) were made, by necessity, across outcrop of moderately welded caprock, in addition to the more densely welded zone in the central portion of the Tiva Canyon Member; fractures are considerably more abundant in densely welded tuff than in moderately welded tuff. Both of these long traverses are represented by two rose diagrams, one for the east half and one for the west half; the east halves of these long traverses were made in regions more extensively underlain by moderately welded tuff. Second, by necessity the long traverses were made in part over zones extensively covered with talus; talus preferentially covers gullies which are zones expected to contain high fracture densities. The observation that north-facing slopes, which have more extensive cover, have only about half the fracture densities of the corresponding south-facing slopes, supports this conclusion (fracture densities are calculated only for the fraction of traverse with good exposure). These factors suggest that the most reliable fracture densities are those measured along short traverses, where both maxiumum exposure and uniform stratigraphic interval can be found. The number of variables that cannot be quantified and are inherent to the measure of fracture densities measured along traverses

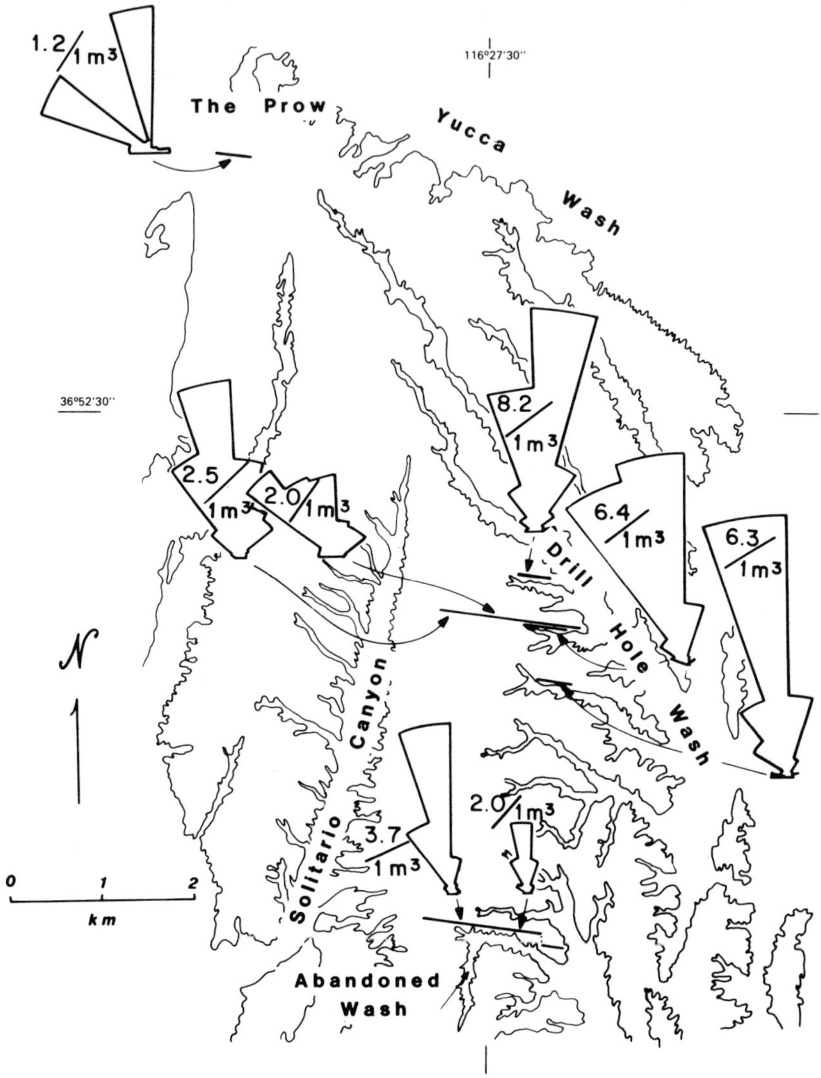


Figure 13. Rose diagrams of the fracture strikes encountered along traverses essentially perpendicular to the average fracture attitude. Five traverses were made in the Tiva Canyon Member, two longer traverses and three shorter traverses. One of the two longer traverses (A) was made south of Drill Hole Wash and the other (B) was made just north of Abandoned Wash. The northernmost of the traverses was made in the tuffaceous beds of Calico Hills. The outline of Yucca Mountain is the Tertiary volcanic rock contact with Quaternary alluvium and colluvium.

suggest that minor differences in fracture densities are not significant.

As mentioned above, the abundance of fractures is a function of both tectonic stresses and cooling stresses. Because nonwelded ash-flow tuffs and bedded tuffs have not been subjected to the stresses related to thermal contraction during cooling, the abundance of fractures in nonwelded tuff might represent exclusively the abundance of tectonic fractures, and the difference between abundance of fractures in welded and nonwelded tuffs might represent the abundance of cooling fractures. This logic breaks down, however, by observation that only a very small portion of the fractures in the rose diagrams of Figure 13 can be assigned to random cooling fractures. The vast majority, therefore, must be tectonically induced. Thus, there is a large difference between the number of tectonic fractures in nonwelded and welded tuffs. Nonwelded tuffs, therefore, have rock mass physical properties that respond to a tectonic stress by formation of relatively few fractures, whereas, welded tuffs have physical properties that respond the the same tectonic stress by formation of more numerous fractures. This conclusion is confirmed by the constant dominant strike of these fractures throughout the mountain, indicating that most of the fractures are of tectonic origin.

5) Dominant fracture strikes are essentially parallel to the north-northwest-striking (about N. 15$^{o}$ W.) faults in the region near Abandoned Wash. Northeast of Drill Hole Wash, however, the north-northwest-striking fractures distinctly intersect the northwest-trending washes and associated strike-slip fault traces. The relationship between the north-northwest-striking fractures and the northwest-striking stike-slip faults is not yet understood, but the 30$^{o}$ angle between them suggests that the fractures may be riedel shears or tension fractures.

6) At numerous localities, slickensides on north-northwest- and northwest-striking faults are nearly horizontal, characteristic of strike-slip motion. In contrast, slickensides on north- to north-northeast-striking faults in the southern and central portions of Yucca Mountain, and particularly those faults that strike in a more north to north-northwest direction, locally concentrated near Yucca Wash, have down-dip attitudes, characteristic of normal fault motion. On two faults with near horizontal slickensides, chattermarks are present. Chattermarks form hairline crescent-shaped fractures [31-32] that are commonly found on glaciated rock surfaces and less commonly on shallow faults. The focus points or noses of the crescents point in the relative direction of movement of the rock face that contains the chattermarks. In both cases on Yucca Mountain,

right-lateral strike-slip motion is unambiguously indicated. Although the magnitude of strike-slip displacement is commonly too small to be measured, at one locality west of drill hole USW-G2, a 10- to 30-m right-lateral offset in the outcrop pattern at the base of the Tiva Canyon Member was measured (Figure 11).

In addition to the structural features descibed above, there are several characteristics of fractures, such as fracture aperture, vein mineralogy, and extent of vein filling that are also very critical for a complete understanding of fracture control on hydrology. Unfortunately, surface weathering processes and drilling disturbances preclude observation of undisturbed fracture apertures. Also the mineralogy and extent of veins in fractures is so highly variable that generalizations should be avoided. Probably, the cumulative effect of both fracture aperture and mineral filling is best measured by empirical hydrologic testing.

## Correlation of Structure with Physical Property Stratigraphy

Based upon fracture attitudes and densities measured downhole from both unoriented cores, oriented cores, and downhole televison video tapes, an impressive positive correlation between degree of welding and density of fractures in core recovered from drill hole USW-GU3/G3 is shown in Figure 14. The density of fractures has been corrected for the attitude relative to the core axis as described above, and is expressed as the number of fractures/unit $m^3$. To depths of 940 m, the more elastic (lower Young's Modulus or stiffness) zeolitized nonwelded and partially welded tuffs [27] have as few as 1 to 3 fractures/unit $m^3$, whereas the more brittle (higher Young's modulus or stiffness) densely welded tuffs typically have 20 to 40 fractures/unit $m^3$. Fracture densities in core from drill holes USW-G2 and -G1 are somewhat lower, but the overall patterns are similar. Below 940 m, the abundance of fractures decreases by nearly an order of magnitude in core from USW-GU3/G3. Further investigations are being conducted to determine whether this decrease in fracture density is related to the more ductile behavior at higher confining pressures, to the somewhat more altered state of these more deeply buried tuffs, or merely to a change in tectonic style, independent of depth. The degree of welding of the more extensively welded tuffs below 940 m falls more into the moderately welded range rather than the densely welded range, and the degree of alteration is significantly greater. Although there is an expected decrease in fracture

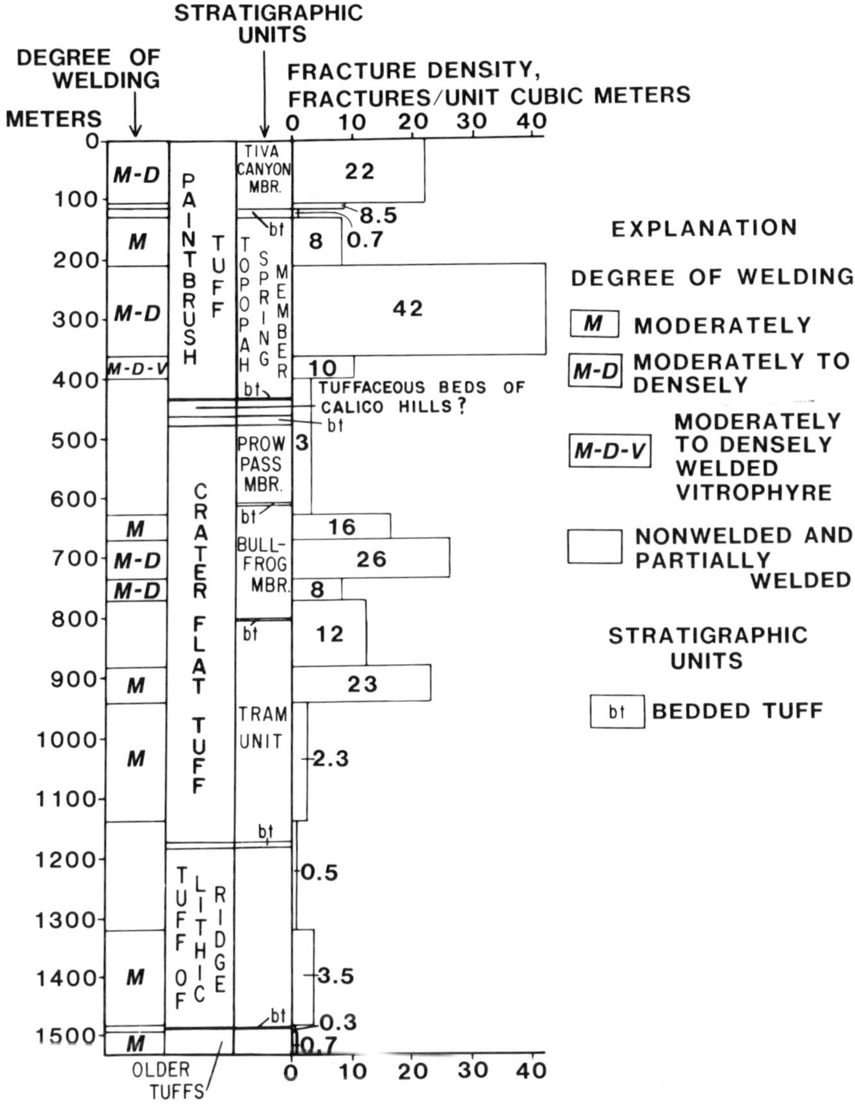


Figure 14. Density of fractures compared to physical-property stratigraphy expressed by differences in degree of welding in drill hole USW-GU3/G3 shown in Figure 4. These fractures include only mineralized cracks or cracks that have evidence of shear such as breccia, gouge, or slickensides. The density of fractures has been calculated per unit $m^3$. The density of fractures at drill holes USW-G1 and -G2 is somewhat lower than at -GU3/G3, but the overall correlation with degree of welding is similar.

density related to these factors, the decrease from 23 fractures/unit $m^3$ in the moderately welded Tram unit above 940 m to 2.3 fractures/unit $m^3$, also in moderately welded Tram unit, below 940 m requires an explanation other than a change in degree of welding.

The densities of fractures in densely welded tuff measured in core (20 to 40/unit $m^3$) do not correlate well with densities measured in well-exposed outcrops of densely welded tuff (6 to 8/unit $m^3$). In all probability, fractures are both more readily seen in core, and the mechanical disturbance of coring enhances parting along abundant hairline aperture fractures not visible on surface exposures. The dominant northwest strike of the fractures measured in oriented core (N. 15° W.,75° SW., Figure 15) from drill hole USW-GU3/G3 (R. B. Scott, USGS and Mayra Castellanos, F&S, written commun., 1982), however, does agree well with those measurements on the surface (Figure 13). Also the average fault attitude in this core (N. 21° W., 73° SW.) is similar to the north-northwest-striking faults mapped on the surface in the vicinity of Abandoned Wash (Figure 11).

## Drill-Hole Deviation

Drill holes on Yucca Mountain tend to deviate naturally from vertical, almost uniformly to the southwest and west (Figure 16). Holes USW-G1, USW-H1, and UE25a-1, all within Drill Hole Wash, trend S. 30° W. to S. 60° W., to a depth of over 1 km, essentially perpendicular to Drill Hole Wash. Holes USW-G2 and USW-G3, in contrast, trend slightly south of west, from S. 80° W. to S. 83° W. Because Drill Hole Wash appears to be controlled, at least in part, by fracture and fault systems that strike nearly parallel to the wash and dip steeply (60°-90°) toward the southwest, the uniform drill-hole deviation direction may also be controlled if the core bit tends to follow the path of least resistance down near-vertical fractures and faults. However, within the Yucca Mountain block, away from influence of the structures in the washes, the trend of dip of dominant fractures and faults is only slightly south of west; apparently the drill-hole deviation follows a similar trend and plunge. Thus, drill-hole deviation directions appear to be sensitive indicators of anisotropic mechanical properties of the rock, in this case, fracture attitudes.

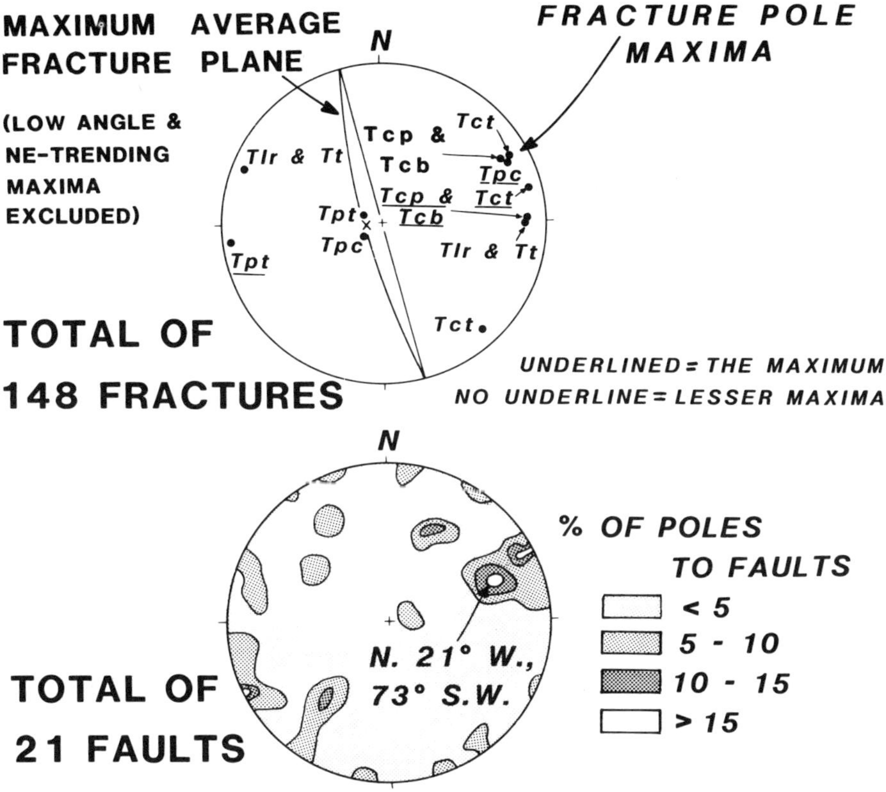


Figure 15. Lower hemisphere poles to attitudes of average fractures and faults measured in oriented core from USW-GU3/G3 and corrected for drilling deviation (R. B. Scott, USGS, and Mayra Castellanos, F&S, written commun., 1982). The average fracture plane (N. 15° W., 81° SW.) was calculated in the upper diagram by averaging the stereonet maxima for all the poles to fracture planes from each stratigraphic unit. Also in the upper diagram, the averaged direction of drill hole deviation is shown by the X. The stereonet maximum for the fault attitude (N. 21° W., 73° SW.) is shown in the lower diagram. The Tiva Canyon Member (Tpc) and Topopah Spring Member (Tpt) are part of the Paintbrush Tuff. The Prow Pass Member (Tcp), Bullfrog Member (Tcb), and Tram unit (Tct) from the Crater Flat Tuff. Tlr symbolizes the tuff of Lithic Ridge and Tt symbolizes the older unnamed tuffs.

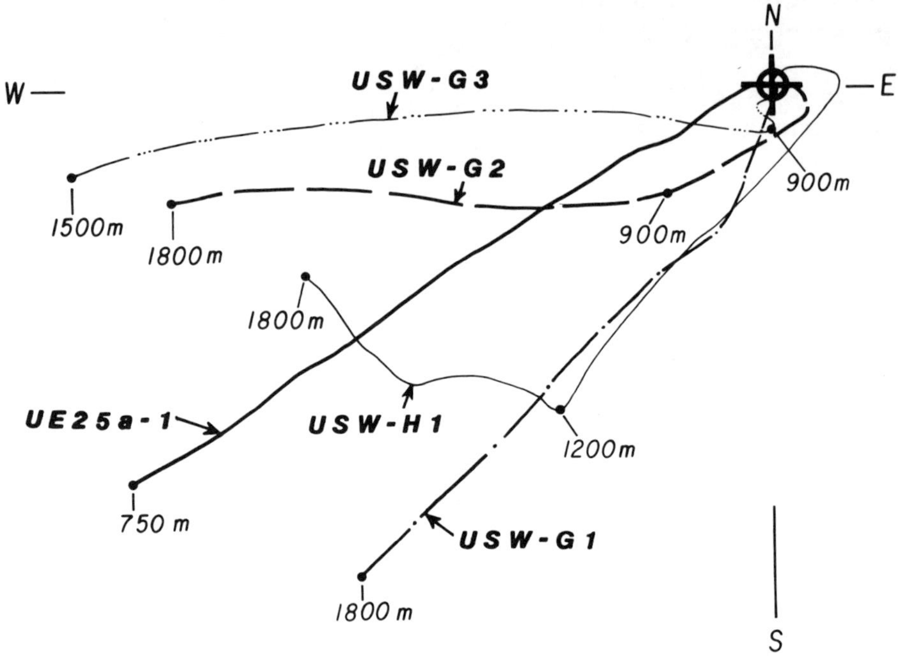


Figure 16. Drill-hole deviation directions on Yucca Mountain. Note that no horizontal scale is given because the amount of lateral deviation varies. Line lengths are normalized to represent percent of horizontal deviation.

## DISCUSSION

### Regional Structural Framework

Although the dominant Tertiary faults in the NTS are north-northeast-striking, Basin and Range style, normal faults [18,21], Carr [11] reported a significant overprint of strike-slip fault displacement, particularly left-lateral displacement on the north-northeast-striking faults. Carr did not observe right-lateral strike-slip displacement on northwest-striking faults elsewhere in the NTS; he concluded that strike-slip displacement is not expected because these faults would be under compression under the current stress regime. However, on Yucca Mountain, evidence of right-lateral displacement exists on north-northwest-striking faults south of Drill Hole Wash and on northwest-striking faults north of Drill Hole Wash. Major movement on these fault systems and the north-northwest-striking fracture system do not appear to affect the Rainier Mesa Member at Yucca Mountain; thus, the probable age of this fracturing and faulting is between the age of the faulted Tiva Canyon Member, about 12.5 m.y., and the age of the Rainier Mesa Member, about 11.3 m.y. From a regional perspective, the general attitude, sense of displacement, and age of these northwest-striking features are very similar to those features of regional strike-slip zones including the Las Vegas Valley and Walker Lane shear zones (Figure 17) [33-34]. A

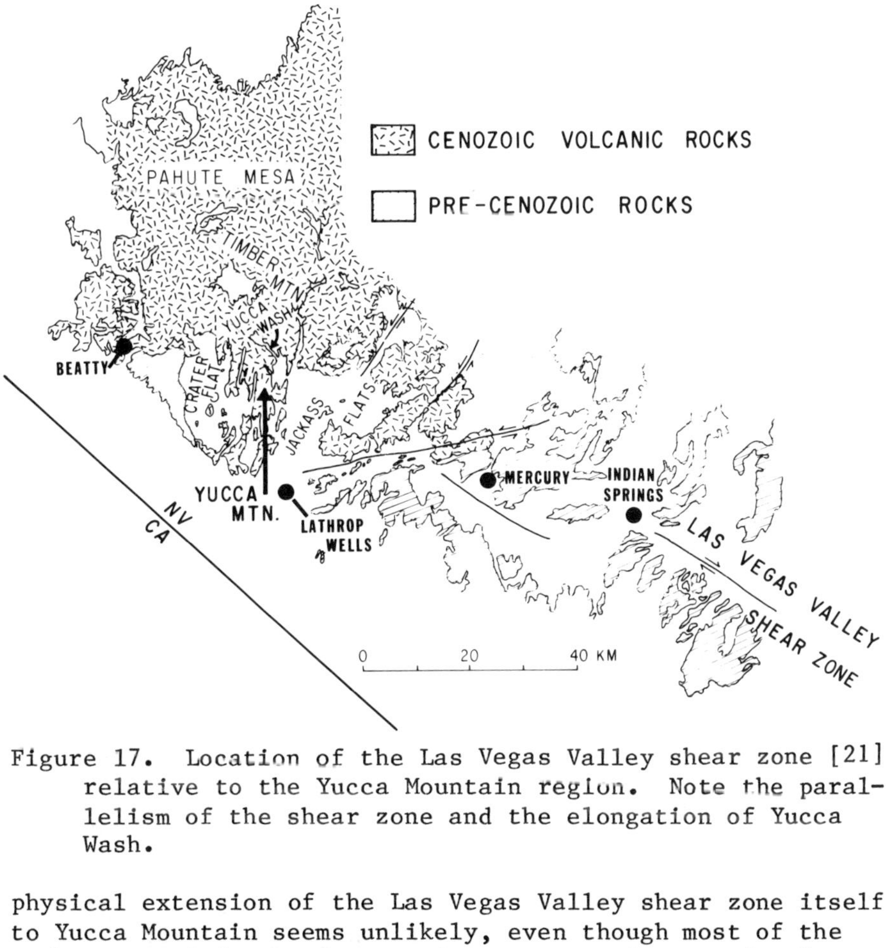


Figure 17. Location of the Las Vegas Valley shear zone [21] relative to the Yucca Mountain region. Note the parallelism of the shear zone and the elongation of Yucca Wash.

physical extension of the Las Vegas Valley shear zone itself to Yucca Mountain seems unlikely, even though most of the rocks exposed in the intervening area postdate regional shear zone activity [36-37]. A diffuse zone of right-lateral stike-slip with small displacements is probably more likely, but very difficult to recognize in the field.

The major observable displacement on faults in the Abandoned Wash area along both the numerous north-northwest-striking faults and the major north-northeast-striking block-bounding fault appears to be dip-slip and, as a result, that area appears to have experienced significant extension. The general increase in eastward dip on the down-faulted blocks west of the major block-bounding fault in this area is very similar to the roll-over features or "reverse drag" observed along growth faults in the Gulf Coast region and normal faults in the Colorado Plateau region [38-39]. If the features at Yucca Mountain are of similar origin, then the dips of major block-bounding faults under Yucca Mountain probably decrease at some as yet unobserved depth in a manner similar to those of listric growth faults on the Gulf Coast, as Wernicke and Burchfiel [40] have recently suggested for extensional systems of the Great Basin. Where large vertical displacements (about $100^+$ m) occur, such as along the eastern margin of Solitario Canyon (Figure 11), chaotic zones of breccia and extremely irregular fault blocks including overturned sequences are present on the hanging-wall side of the fault. Presumably these chaotic zones develop where movement along the listric fault creates a significant gap at the head of the fault and blocks collapse in a disorganized fashion to fill a developing gap. Abandoned Wash is an example of smaller vertical displacement between major blocks (about 25 m), and therefore, smaller gap formation on the hanging-wall of the listric system. In this case, the fractured rock in the roll-over structure has retained coherent block-to-block contact with the major normal fault. Drag on these surfaces produces the offset observed; in essence, this is a brittle response to drag along the major fault (Figure 12). These features further suggest that the brittle behavior seen in the Tiva Canyon and Topopah Spring Members at the surface probably goes through a transition at depth to more ductile behavior where these faults begin to flatten. The abundant fractures characteristic to depths of about 1 km depth and the mechanical contrast that produces uniform drill-hole deviation both reinforce the concept that the exposed portion of Yucca Mountain is within the upper brittle behavior region. Data do not exist at present to predict the depth of the transition to more ductile behavior, but the presence of "reverse drag" dips should persist to at least 1 km.

An alternative interpretation to the listric fault hypothesis given above is that the steeper foliation attitudes observed in Abandoned Wash are related to rotation by a component of strike-slip displacement (W. J. Carr, oral commun., 1982). In the absence of more discriminatory evidence, the listric fault explanation appears most likely.

Conceptual Hydrologic Model

Direct hydrologic investigations of the unsaturated zone at Yucca Mountain are underway but are too preliminary to construct an an empirical model of the hydrology at this early stage. However, knowledge of the gross physical property stratigraphy combined with a three-dimensional model of fracture and fault attitudes and densities, in conjunction with regional hydrologic observations and regional conceptual models of the hydrology of ash-flow tuffs [5,7,22], does allow construction of a very simple conceptual hydrologic model at Yucca Mountain.

Limited physical-property data [25] are available from core samples from the unsaturated zone at Yucca Mountain. These data include matrix hydraulic conductivities that fall within the ranges listed in Table I, for the most part ($2 \times 10^{-10}$ to $3 \times 10^{-9}$ cm/s for the densely welded Tiva Canyon Member, $8 \times 10^{-10}$ to $2 \times 10^{-7}$ cm/s for the densely welded Topopah Spring Member, and $4 \times 10^{-9}$ to $1.5 \times 10^{-6}$ cm/s for the underlying zeolitized nonwelded ash-flow and bedded tuffs). Also these data include natural, saturated, and dry bulk densities, from which the degree of saturation was calculated. The range of saturation of the welded Tiva Canyon Member samples is 33 to 50 percent, the underlying nonwelded ash-flow and bedded tuffs is 61 to 90 percent, the welded Topopah Spring Member is 17 to 91 percent, and the underlying nonwelded ash-flow and nonwelded tuffs is 82 to 100 percent. Even below the static water level, within the deepest interval listed above, the range of saturation is between 82 and 100 percent. The lower densities, higher porosities, higher degree of saturation, and higher surface areas of the nonwelded tuffs suggest that they may act as temporary capillary barriers (W. E. Wilson, USGS, oral commun., 1982), but the unsaturated zone has probably reached a steady-state flow such that each layer is as saturated as possible with a balance of capillary forces, gravity, and back pressure of trapped gases. Thus, any pulse of water migrating through the rock would be in excess of what the rock can absorb, and therefore, would eventually pass through these nonwelded layers.

The generalized and greatly simplified cross section drawn through the unsaturated zone in Yucca Mountain in Figure 18 assumes the more welded portions of the Tiva Canyon and Topopah Springs Members to be highly transmissive units with high effective conductivities. Because the fracture densities in lithophysal zones in the densely welded tuffs of the Tiva Canyon Member (14 to 22 fractures/unit $m^3$) do not appear to differ appreciably from those within nonlithophysal densely welded zones (22 to 26 fractures/unit

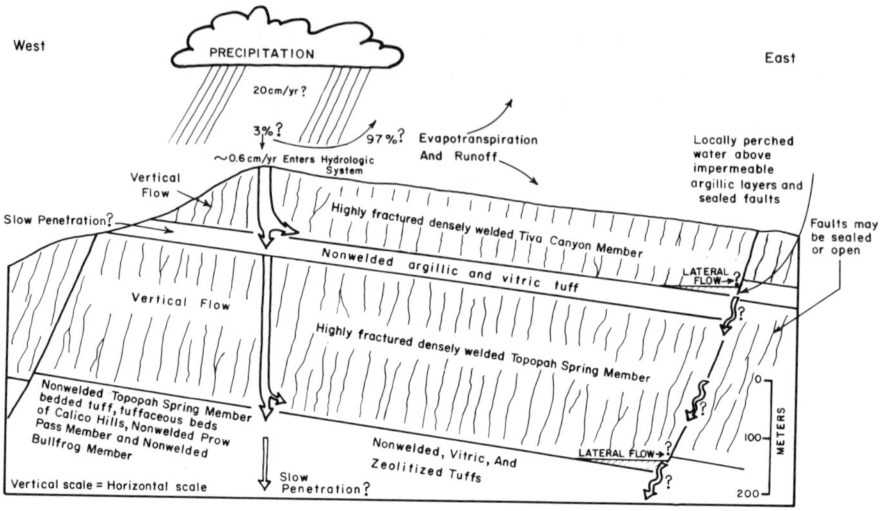


Figure 18. Conceptual hydrologic model of Yucca Mountain, assuming a positive net recharge for the mountain. Determination of vertical and possible horizontal rates of transmission of recharge through densely and moderately welded Tiva Canyon and Topopah Spring Members and through less welded vitiric, argillic and zeolitized tuffs awaits hydrologic testing specifically designed for the unsaturated zone. If faults are locally sealed, they may create local perched water tables, but if they are open, the faults may rapidly transmit water.

$m^3$), it is assumed that lithophysal zones will have hydrologic properties similar to densely welded nonlithophysal zones. The abundance of cavities in lithophysal zones may increase the conductivities locally where the fractures interconnect with cavities. Between and below these more welded portions of the Tiva Canyon and Topopah Spring Members are bedded tuffs and nonwelded to partially welded ash-flow tuffs that have few fractures. Where these less welded units are vitric, they may act as "leaky aquitards", the term used by Winograd and Thordarson [5], because they have relatively high matrix conductivities. However, where these bedded tuffs and less welded tuffs are altered to clays and zeolites [29], their matrix conductivities are considerably lower than those of vitric tuffs, and they have low hydraulic conductivities because of the low fracture densities. These less welded altered tuffs may have the capacity to act as capillary barriers because of the lower hydraulic conductivities. Observations by Thordarson [22], at Rainier Mesa northeast of Yucca Mountain in a similar setting on the NTS, indicate that perched water does occur above locally saturated zeolitized tuffs within the unsaturated zone.

Of the small annual precipitation that occurs in this arid region (less than 20 cm/yr in the Jackass Flats drainage basin), less than about 3 percent is estimated [40] to enter the hydrologic system as recharge on a basin-wide average. Because Yucca Mountain stands roughly 400 m higher than Jackass Flats, it is assumed here that a positive recharge does occur at 3 percent of 20 cm/yr (actual measurements of precipitation or recharge on Yucca Mountain have not been made). Thus, roughly 0.6 cm of water would enter the system annually and would be transmitted through the highly fractured zone of the densely welded portion of the Tiva Canyon Member. Where the water encounters vitric tuffs below the welded Tiva Canyon Member, the water may tend to be temporarily absorbed in the porous nonwelded tuffs above the Topopah Spring welded tuffs; however, pore sizes in vitric tuff are considerably greater than those in zeolitized tuffs, and eventually water would penetrate these vitric layers. On the other hand, where these nonwelded vitric layers are clay-rich and have low conductivities, they may become essentially water saturated, and may inhibit vertical flow sufficiently to locally or temporarily pond water, producing perched water tables above the Topopah Spring Member. After vertical passage through densely welded zones and the vitrophyre of the Topopah Spring Member, the water would encounter the deeper zone of nonwelded tuffs. Although these tuffs are extensively zeolitized, the zeolitization is discontinuous, particularly toward the southern portion of Yucca Mountain where the water table is deeper and the tuffs are still vitric (R. B. Scott, USGS, and Mayra Castellanos, F&S, written commun., 1982). Whether zeolitization is a requirement and whether argillization can create similar conditions conducive to perched water tables in ash-flow tuffs on Yucca Mountain is not known at the present time. Direct evidence to suggest that perched water exists within Yucca Mountain has not been found, but wet areas on the surface exist above zeolitized tuffs along the steep walls of Yucca Wash, indirectly suggesting that some perched water may be present along this horizon. Such hydrologic barriers may also cause limited lateral movement of water down dip (east and southeast) of the stratigraphic boundary and along the major fracture attitude (south-southeast and southeast). This general southeast direction would be essentially parallel to flow directions down the hydrologic gradient (F. E. Rush, USGS, oral commun., 1982).

Faults may either be sealed by formation of fault gouge or secondary mineralization, forming relatively nontransmissive barriers and creating local perched water, or they may be open, highly transmissive paths through both nonwelded tuffs and the highly transmissive Topopah Spring Member. Eventually that portion of the recharge that moves vertically must permeate the nonwelded base of the Topopah Spring Member, the nonwelded, zeolitized tuffaceous beds of Calico Hills, various zeolitized and argillized bedded tuffs, and the nonwelded to partially welded zeolitized Prow Pass Member before reaching the static water level, unless a fault acts as a short-circuit conduit. Faults may be highly impermeable in general, especially in nonwelded zones but observations of open faults in core to depths greater than the static water level suggest that unless detailed empirical data exist to support the contention that a specific fault is impermeable, high permeabilities should be considered as possible, at least locally.

The validity of the simplified conceptual model shown in Figure 18 will be tested as empirical data are gathered. At present the model can best be used to design future hydrologic testing and to recognize areas where additional structural and hydrologic data are needed. Obviously, the reality of hydrologic flow within the unsaturated zone under Yucca Mountain is far more complex than the model is capable of suggesting; a number of unknown factors will have to be investigated (R. K. Waddell, USGS, oral commun., 1982): What is the annual average precipitation and recharge at Yucca Mountain? What is the relation between recharge flux and lithophysal zones? How does water flow through layers of variable degrees of saturation, degrees of welding, and variable densities of fracturing? Do fractured densely welded tuffs have fracture conductivity as the dominant mode and nonwelded tuffs have matrix conductivity as the dominant mode of water flow? How much smaller are matrix and effective conductivities in the unsaturated zone than in the saturated zone? At what rates does water flow through rocks with these variable characteristics? To what degrees do porous layers act as capillary barriers? Until these and other problems are addressed, a more realistic and complex model is difficult to produce. The viability of the unsaturated zone under Yucca Mountain as a nuclear waste repository will depend, in part, upon reaching an understanding of the complex physical process of fluid transmission in the unsaturated zone of Yucca Mountain.

CONCLUSIONS

Determination of the viability of nuclear waste disposal in the unsaturated zone of ash-flow tuffs at Yucca Mountain is highly dependent upon a three-dimensional evaluation of physical, structural, and hydrologic properties of the rock mass.

A) Listed below are four major conclusions reached from the investigation of rock mass physical properties.

A1) The physical-property stratigraphy in tuffs at Yucca Mountain is defined principally by layers of contrasting degrees of welding, bulk density, and porosity. Physical-property stratigraphy, rather than conventional petrologic stratigraphy, must be used to characterize the inhomogeneous and layered rock body.

A2) Within densely welded portions of the Tiva Canyon and Topopah Spring Members of the Paintbrush Tuff, the density of fractures measured on core 6.25 cm in diameter is commonly 20 to 40 fractures/unit $m^3$. Detailed surface traverses indicate a lower fracture density for this rock type, in the range of 6 to 8 fractures/unit $m^3$.

A3) Partially welded and nonwelded tuffs have considerably fewer fractures than welded tuffs, commonly as few as 1 to 3 fractures/unit $m^3$. Also they are commonly altered to clays and zeolites.

A4) Although a number of surface characteristics that create mappable zonations (such as the clinkstone, hackly, and columnar zones) appear to be related to detailed devitrification fabrics, the fracture density appears to be relatively independent of this variable. Also, the lithophysal content and grain density do not appear to appreciably affect the fracture density.

B) Listed below are four major conclusions reached from the investigation of the structural geology of the unsaturated zone.

B1) The dominant strike of fractures in both welded and nonwelded tuffs, as measured in both outcrop and core, is about N. $15^\circ$ W.. This indicates that most of the observed fractures are tectonically induced, as a random distribution of fracture strikes should result from contraction-induced jointing.

B2) The small-displacement north-northwest strikes of faults in the Abandoned Wash vicinity in the southern portion of Yucca Mountain are parallel to the dominant strike of fractures. However, north of Drill Hole Wash, faults appear to be more nearly parallel to the washes, about N. $25^\circ$ W. to N. $40^\circ$ W.

B3) Several of the northwest-striking minor faults have slickensides and chattermarks compatible with right-lateral, strike-slip motion. This attitude and sense of displacement suggests that a genetic relation between the regional Las Vegas Valley and Walker Lane shear systems and these local Yucca Mountain structures may exist. This movement appears to postdate the Tiva Canyon Member of the Paintbrush Tuff (12.5 my old) but predate the Rainier Mesa Member (11.3 my old) of the Timber Mountain Tuff.

B4) On the hanging-wall side of several major north-northeast-striking normal faults in Yucca Mountain, the increase in foliation dip may be analogous to roll-over structures associated with listric faults. Also, the closely-spaced, small-displacement normal faults in Abandoned Wash may result from drag along the major faults. The major movement on normal faults on Yucca Mountain occurred within the same time interval as the strike-slip faults mentioned above.

C) The conclusions reached from physical-property and structural investigations of Yucca Mountain allow the construction of the following preliminary conceptual hydrologic model based upon the fundamental assumption that a net positive recharge occurs on the mountain.

C1) The uniform high fracture density and brittle character of densely welded tuffs indicate that, in spite of their low matrix conductivities, their effective hyraulic conductivities are uniformly high; this conclusion is supported by limited in situ hydraulic conductivity measurements within the saturated zone of the densely welded Topopah Spring Member near Yucca Mountain [5]. Thus, in the unsaturated zone, the fractures in densely welded tuffs should readily transmit fluids, provided a maximum or steady-state degree of saturation of the rock matrix exists; that is, assuming the rock matrix has already absorbed all the water it can retain in the unsaturated zone.

C3) The low fracture density and low matrix hydraulic conductivities of zeolitized or argillized nonwelded tuffs suggests that they should serve as potential aquitards below the water table [5] and as infiltration barriers within the unsaturated zone.

C4) The relatively high matrix conductivities of nonwelded vitric tuffs indicates that these tuffs should serve as relatively permeable zones, even in the absence of fracturing [5], but the rate of transmission of flow in the unsaturated zone may be inhibited to a degree by possible capillary action of these porous tuffs.

C5) Although the vast majority of recharge probably moves vertically, limited lateral flow associated with perched water may occur above relatively impermeable zeolitized or argillized nonwelded zones. Even though Thordarson [22] did not find evidence to support significant lateral flow in a similar setting in the unsaturated zone under Rainier Mesa northeast of Yucca Mountain, he did find perched water above highly zeolitized tuffs there. The east to southeast dip and north-northwest-striking fractures on Yucca Mountain, absent at Rainier Mesa, would be conducive to lateral flow of perched water in an east or southeast direction.

C6) Development of this hydrologic model has identified several areas where important fundamental data are needed for an understanding of the hydrology of the unsaturated zone: a) What is the range of annual precipitation on Yucca Mountain and what is the range of annual recharge? b) Is the dominant mode of transmission of fluids along fractures in densely welded tuffs, but within the rock matrix in zeolitized or argillized nonwelded tuffs and in vitric nonwelded tuffs? c) How do rates of fluid transmission in the unsaturated zone differ from those in the saturated zone within similar rocks? d) Do porous vitric tuffs act as capillary barriers in a steady state system? e) Does limited lateral flow occur and does perched water exist? f) Do specific faults within Yucca Mountain act as hydrologic barriers or as highly permeable hydrologic short circuits?

## ACKNOWLEDGMENTS

As in any attempt to correlate several disciplines, much credit belongs to numerous colleagues: In addition to formal USGS technical reviews, numerous USGS hydrologists including Ike Winograd, Bill Wilson, Rick Waddell, Merrick Whitfield, and Jim Robison, have freely given advice. Mike Carr has also provided a particularly valuable informal geologic review. Barry Swartz of SNL transmitted the physical-property measurements for this work and Larry Teufel, also of SNL, gave advice concerning the behavior of these ash-flow tuffs under stress. Judy Brandt and Byron Cork, among others of Fenix and Scisson, provided field assistance. Finally Jim Mercer's understanding of interminable delays in submitting the manuscript is deeply appreciated.

## REFERENCES

1. Eckel, E. B., ed. "Nevada Test Site," *Geol. Soc. Amer. Mem. 110*: 290p. (1968).

2. Winograd, I. J. "Radioactive waste disposal in thick unsaturated zones," *Science 212*: 1457-1464 (1981).
3. Christiansen, R. L., and Lipman, P. W. "Geologic map of the Topopah Spring NW quadrangle, Nye County, Nevada," *U. S. Geol. Survey Geol. Quad. Map GQ-444*, scale 1:24,000 (1965).
4. Lipman, P. W., and McKay, E. J. "Geologic map of the Topopah Spring SW quadrangle, Nye County, Nevada," *U. S. Geol. Survey Geol. Quad. Map GQ-439*, scale 1:24,000 (1965).
5. Winograd, I. J., and Thordarson, William " Hydrogeologic and hydrochemical framework, south-central Great Basin, Nevada-California, with special reference to the Nevada Test Site," *U. S. Geol. Survey Prof. Paper 712-C*: C1-C126p. (1975).
6. Smith, R. L. "Zones and zonal variations in welded ash-flows," *U. S. Geol. Survey Prof. Paper 345-F*: F149-F154 (1960).
7. Winograd, I. J. "Hydrogeology of ash flow tuff: a preliminary statement," *Water Resources Research 7*: 994-1006 (1971).
8. Byers, F. M., Jr., Carr, W. J., Orkild, P. P., Quinlivan, W. D., and Sargent, K. A. "Volcanic suites and related cauldrons of Timber Mountain-Oasis Valley caldera complex, southern Nevada," *U. S. Geol. Survey Prof. Paper 919*: 70p. (1976).
9. Christiansen, R. L., Lipman, P. W., Orkild, P. P., and Byers, F. M., Jr. "Structure of the Timber Mountain caldera, southern Nevada, and its relation to Basin-Range structure," *U. S. Geol. Survey Prof. Paper 525B*: B43-B48 (1965).
10. Christiansen, R. L., Lipman, P. W. Carr. W. J., Byers, F. M., Jr., Orkild, P. P., and Sargent K. A. "Timber Mountain-Oasis Valley caldera complex of southern Nevada," *Geol. Soc. Amer. Bull. 88*: 943-959 (1977).
11. Carr, W. J. "Summary of tectonic and structural evidence for stress orientation at the Nevada Test Site," *U. S. Geol. Survey Open-File Rpt. 74-176*: 53p. (1974).
12. Spengler, R. W., Byers, F. M., Jr., and Warner, J. B. "Stratigraphy and structure of volcanic rocks in drill Hole USW-G1, Yucca Mountain, Nye County, Nevada," *U. S. Geol. Survey Open-File Rpt. 82-1349*: 36p. (1981).
13. Spengler, R. W., Muller, D. C. and Livermore, R. B. "Preliminary report on the geology of drill hole UE25a-1, Yucca Mountain, Nevada Test Site," *U. S. Geol. Survey Open-File Rpt. 79-1244*: 43p. (1979).

14. Spengler, R. W., and Rosenbaum, J. G. "Preliminary interpretations of results obtained from boreholes UE25a-4, -5, -6, -7, Yucca Mountain, Nevada Test Site," *U. S. Geol. Survey Open-File Rpt. 80-929*: 35p. (1980).
15. Snyder, D. B., and Carr, W. J. "Preliminary results of gravity investigations at Yucca Mountain and vicinity, southern Nye County, Nevada," *U. S. Geol. Survey Open-File Rpt. 82-701*: 36p. (1982).
16. Carr, W. J. "Volcano-tectonic history of Crater Flat, southwestern Nevada, as suggested by new evidence from drill hole USW-VH-1 and vicinity," *U. S. Geol. Survey Open-File Rpt. 82-457*: 23p. (1982).
17. Lipman, P. W., Christiansen, R. L., and O'Connor, J. T. "A compositionally zoned ash-flow sheet in southern Nevada," *U. S. Geol. Survey Prof. Paper 524-F*: F1-F47 (1966).
18. Stewart, J. H., and Carlson, J. E. "Geologic map of Nevada," *U. S. Geol. Survey: scale 1*:500,000 (1978).
19. Armstrong, R. L. "Sevier orogenic belt in Nevada and Utah," *Geol. Soc. Amer. Bull. 79*: 429-458 (1968).
20. Barnes, Harley, and Poole, F. G. "Regional thrust fault system in Nevada Test Site and vicinity" *in* Nevada Test Site, *Geol. Soc. Amer. Memoir 110*: 233-238 (1968).
21. Ekren, E. B., Rogers, C. L., Anderson, R. E., and Orkild, P. P. "Age of Basin and Range normal faults in Nevada Test Site and Nellis Air Force Range, Nevada," *in* Nevada Test Site, *Geol. Soc. Amer. Memoir 110*: 247-250 (1968).
22. Thordarson, W. "Perched ground-water in zeolitized-bedded tuff, Rainier Mesa and vicinity, Nevada Test Site, Nevada," *U. S. Geol. Survey Open-File Rpt. TEI-862*: (1965).
23. Blankennagel, R. K., and Wier, J. E. "Geohydrology of the eastern part of Pahute Mesa, Nevada Test Site, Nye County, Nevada," *U. S. Geol. Survey Prof. Paper 712-B*: B1-B34 (1973).
24. Young, R. A. "Water supply for the Nuclear Rocket Development Station at the U. S. Atomic Energy Commission's Nevada Test Site," *U. S. Geol. Survey Water Supply Paper 1938*: 19p. (1972).
25. Anderson, L. A. "Rock property analysis of core samples from the Yucca Mountain UE25a-1 borehole, Nevada," *U. S. Geol. Survey Open-File Rpt. 81-1338*: 36p. (1981).

26. Lappin, A. R., Van Buskirk, R. G., Enniss, D. O., Butters, S. W., Prater, F. M., Muller, C. B., and Bergosh, J. L. "Thermal conductivity, bulk properties, and thermal stratigraphy of silicic tuffs from the upper portions of hole USW-G1; Yucca Mountain, Nye County, Nevada," *SAND81-1873, Sandia National Laboratories, Albuquerque, NM*: 78p. (1982).
27. Olsson, W. A., and Jones, A. K. "Rock mechanics properties of volcanic tuffs from the Nevada Test Site," *SAND80-1453, Sandia National Laboratories, Albuquerque, NM*: 54p. (1980).
28. Lappin, A. R. "Bulk and thermal properties of the functional tuffaceous beds in holes USW-G1, UE25A#1, and USW-G2, Yucca Mountain, Nevada," *SAND82-1434, Sandia National Laboratories, Albuquerque, NM*: 62p. (1982).
29. Caporuscio, F., Vaniman, D., Bish, D., Broxton, D., Arney, B., Heiken, G., Byers, F., Gooley, R., and Semarge, E. "Petrology studies of drill cores USW-G2 and UE25b-1H, Yucca Mountain, Nevada," *LA-9255-MS, Los Alamos National Laboratory, Los Alamos, NM*: 111p. (1982).
30. Terzagi, R. D. "Sources of error in joint surveys," *Geotechnique 15*: 287-304 (1965).
31. Chamberlin, T. C. "The rock scorings of the great ice invasions," *U. S. Geol. Survey 7th. Ann. Rpt.*: 218-223 (1885).
32. Johnson, B. C. "Characteristics and mechanics of formation of glacial arcuate abration cracks," *unpubl. Ph.D. dissertation, Penn. State Univ.*: 253p. (1975).
33. Stewart, J. H., Albers, J. P., and Poole, F. G. "Summary of regional evidence for right-lateral displacement in the western Great Basin," *Geol. Soc. Amer. Bull. 79*: 1407-1414 (1968).
34. Guth, P. L. "Tertiary extension north of the Las Vegas Valley shear zones, Sheep and Desert ranges, Clark County, Nevada," *Geol. Soc. Amer. Bull. 92*: 763-771 (1981).
35. Fleck, R. J. "Age and possible origin of the Las Vegas Valley shear zone, Clark and Nye Counties, Nevada," *Geol. Soc. Amer. Abstracts with Programs 2*: 333 (1970).
36. Sargent, K. A. and Stewart, J. H. "Geologic map of the Specter Range NW quadrangle, Nye County, Nevada," *U. S. Geol. Survey Quad. Map GQ-884*: scale 1:24,000 (1971).
37. McKay, E. J. and Williams, W. P. "Geologic map of the Jackass Flats quadrangle, Nye County, Nevada," *U. S. Geol. Survey Quad. Map GQ-368*: scale 1:24,000 (1964).

38. Bruce, C. H. " Pressured shale and related sedimentary deformation: mechanisms for development of regional contemporaneous faults," *Amer. Assoc. Petrol. Geol. Bull. 57*: 878-886 (1973).
39. Hamblin, W. K. "Origin of "reverse drag" on the downthrown side of normal faults," *Geol. Soc. Amer. Bull. 76* 1145-1164 (1965).
40. Wernicke, B. and Burchfiel, B. C. "Modes of extensional tectonics," *Jour. Structural Geol. 4*: 105-115 (1982).
41. Rush, F. E. "Regional groundwater system in the Nevada Test Site area, Nye, Lincoln, and Clark Counties, Nevada," *Nevada Dept. Conservation Natural Resources Reconnaissance Series, Rpt. 54*: 25p (1970).

acetic acid 52-53
acetone 52-53,192
adsorption 86,198
air entry potential 167-168
aniline 52-53
arsenic 216,219-221
atmospheric diffusion 30
attenuation 86
Atterberg Limits 103

Barnwell Formation 261,264
Barnwell, South Carolina 27, 42,44-46
Basin and range geomorphic province 196,197,199, 200,206,208
benzene 216-222
bis-2 ethyl hexyl phthalate 218,226
Boussinesq's momentum correction factor 29
Brooks-Corey method 140

cadmium 216,221
Cape May Formation 214-215
capillary pressure 75,109, 138
capillary retention curves 127,145
Childs model 267
Chino clay 41
chloroform 216-222
chromium 216,219,221-222
clay dissolution 67
Cohansey Formation 211,214, 227
Collis-George model 267
compaction 102,104
compressibility 131
Congaree Formation 261
Cooper-Jacob equation 50,53
copper 219
co-precipitation 86
Corioli's energy correction factor 29

Darcy's law 85,87,238-239, 265-266,270
decay 86
desaturation curves 106
1,2 dichloroethane 216-222

electrical resistivity 179
electromagnetic method 179
Ellenton Formation 261
entrapped air 121,147
ethylene glycol 52-53
evaporation 40-41
evaporation potential 32, 42-43

Fairgrange Till 103-104, 106,108
falling head permeability test 103,233
Fick's law 31,37
field permeameter 236
finite element method 238
fractured rock 165-167,174

geothermal gradient 204
gravity water 34-36,39,42,45
Green-Ampt infiltration equation 90
ground penetrating radar 180, 182-189,192
ground water mound 85

Hamon equation 32
Hanford Location, Washington 77
Hawthorn Formation 261,264
hazardous waste 49,81,98,211
heptane 53
high-level radioactive waste disposal 1,2,147,165
Houston black clay 51-52
hydraulic conductivity 17-18, 23,30,33,36,44,50,52-53, 71,73-75,78-79,87-88, 91-92,101-102,104-105, 107-109,113,115,117,121, 124-125,127-128,166,214, 240,243,246,250-252,256, 259-260,265-267,274,276, 278-282,284-285
hydraulic diffusivity 36
hydraulic fabric 131,135-136

hydraulic head 24,51,118-119, 241
hygroscopic water 34-35,37,45

infiltration 11,13,20,27-28, 39-40,42,44,45-46,55, 115-116,118=120,123-124, 264,269,274-277
intrinsic permeability 51, 54-55,58,60,66-67,72

Jaeger's hydrostatic pressure correction factor 29

Kelvin capillary model 134
Kirkwood Formation 208, 214-215

Laplace equation 160
leachate 49-51,211-212,219, 223,226
lead 216,219-221
low-level radioactive waste 1-2,5-6,11,27,42,46, 255-256
Lufkin clay 51-52
lysimeter 116,230,233,235

Maddock sandy loam 75-76
Manning's coefficient 29
Maxey Flats, Kentucky 6,8
McBean Formation 264
mercury 216,219-221
methyl chloride 216-218
methylene chloride 216-221
mill tailings 91,130-132
modified Poiseuille equation 171-172
moisture characteristic curve 241-243
moisture content (see water content)
moisture content-capillary pressure relationship 102
moisture content-capillary pressure test 103,105
molecular diffusion 205
Mt. Simon sandstone 71-72
Muamlem method 13

natural wetness distribution curve 34
neptane 52
Nevada Test Site 1,3-4,196, 289
Newton-Raphson method 55,58
nickel 216,221
non-water impoundment 82

oil shale 138-139
osmotic suction 128,138
overland flow 28,32-33,39,46

Panoche silty clay loam 75-76
pellicular water 34-37,41-43, 45
Peoria Loess 13-16,18,20-21, 103-104,106,108
permeability (see intrinsic permeability)
phenols 216-218
phreatic surface 85
plasticity 128
plutonium 256
Poiseuille equation 169-170
pore-size distribution 92, 102,127
porosity 36-37,44,72,92,101, 109,115,127-129,214,266, 271,274
Prandtl's mixing length coefficient 31
precipitation 86,198
pressure head 13-14,17-22, 87-88,91-92,106,117,121, 140-142
Price Landfill 211,220,222, 225

radionuclides 195-196, 198-199,207-208,255
relative hydraulic conductivity 54,116,241
Richard's equation 12,50,53

salt content 128
saturation 54,57-59,64,87,
106,108-109,116,119,133,
140,152,157-158,160,163,
203,207,232,241,244-245,
249,278
Savannah River Laboratory/
Plant 255-262,264,267,
271,274,279
seepage 81-89,93-97,229,
243,248
seepage face 19-19,21-22,24
Sevilleta dune sand 73
shear strength 131
smectite 51
sodium adsorption ratio 128
sodium chloride 186,191
soil moisture retention
curves 106
soil tension 76-77
Soret effect 176
sorption (see adsorption)
specific heat 201
specific moisture
capacity 54
specific yield 92
Stanford watershed model 28
suction 131,265-266,279

tensiometer 116-118,269
thermal conductivity
201-205,207
Tipperary Sand 151-156,159,
161-163
toluene 216-222
tortuosity 129
trans-1,1 dichloroethylene
216-217,219-221
1,1,1 trichloroethane 216
1,1,2 trichloroethane 216
trichloroethylene 216-218,
220-221
trench cap
(see trench covers)
trench covers 11,28,33-34,
37,42,44,101
tuff welded 204,206,289
non-welded 204,289
tuffaceous rocks 196-197,199
Tuscaloosa Formation 261,264

uranium mill tailings
(see mill tailings)
uranium recovery wastes
1-2,9
U.S. Department of Energy 2
U.S. Environmental Protection
Agency 2
U.S. Nuclear Regulatory
Commission, Division of
Waste Management 1

vadose zone 113,195,199
vinyl chloride 216-222

water content 13-17,19-20,
22,88,106-107,109,115,
117,124,127,133,137-138,
141-142,151,156-158,176,
203,243,272,280
water impoundment 82-83
water retention curve 116
Wick effect 1,11-12,78

xylene 52-53

Yucca Mountain 3,289

zinc 216,221